Intelligent Robots and Cobots

Scrivener Publishing
100 Cummings Center, Suite 541J
Beverly, MA 01915-6106

Industry 5.0 Transformation Applications

Series Editors: Dr. S. Balamurugan (sbnbala@gmail) and Dr. Sheng-Lung Peng

The increase in technological advancements in the areas of artificial intelligence (AI), machine learning (ML) and data analytics has led to the next industrial revolution "Industry 5.0". The transformation to Industry 5.0 collaborates human intelligence with machines to customize efficient solutions. This book series covers various subjects under promising application areas of Industry 5.0 such as smart manufacturing, intelligent traffic, cloud manufacturing, real-time productivity optimization, augmented reality and virtual reality, etc., as well as titles supporting technologies for promoting potential applications of Industry 5.0, such as collaborative robots (Cobots), edge computing, Internet of Everything, big data analytics, digital twins, 6G and beyond, blockchain, quantum computing and hyper-intelligent networks.

Publishers at Scrivener
Martin Scrivener (martin@scrivenerpublishing.com)
Phillip Carmical (pcarmical@scrivenerpublishing.com)

Intelligent Robots and Cobots

Industry 5.0 Applications

Edited by

V. Ramasamy
Dept. of Computer Science and Engineering, Vel Tech Rangarajan Dr. Sagunthala R&D Institute of Science and Technology, Chennai, Tamilnadu, India

S. Balamurugan
Intelligent Research Consultancy Services (iRCS), Coimbatore, Tamilnadu, India

and

Sheng-Lung Peng
Dept. of Creative Technologies and Product Design, National Taipei University of Business, Taiwan

Scrivener Publishing

WILEY

This edition first published 2025 by John Wiley & Sons, Inc., 111 River Street, Hoboken, NJ 07030, USA and Scrivener Publishing LLC, 100 Cummings Center, Suite 541J, Beverly, MA 01915, USA

For more information about Scrivener publications please visit www.scrivenerpublishing.com.

Wiley Global Headquarters
111 River Street, Hoboken, NJ 07030, USA

For details of our global editorial offices, customer services, and more information about Wiley products visit us at www.wiley.com.

Library of Congress Cataloging-in-Publication Data

ISBN 978-1-394-19817-7

Front cover image courtesy of Adobe Firefly
Cover design by Russell Richardson

Set in size of 11pt and Minion Pro by Manila Typesetting Company, Makati, Philippines

Printed in the USA

10 9 8 7 6 5 4 3 2 1

The editorial team extends heartfelt dedication to their families—parents, spouses, and children—for their boundless support throughout the creation of this work. Additionally, the editors express sincere gratitude to their colleagues at their esteemed institution for their unwavering love, blessings, and encouragement. Finally, this book is respectfully dedicated to the entire research community, whose collective efforts continue to enrich our understanding of the world.

Contents

Preface

Advancements in science and technology are rapidly modernizing human lifestyles, especially in the 21st century. Industries are increasingly leveraging automation and cutting-edge technology to streamline tasks and meet evolving demands. Industry 1.0 was built on "water and steam," Industry 2.0 harnessed "advanced electrical technology," Industry 3.0 introduced automation through electronics and IT, and Industry 4.0 brought forth intelligent, autonomous robots. Now, Industry 5.0 focuses on using collaborative robots, or cobots, enabling users to design with greater freedom. To fully realize this potential, the IT industry must contribute more creatively to future industrial needs. This book, structured into 18 chapters and three sections - Fundamentals; Applications; and Challenges - reflects the current and emerging market trends that shape industrial growth. Each chapter explores how businesses incorporating hardware and software—like AI, cognitive computing, blockchain, IIoT, and more—are capitalizing on these innovations to maintain a competitive edge.

The research and development in the area of technology has increased the cost-effectiveness and acceptance of these IoT-enabled devices in many different industries. Various sectors including manufacturing, healthcare sectors, transportation sector, and the agriculture sector have begun incorporating robots and cobots into their operations. They are aiming to increase their productivity, reduce the downtime of their equipment, and optimize resource utilization. Chapter 1 offers a comprehensive review of how robots and cobots in the Industrial Internet of Things (IIoT) have the potential to revolutionize businesses, serving as a valuable resource for researchers interested in this transformative technology.

Chapter 2 focuses on the role of cobots in the successful implementation of Industry 5.0, examining their potential applications across various industries. It compares the functions of robots and cobots, highlighting their key differences, and introduces a conceptual cobot-based cyber-physical system. The chapter also investigates the explicit use of human intelligence in different applications and discusses the risks and security threats to cobots, along with mitigation techniques.

Chapter 3 explores how cobots offer the most significant value in scenarios where human proximity is essential. The direct interaction between humans and robots can be either the greatest advantage or limitation of collaborative systems, depending on its impact on human factors like ergonomics and psychological stress. Recent case studies have demonstrated the successful implementation of cobots across various industries, establishing their role in the Industry 5.0 revolution, especially in the health sector.

Chapter 4 offers a comprehensive exploration of the dynamic interaction between robots, cobots, and smart transportation, highlighting their transformative potential. The use of robotic technology, particularly collaborative robots, or cobots, has made a profound impact on the evolution of transportation within smart cities. It also examines various machine learning methods and IoT applications that enhance synergy within Intelligent Transportation Systems (ITS).

Chapter 5 explores low/no-code development, a method that enables application design and development using interactive graphical interfaces, eliminating the need for traditional coding. Low/No-code development platforms established a new paradigm in application development, offering ease of use around the environment that allowed non-IT professionals to build applications with very minimal or no coding experience. This approach allows non-IT professionals to create applications without writing a single line of code.

Chapter 6 provides critical insights into the benefits that Industry 5.0 offers to the workforce, as well as the challenges workers will face during this transition to the Fifth Industrial Revolution. It outlines the skills workers need to be equipped for Industry 5.0 and addresses the issues surrounding the integration of technology with the human workforce. The chapter also offers considerations for managers when merging technology with human resources and provides recommendations for reskilling and upskilling workers to prepare them for the future of work.

Chapter 7 delves into the concepts and applications of intelligent robots and cobots within the context of Industry 5.0 transformation. It provides an overview of Industry 5.0, emphasizing its goal of merging human capabilities with advanced technologies, and examines the characteristics, functionalities, and potential benefits of intelligent robots and cobots.

Chapter 8 reviews multiple human-machine collaboration strategies studied to guide businesses and researchers in making informed decisions in Industry 5.0, where human-machine collaboration is the primary focus.

The objective of intelligent process automation is to assist cobots in establishing both production objectives and industrial plant safety measures. The chapter emphasizes how Industry 5.0 aims to transform human-technology interactions through Artificial Intelligence.

Chapter 9 delves into the development of a vision-guided robotic device, also known as a cobot, designed for performing a variety of precise tasks. Cobots can be used for hygienic and precise manipulation and assembly of medical equipment or implants, reducing the danger of human contamination, penetrating illnesses, and maintaining clean environments. The chapter highlights how cobots' ease of programming, installation, and collaboration helps maximize throughput and maintain consistent quality.

Chapter 10 covers the design, construction, applications and role of robotic arms in the industry. A computer-aided design (CAD) model of a robotic arm is developed using CREO/SolidWorks and analyzed with ANSYS, demonstrating the crucial role of material properties under various loading conditions. The chapter aims to create a robotic arm that is highly accurate, precise, lightweight, and easily transportable.

Chapter 11 explores the potential of Industry 5.0, which aims to overcome challenges by fostering collaboration between robots and humans, rather than competition. Industry 5.0 is expected to profit from different promising advances and applications empowering expanded creation and unconstrained conveyance of modified items. It thoroughly discusses the various applications of Industry 5.0, including intelligent healthcare, cloud manufacturing, supply chain management, and production.

Collection of proper aviation resources determines how the chatbots and robots are going to perform and up to what accuracy. Translation analysis determines the accuracy of translations and transliterations, whereas other parameters determine how that particular aviation bots interact with passengers. Chapter 12 focuses on effectively managing the development of various bots related to the aviation industry.

Chapter 13 explores hydroponics, a soil-less plant cultivation method within horticulture, and examines how cobots can streamline this process and enhance plant survival. Existing traditional farming systems can lead to soil degradation, reduced crop yields, and water scarcity due to high water usage for irrigation. Together, hydroponics and cobots create a more efficient and effective system for plant growth.

Chapter 14 highlights the latest advancements in big data analytics for cobots, addressing the challenges researchers face within IoT environments. Data analytics for IoT within the realm of cobots involves collecting, processing, and analyzing the copious data generated by IoT devices to extract insights and inform decision-making. Cobots and other IoT devices produce various data types, ranging from structured to unstructured, originating from sources like sensors and wearables. It discusses gaps in analytics, research methodologies, and data management techniques.

Chapter 15 focuses on the design and control of soft robots, reviewing commonly used manufacturing materials and their characteristics. Soft robots are well-known for their ability to control the range of force in which they act such as the ones used in material prehension. Also, soft robots are smart robots. It highlights the diverse applications of soft robots in everyday life, such as wearables, mechanical motion, surgeries, and more.

Chapter 16 introduces an innovative model where collaborative robots (cobots) leverage microservices and swarm intelligence within a shared industrial workspace. Evaluating the deployment of the collaborative robot service is necessary in the industrial sectors as well as for researchers. This chapter also addresses deployment constraints, evaluation processes, and service performance when applying swarm intelligence via microservices. The work will also highlight the deployment constraint and performance aspects of service while utilizing the swarm intelligence through microservices.

Chapter 17 examines the security of industrial robots and cobots, beginning with an introduction to the Fourth Industrial Revolution and its various facets. Because, when dealing with a cobot, it is recommended to have a heightened knowledge of the many areas of cyber security and how they relate to the safety of the system. It then explores the vulnerabilities and attacks that industrial robots and cobots may encounter.

The integration of cobots into interconnected networks and their reliance on digital communication channels raise significant cybersecurity concerns. Traditional security measures such as firewalls and encryption mechanisms may not be sufficient to safeguard cobots against evolving cyber threats. Chapter 18 examines how blockchain technology can address cybersecurity issues related to cobots. The blockchain-based smart contracts can play a vital role in securing interactions between cobots and their human operators or other entities. Smart contracts enable the establishment of predefined rules and protocols, ensuring that cobots execute only authorized actions.

We hope that readers will find this book beneficial. The editors are grateful to the reviewers who have contributed to improving the quality of the book through their constructive comments. The editors also thank Martin Scrivener and Scrivener Publishing for their support and publication.

Editor(s):
Dr. V. Ramasamy
Associate Professor in the Department of CSE at Vel Tech Rangarajan Dr. Sagunthala R&D Institute of Science and Technology (Deemed to be University), Chennai, Tamilnadu, India
datasciencerams@gmail.com
Dr. S. Balamurugan
Director-Research and Development, Intelligent Research Consultancy Services (iRCS), Coimbatore, Tamilnadu, India
sbnbala@gmail.com
Dr. Sheng-Lung Peng
Professor and the Director of the Department of Creative Technologies and Product Design, National Taipei University of Business, Taiwan
slpeng@ntub.edu.tw
November 2024

Acknowledgement

The editorial team extends heartfelt gratitude to our institution for fostering an encouraging research environment that laid the foundation for this proposal. We are deeply appreciative of the diverse group of contributors from various nations and offer special thanks to the reviewers worldwide who have diligently examined each chapter to maintain the book's high standards. Their insightful feedback has been invaluable. We sincerely thank all parties involved for their dedication and willingness to take on tasks that pushed them beyond their usual comfort zones. We look forward to reuniting with you in the next edition of our publication.

Part 1
FUNDAMENTALS

1

Cobots for Industry 5.0 Transformation

Ahmed F. Siddiqui[1], Aditya J. Paul[1], Sushruta Mishra[1]* and S. Balamurugan[2]

[1]Kalinga Institute of Industrial Technology, Deemed to be University, Bhubaneswar, Odisha, India
[2]Albert Einstein Engineering and Research Labs (AEER Labs), Vice Chairman-Renewable Energy Society of India (RESI), Tamil Nadu, India

Abstract
The potential of cobots in the Industrial Internet of Things (IIoT) to transform the industries has been examined in this research article. The research and development in the area of technology has increased the cost-effectiveness and acceptance of these IoT-enabled devices in many different industries. Various sectors including manufacturing, healthcare sectors, transportation sector, and the agriculture sector have begun incorporating robots and cobots into their operations. They are aiming to increase their productivity, reduce the downtime of their equipment, and optimize resource utilization. This article looks into the different ways that cobots are being used in these sectors. It highlights their role in enhancing industrial processes. Also, the article addresses potential challenges in deploying robots and cobots in businesses. These challenges such as acceptance issues and compatibility issues need to be addressed. The advantages of deploying cobots in IIoT settings are also demonstrated, which also showcases how this technology can lead to significant improvements across various industries. Overall, this article provides us a comprehensive review of how robots and cobots in IIoT have the potential to revolutionize businesses and serves as a valuable resource for other researchers who are interested in this transformative technology.

Keywords: Robots, cobots, Internet of Things (IoT), Industrial Internet of Things (IIoT), Industry 4.0, Industry 5.0, cyber-physical systems (CPS)

**Corresponding author*: sushruta.mishrafcs@kiit.ac.in

V. Ramasamy, S. Balamurugan and Sheng-Lung Peng (eds.) Intelligent Robots and Cobots: Industry 5.0 Applications, (3–18) © 2025 Scrivener Publishing LLC

1.1 Introduction

The Internet of Things, also known as IoT, intelligently connects our surrounding items and raises the quality of life by using embedded sensors and networks. It is a relatively new global networking infrastructure that connects various networks to enable the process of gathering, processing, managing, and distributing data via actual physical devices and appliances using pervasive sensory, communication, networking, and information processing technologies. It is rapidly developing in the field of contemporary wireless telecommunications [1–3]. This infrastructure is made up of a sizable number of physical items that are controlled and monitored online in addition to internet-connected equipments and devices. Their value comes from the ability of the equipments and devices to communicate with one another and their surroundings to achieve common objectives [4]. In domain-specific applications, smart devices and the activities that they are designed to perform can produce dependable, effective, and secure services [1, 3]. The term "Things" in the Internet of Things refers to both physical and non-physical objects. It includes robots, cobots, and other machines; it may also include human and animal bodies and data collected from sensors. IoT can capture the data, share it, and also analyze it. It is of great use for the industry and can result in several advantages. Adopting the IoT-based technologies in order to track and monitor products can improve the efficiency of the industry when it comes to manufacturing and also reduce the time it takes to deliver the goods. Development of robots and cobots is made possible by IoT-engineered mechanisms and software systems, which greatly minimizes and eradicates equipment breakdown, process defects, and human errors across the entire process. IoT enhances operations and offers a variety of creative solutions, such as creating a new business model [5]. During the implementation of the IoT in large-scale industrial applications, numerous challenges have been faced, encompassing issues related to power consumption, expandability, linkages, and data-related matters (communication, data rate, time delay, and standardization), as well as the aspects of safety and protection. Nonetheless, certain industries have already successfully used IoT technologies, such as the healthcare services, supply chain, and infrastructure monitoring [2, 6, 7]. Industries all around the world are looking for robotic solutions aspiring to bring a revolution in their production and manufacturing processes; they are aiming to boost efficiency and throughput and also make it cost-effective while minimizing the need of human labor. Currently, conventional industrial robots are primarily utilized by large manufacturers engaged in high-volume production. Small- and medium-sized businesses (SMEs) constitute a substantial 98% of the European manufacturing sector and have yet to catch up with advanced robotic automation for their operations. SMEs are trying to use robots to boost their productivity and also to keep up with the market demands. Robotic automation can help the SMEs to become more cost effective in their manufacturing processes; this also allows them to compete with others [8]. The swift integration of cobots is completely transforming the manufacturing procedures of small- and mid-sized businesses (SMBs). These cobots offer ease of use; they also occupy very less space and collaborate with human operators easily in shared workspaces. The ability of cobots to work along with human workers makes them very useful. An increasing number of SMEs are trying to use cobots in activities where they can integrate it. There are vast possibilities for better efficiency and better production processes [9].

The following points will be discussed in the paper:

- Discuss how IoT is essential for industries.
- Discuss the issues with acceptance of cobots.
- Discuss how cobots function in the industries.
- Discuss how cobots collaborate and result in automation.

1.2 Related Works

Industry 4.0's foundational idea of intelligent manufacturing depicts a system that can adapt to a variety of shifting circumstances and goods. Flexible lines autonomously modify manufacturing procedures to accommodate various product types and varying environmental circumstances [10]. Consequently, this results in heightened excellence, productivity, and adaptability, alongside extensive and eco-friendly production of personalized goods, as fewer resources are utilized [11].

In 2011, a German initiative led by the Federal government in collaboration with academic institutions and businesses had introduced the concept of "Industry 4.0." A significant strategic endeavor of this initiative was to develop innovative manufacturing techniques, aimed at boosting the productivity and efficiency of local industries [12].

A burgeoning innovation known as a cyber-physical system (CPS) focuses on integrating computational applications as a network of interconnected physical and digital elements that oversee real-time automation in industrial infrastructures [13].

Researchers of [14] developed a smart collaborative balancing method for ruthlessly altering the instrumentation of system operation and successfully optimizing the job pattern. The relevant problems are addressed by matrix operation using a single congestion interval. The invasion defense system is also covered and displayed. Then, concrete suggestions for coordinating different network tactics are made.

In an architecture developed by researchers of [15], it integrates deep convolutional neural networks with actual network information for early identification of distributed-denial-of-service (DDoS) assaults organized by a botnet overseeing malevolent gadgets. These surrogate devices autonomously execute SMS flooding, quiet calls, signaling, or a combination of these strategies, with the goal of orchestrating DDoS attacks within a cluster, potentially disrupting the functioning of CPSs. These attacks target various services including the Internet, calls, and SMS, or it may be a combination of these. This poses significant threats to the system's functionalities and makes it compromised. Gushev [16] focuses on the computation framework, which uses a particular framework to develop these concepts.

The modern business world is made up of a network of connected companies and organizations that actively participate in the extensive supplier network [17]. In this dynamic world, supply chain management and logistics have experienced significant paradigm and systemic changes [18] and are vulnerable to a wide range of threats. This process is now an essential component of how any organization functions and efficient administration of this procedure is now undeniably essential for the sustainability of the enterprises in an ever-changing and fiercely competitive setting [2]. However, technological advancements have improved business-to-business communication and made it easier to transport and monitor physical goods throughout the network of supply chain.

Smart household appliances are an Industry 4.0 enabler that have the ability to improve consumer happiness, and they help with energy economy, better personalization, and big data analytics. The interconnected network of objects elevates instant communication, data gathering, and analytical capabilities, revolutionizing traditional enterprises into a digital model [19]. Now, it is that data is a valuable resource to help consumers demand cheap smart products and systems with higher levels of personalization than ever before [20]. The household device industry has been one of the forerunners in implementing cutting-edge technologies, such as IoT and Cloud, since the early 2010s [21]. Through a seamless cooperation, IoT enables sensing, acting, and real-time data transmission skills [22, 23]. The crucial IoT benefit depends on making it possible to handle vast amounts of data by using big data analytics [24]. With the help of IoT, it can be made possible for the different types of sensors to share data in real-time for a variety of purposes. This may include managing pollution and tracking traffic flow, and they can also help in managing road intersections [25]. A separate network link like Bluetooth or GSM is used to assist in the development of IoT-enabled smart homes. This enables remote tracking and management of household appliances. Also, people have the option to use a virtual helper such as Amazon Alexa, Google Nest, Apple Siri, or Microsoft Cortana. This enables effortless voice command over intelligent home devices [26]. An increasing array of commonplace objects is undergoing transformation into intelligent, IoT-enabled merchandise, and collaborative robots. The aim is to establish an extensive array of interconnected products utilizing suitable Industry 5.0 technologies, encompassing, among others, fridges, laundry machines, automated dishwashers, stoves, drying appliances, air conditioning units, and heating systems [27]. The advent of innovative Industry 5.0 advancements has ignited a revolution in intelligent home applications, although there are indications that affordable customized items may not be readily accessible in the near future. The primary value of creating smart products lies in utilities, intelligent urban centers, and interconnected networks [28].

Research predicts an exponential growth in the global market sales of cobots, projected to reach $11.5 billion in 2025 from $116 million in 2015, with around 700,000 units sold [29]. The researcher made it very clear that cobots revolutionize human ergonomics and revolutionize manufacturing productivity. However, a triumphant implementation of cutting-edge technology demands more than just financial investment; the pivotal role of employee attitudes and acceptance cannot be underestimated. Their unwavering support and acceptance are paramount to the technology's triumphant integration and ultimate success [30].

1.3 IoT for Industries

The current collection of literature concerning Industry 5.0 predominantly centers on futuristic and interdisciplinary viewpoints. Unlike Industry 4.0, which primarily concentrates on utilizing digital tech to tackle particular manufacturing obstacles and boost efficiency, Industry 5.0 adopts a wider and all-encompassing methodology. This new paradigm aims to create a more customer-centric environment, embracing a holistic purpose that goes beyond mere technological advancements. Industry 5.0 envisions a future where manufacturing processes seamlessly integrate with human expertise and creativity to produce goods and services that cater precisely to the needs and preferences of customers. Rather than solely relying on automation and smart technologies, Industry 5.0 emphasizes the harmonious coexistence of humans and machines, fostering a collaborative and inclusive work environment. Moreover, Industry 5.0 extends its reach beyond manufacturing sectors, seeking to revolutionize and optimize processes across various industries. This cross-sectoral approach envisions a synergy between sectors, with shared knowledge and innovation propelling a more interconnected and efficient economy [31].

1.4 Issues with Cobots in Industry 5.0

Collaborative robots (cobots) are increasingly utilized in Industry 5.0, yet employee acceptance remains a challenge. The conventional technology acceptance model falls short in explaining this phenomenon for cobots integrated with artificial intelligence (AI). Anthropomorphism, attributing human-like traits to cobots, also fails to elucidate the issue. The issues that employees have come from factors like trust; they also have a fear of job displacement and uncertainty about working alongside intelligent machines. A proper study of human–cobot interaction is very necessary to understand employee attitudes in a better way. Researchers should consider psychological aspects as well as socio-cultural aspects that may affect the employee's behavior. They should also look into usefulness, the ease of using cobots, and compatibility with other already existing workflows. Also, the ethical concerns in regard to privacy and data security must be addressed to build trust. To improve acceptance, we need industries to focus on educating the employees, training them, and involving them in the technology adoption process. Addressing the issues that the employees face can lead to a more harmonious and productive human–cobot collaboration.

Anthropomorphism involves attributing human-like attributes such as emotions, intentions, motivations, imagination, or actual behavior to non-human entities. Interestingly, robot acceptance indicates a clear willingness of the user group to start using and employing the technology for its intended purposes. This tells us that the group is open and ready to start using robots and consider them as important tools that can help save some time and reduce human effort as well [32].

Interactive human–robot interaction (IHRI) stands as a vital domain within cognitive ergonomics. Its primary objective lies in ensuring optimal compatibility between human cognitive abilities and robots during interactions. By employing psychological models and concepts from cognitive science, IHRI aims to identify key factors influencing cognitive functions. The ultimate goal is to minimize errors, reduce unnecessary cognitive burden, and alleviate psychological pressure experienced by employees. By doing so, IHRI seeks to create a workplace environment that fosters safety, well-being, and high performance. In essence, this research embraces the cognitive ergonomics approach to delve deeper into understanding and enhancing the dynamics of human–robot interactions for the benefit of both human workers and robotic systems [33].

As a solution to address the issue of low cobot acceptance at the mechanism level, enhancing employees' robot use self-efficacy through diverse training methods proves promising. Prior to implementing cobots in the factory, comprehensive training sessions can be conducted. The comprehensive training programs involve various activities, including educating employees on cobot principles, providing surrogate experiences through interactions with small toy robots, and engaging in collaborative work under professional guidance to create successful experiences. By empowering employees with essential knowledge and practical skills, the apprehension and concerns related to cobots and any perceived threat they may pose can be effectively minimized.

This systematic approach allows employees to develop a greater sense of confidence and competence in utilizing cobots effectively. Through hands-on training and real-life application, the training process empowers employees to overcome initial hesitations and uncertainties, paving the way for increased acceptance and integration of cobots into their daily work routines. Consequently, the successful implementation of such training initiatives fosters a harmonious human–robot collaboration, ultimately contributing to improved productivity, safety, and job satisfaction in the workplace.

1.5 Cobots in Industries

Industry 4.0 was contemplated as a way out to tackle both technical and societal challenges, harnessing the remarkable advancements in IoT and CPSs. This transformative idea led to the rise of the Work 4.0 model in Germany, initiating vital discussions about the substantial effects of Industry 4.0 on everyday work and society as a whole [34]. However, it seems Industry 4.0 has mainly focused on technological advancements, possibly overlooking its broader impact on society. To accomplish a complete and beneficial transformation, it is crucial to find equilibrium, recognizing the ethical and societal aspects of this industrial revolution. Adopting inclusivity and sustainability can lead Industry 4.0 toward a brighter and more prosperous future, where technology and society coexist in ideal harmony. Industry 5.0 complements Industry 4.0 with a focus on human-centric design, sustainability, and resilience.

By making use of advanced materials including smart properties with integrated sensors makes industrial workplaces safer and improves how people work together. Simultaneously advanced technologies like cutting-edge virtual reality and cooperative automation help things run more smoothly and efficiently. Sustainability is prioritized through environmentally friendly practices, aiming for a resilient and ecologically conscious industry. By emphasizing human requirements and promoting adaptability, Industry 5.0 visualizes a future where technology brings about positive change for both individuals and the planet. Cobots, also referred to as collaborative robots, are provided with various sensors, taking advantage of advanced industrial robotics technology. These exceptional machines are implemented in factories and various industrial settings where they are provided with robust vision systems. In contrast to conventional industrial robots, cobots are designed with safety features, to assure effective collaboration with human workers. The primary aim of cobots is to place humans at the heart of the manufacturing process and alleviate them from performing arduous, hazardous, or monotonous tasks. Importantly, cobots are not meant to replace human workers; instead, they work alongside operators, complementing their skills and capabilities. Once installed, cobots are intended to remain stationary at their designated work site, streamlining and enhancing the overall production process [35–37].

Businesses are growing more attracted to the benefits of utilizing these robots in various industries, which in result making the cobot sector a promising area of rapid growth. Due to their versatility and effectiveness, cobots are being used across diverse production settings, leading notable advancements in the industrial robotics industry. Through seamless collaboration with human operators, these transformative robots are reshaping manufacturing and improving previously labor-intensive or dangerous tasks. The increasing popularity of cobots in the business world is evidence of their capabilities for transforming work processes and raising productivity [38, 39].

Cobots are cutting-edge machines engineered with an unwavering focus on safety. Their advanced collision sensors ensure the utmost protection of human workers during any interaction. Cobots, armed with collaborative arms incorporating force boundaries, ensure that unintended collisions are harmless to human workers. Moreover, these remarkable machines showcase a hand-guided tool, empowering operators to assume direct control, even in automation, and responding exclusively to the operator's instructions. The soaring popularity of cobots knows no bounds, transcending traditional and unconventional sectors alike. However, successful integration demands meticulous management during the initial deployment stages. Sporting a distinctive design featuring force-sensing technology, cobots stand head and shoulders above other robots in the field. Their collaborative nature, working side by side with humans, offering assistance or acting as guides, showcases their exceptional responsiveness to human commands and actions, setting them apart from fully autonomous counterparts. In a realm where innovation meets safety, cobots reign supreme [40, 41].

1.6 Automation and Cobots

The transformation of the innovative collaborative machines of the manufacturing process is done by the laborious activities. These activities include gathering components, conducting quality checks, and feeding machines. All these tasks help humans to shorten the prolonged duration of work. Because these robots turn out to be the value assistants, one would be able to witness them working with the human operators with harmony. This, in turn, enhances the efficiency and productivity of the manufacturing process. Improved production outcomes can only be expected via the mixture of human ingenuity and robotic precision. The cobots are the machines, which are significantly designed to carry out specific functions. These can also help in moving the necessary objects while working. The cobots are well innovated to open new avenues for streamlined and efficient manufacturing processes. If the goal is to achieve a balance between human expertise and technological advancement, then cobots prove to be the best things to embrace as it helps in optimizing operations making the work easy for the manufacturers.

The industrial segment has been highly benefited from the cobot applications in the manufacturing process. The benefits include cost effective designing, flexibility of work, and easy work. However, having an enhanced and swift flexibility in the manufacturing process proves to be somehow not so valuable as it does not impact on performance enhancement and adaptability to the production lines. The cobots have been very widely recognized in the workstation areas where they are catering to the needs of the SMBs. Currently, with the advancement in technology with the cobots, the world has seen advancements in securing human operators, effortless configurability, and friendly user interface (UI). With this, we have seen high acceptance rates with positive feedback and improved performances in the collaborative tasks. There are some other factors as well, which might need an extensive analysis. These include giving emphasis on dependability, operator acceptance, successful interaction with humans, and many more. The cobots have been developed in such a way that it can cater to the needs of the coworkers alongside lowering their work stress, which, in turn, helps in alleviating work pressure [42].

Although the robotics has been here for over a decade, the current era has been some great advancements. This advancement has enhanced the versatility in the manufacturing sector. As well-known by now, the cobots are extensively designed to be artificially intelligent machines that are autonomous and are capable of doing independent work assigned. This is now possible by the direct interaction with the humans in the workstation. It also maintains a physical demeanor at the place. The past decade has seen substantial increase in the use of robots because the population has accepted it to share the workload and safety purpose. The concept of the cobots or collaborative robots enables better human decision making and flexible environments. The best advantage to the advancement in cobots is their factory automation contribution alongside the lightweight designs that they receive to get accommodated to any place.

In Figure 1.1, there is a comprehensive image provided. This showcases the application of cobots within the industrial landscape in relation to degrees of collaboration, relationships, interdependencies, etc. The illustrative description throws light on the important practices or collaboration degrees. The practices include assistive modes of team work, concurrent, consecutive, and autonomous. These provide some distinctive insights to

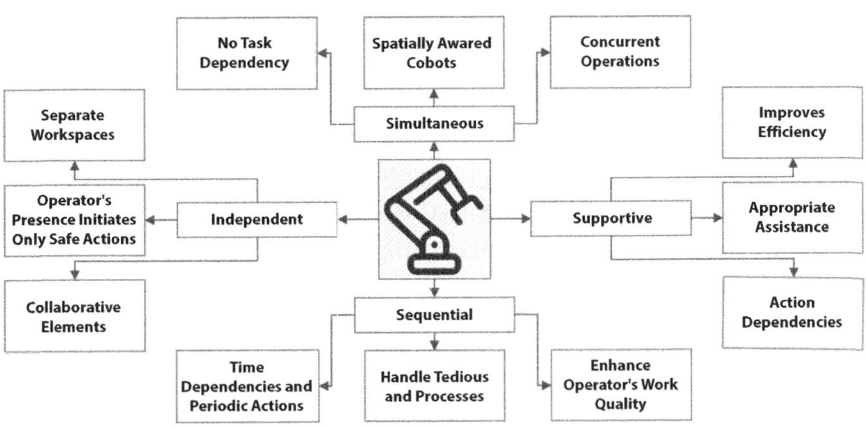

Figure 1.1 Relationships and dependencies related to automation using cobots.

the dynamic interactions between the humans and the collaborative robots in the industrial setting [43, 44].

The popularity of the cobots is rising, and they have now become a part of the advanced Industry 5.0. The cobots have well positioned themselves at the vanguard of the transforming era of technological advancements. The best use of their ability (which is collaboration with humans) has no doubt helped in enhancing factory automation and more flexibility in the manufacturing that leads to a great innovative shift in collaborative manufacturing. With this, one can say that a perfectly synced human–robot synergy will help in boosting efficiency, productivity, and overall manufacturing excellence [45–47].

1.7 Conclusion

The increase in the advancement of the robotic technologies has greatly impacted the manufacturing, expanding of the horizons of potentials, and adaptability within the industries into a more productive and new era. The very forefront of the transformation of this new developmental technology includes the collaborative robot transformation or one can say "cobots" that has highly exalted and exemplified the interactions between human and intelligence and technological innovation. What is more, the cobots are different from their predecessors. These cobots are highly capable of independent functioning, possess extreme AI ability, and are autonomous in nature. Apart from this, the cobots are also found to be engaged in collaborative tasks with the human counterparts in the shared workspaces.

The transformation of Industry 4.0 model to the new Industry 5.0 has embodied a new transformation in the technology, which can be seen with a comprehensive perspective. Customer-centricity and eco-conscious practices can be envisioned into a manufacturing landscape where the two take the center of the stage. With this, it aims to harmonize the creativity and ingenuity of humans in relation with the machine efficiency, which further results in well-tailored goods and services that will align with the preference of the consumers. The main focus of the new industry 5.0 has been shifted toward the convergence of human expertise and technological prowess serving as a cornerstone. This also encourages an act of collaborative as well as inclusive work environment.

Speaking about the cobots, they play the pivotal roles in the Industry 5.0 narrative. Cobots have found their place in the industrial sector and have possessed an array of sensors and technological advancements. This has further enhanced their function to seamlessly operate in the grounds of factories and industries. Unlike the past robots, these new cobots have been made to have learnt the human safety and interaction processes. They function as valuable assistants to their human friends that also complement human skills. So, there seems to be less or no chance of replacing them.

Although the cobots have advanced, integration of the cobots in the Industry 5.0 has also faced challenges. The machines do face challenges like efficiency and innovations; also, they are prone to face challenges of acceptance and adoption by the employees. This situation arises due to the pivotal concerns of the employees such as job displacement and uncertainty and necessitates a comprehensive exploration of the interaction between human and robots. The attitude of the employees also depends on the psychological and social cultural factors that come into the scene once the cobots appear. Hence, this has urged the researchers to delve deeper into the ergonomics and reduce the boundaries to ensure a happy and well-settled human–cobot collaboration.

Lastly, Industry 5.0 has definitely pioneered a big transformation that transcends the mere technological advancement. The main focus of this Industry 5.0 has been on maintaining the synergy between human creativity and technological prowess where the collaboration between humans and the machines in the future will be paramount. Cobots are believed to be the embodying spirits, which will bridge the gaps between human expertise and innovation. However, addressing the challenges in making cobots accepted into society will help the researchers to unlock the machine's full potential. This will again help the human race to enter into a more profitable and productive era where a harmonious collaboration between human and robot will be seen.

References

1. Lin, J., Yu, W., Zhang, N., Yang, X., Zhang, H., Zhao, W., A survey on Internet of Things: Architecture, enabling technologies, security and privacy, and applications. *IEEE Internet Things J.*, 4, 5, 1125–1142, 2017.
2. Da Xu, L., He, W., Li, S., Internet of Things in industries: A survey. *IEEE Trans. Ind. Inform.*, 10, 4, 2233–2243, 2014.
3. Al-Fuqaha, A., Guizani, M., Mohammadi, M., Aledhari, M., Ayyash, M., Internet of Things, A survey on enabling technologies, protocols, and applications. *IEEE Commun. Surv. Tutor.*, 17, 4, 2347–2376, 2015.
4. Atzori, L., Iera, A., Morabito, G., The Internet of Things: A survey. *Comput. Netw.*, 54, 15, 2787–2805, 2010.

5. Bughin, J., Chui, M., Manyika, J., An executive's guide to the Internet of Things. *McKinsey Q.*, Mckinsey&Company, 2, 9, 89–105, 2015.
6. Gubbi, J., Buyya, R., Marusic, S., Palaniswami, M., Internet of Things (IoT): A vision, architectural elements, and future directions. *Future Gener. Comput. Syst.*, 29, 7, 1645–1660, 2013.
7. Kim, S. and Kim, S., A multi-criteria approach toward discovering killer IoT application in Korea. *Technol. Forecasting Social Change*, 102, 143–155, 2016.
8. Pieskä, S., Kaarela, J., Mäkelä, J., Simulation and programming experiences of collaborative robots for small-scale manufacturing. *2018 2nd International Symposium on Small-Scale Intelligent Manufacturing Systems, SIMS*, IEEE, pp. 1–4, 2018.
9. Holm, M. and Schnell, M., Challenges for manufacturing SMEs in the introduction of collaborative robots.
10. Wern, C.W., Ramulu, M., Shukla, A., Investigation of stresses in the orthogonal cutting of fiber-reinforced plastics. *Exp. Mech.*, 36, 1, 33–41, 1996.
11. El-Gallab, M. and Sklad, M., Machining of Al/SiC particulate metal matrix composites. *J. Mater. Process. Technol.*, 83, 1-3, 277–285, 1998.
12. Takeyama, H. and Iijima, N., Machinability of glassfiber reinforced plastics and application of ultrasonic machining. *CIRP Ann.*, 37, 1, 93–96, 1988.
13. Mahesan, K.V., Wongchai, A., Tamizhselvan, C., Manoj Kumar, S., Vijayakumar, P., Singh, B., Automation industry in cyber physical systems based on 4G IoT networks for operation controlling and monitoring management. *Optik*, 272, 170308, 2023, ISSN 0030-4026.
14. Song, F., Ai, Z., Zhang, H., You, I., Li, S., Smart collaborative balancing for dependable network components in cyber-physical systems. *IEEE Trans. Ind. Inform.*, 17, 10, 6916–6924, 2020.
15. Hussain, B., Du, Q., Sun, B., Han, Z., Deep learning-based DDoS-attack detection for cyber–physical system over 5G network. *IEEE Trans. Ind. Inform.*, 17, 2, 860–870, 2020.
16. Gushev, M., Dew computing architecture for cyber-physical systems and IoT. *Internet Things*, 11, Article 100186, 2020.
17. Li, M. and Wang, Z., An integrated robust replenishment/production/distribution policy under inventory inaccuracy. *Int. J. Prod. Res.*, 56, 12, 4115–4131, 2018.
18. Strange, R. and Zucchella, A., Industry 4.0, global value chains and international business. *Multinatl. Bus. Rev.*, 25, 3, 174–184, 2017.
19. Aheleroff, S., Xu, X., Lu, Y., Aristizabal, M., Velásquez, J.P., Joa, B., Valencia, Y., IoT-enabled smart appliances under industry 4.0: A case study. *Adv. Eng. Inf.*, 43, 101043, 2020.
20. Aheleroff, S., Philip, R., Zhong, R.Y., Xu, X., The degree of mass personalisation under industry 4.0. *Procedia CIRP*, 81, 1394–1399, 2019.
21. Zhong, R.Y., Xu, X., Aheleroff, S., Smart manufacturing systems for industry 4.0: a conceptual framework. *Proceedings of International Conference on Computers and Industrial Engineering, CIE*, pp. 10–31, 2017.

22. Yang, C., Lan, S., Shen, W., Huang, G.Q.G.Q., Wang, X., Lin, T., Towards product customization and personalization in IoT-enabled cloud manufacturing. *Cluster Comput.*, 20, 2, 1717–1730, 2017.
23. Gorecky, D., Schmitt, M., Loskyll, M., Zühlke, D., Human-machine-interaction in the industry 4.0 Era, in: *Proceedings - 2014 12th IEEE International Conference on Industrial Informatics, INDIN*, pp. 289–294, 2014.
24. Zhong, R.Y., Wang, L., Xu, X., *An IoT-enabled real-time machine status monitoring approach for cloud manufacturing*, vol. 63, pp. 709–714, Elsevier B.V., Procedia CIRP, 2017, ISSN 2212-8271, https://doi.org/10.1016/j.procir.2017.03.349
25. Datta, S.K., Vehicles as connected resources. *IEEE Veh. Technol. Mag.*, 12, 2, 26–35, 2017.
26. Mtshali, P. and Khubisa, F., A smart home appliance control system for physically disabled people. *2019 Conf Inf. Commun. Technol. Soc. ICTAS*, pp. 1–5, 2019.
27. Zhong, R., Dai, Q., Qu, T., Hu, G., Huang, G., RFID-enabled real-time manufacturing execution system for mass-customization production. *Robot. Comput.*, 29, 2, 283–292, 2013.
28. Zheng, P., Yu, S., Wang, Y., Zhong, R.Y., Xu, X., User-experience based product development for mass personalization: a case study. *Procedia CIRP*, 63, 2–6, 2017.
29. Stettler, J., Why Is China Going to Chicago?, 2017.
30. Colim, A., Morgado, R., Carneiro, P., Costa, N., Faria, C., Sousa, N. *et al.*, Lean manufacturing and ergonomics integration: defining productivity and wellbeing indicators in a human–robot workstation. *Sustainability*, 13, 4, 1931, 2021.
31. Verma, A., Bhattacharya, P., Madhani, N., Blockchain for Industry 5.0: Vision, Opportunities, Key Enablers, and Future, Directions. *IEEE Access*, 10, 69160–69199, 10 June 2022.
32. Epley, N., Waytz, A., Cacioppo, J.T., On seeing human: a three-factor theory of anthropomorphism. *Psychol. Rev.*, 114, 4, 864–886, 2007.
33. Lenz, C. and Knoll, A., Mechanisms and capabilities for human robot collaboration. *The 23rd IEEE International Symposium on Robot and Human Interactive Communication*, pp. 666–671, 2014.
34. White Paper Work 4.0 L., White paper work 4.0: White paper, Federal Ministry of Labour and Social Affairs, Berlin, 2017.
35. Simões, A.C., Soares, A.L., Barros, A.C., Factors influencing the intention of managers to adopt collaborative robots (cobots) in manufacturing organisations. *J. Eng. Tech. Manage.*, 57, Article 101574, 2020.
36. Galin, R., Meshcheryakov, R., Kamesheva, S., Samoshina, A., Cobots and the benefits of their implementation in intelligent manufacturing. *IOP Conf. Ser. Mater. Sci. Eng.*, 862, 3, Article 032075, 2020.

37. Frijns, H.A. and Schmidbauer, C., Design Guidelines for Collaborative Industrial Robot User Interfaces. *IFIP Conference on Human-Computer Interaction*, Springer, Cham, pp. 407–427, 2021.
38. Yuvethieka Sri, G.V., Balancing Assembly Line Using Collaborative Robots in Modern Manufacturing Industry under Improvements of Efficiency and Ergonomics Study. *Turk. J. Comput. Math. Educ. (TURCOMAT)*, 12, 12, 538–546, 2021.
39. Mohapatra, S.K., Mishra, S., Tripathy, H.K., Energy Consumption Prediction in Electrical Appliances of Commercial Buildings Using LSTM-GRU Model, in: *2022 International Conference on Advancements in Smart, Secure and Intelligent Computing (ASSIC)*, 2022, November, IEEE, pp. 1–5.
40. Sadik, A.R. and Urban, B., An ontology-based approach to enable knowledge representation and reasoning in worker–cobot agile manufacturing. *Res. Rep. - NIFS Ser.*, 9, 4, 90, 2017.
41. Djuric, A.M., Urbanic, R.J., Rickli, J.L., A framework for collaborative robot (CoBot) integration in advanced manufacturing systems. *SAE Int. J. Mater. Manuf.*, 9, 2, 457–464, 2016.
42. Matheson, E., Minto, R., Zampieri, E., Faccio, M., Rosati, G., Human–Robot Collaboration in Manufacturing Applications: A Review. *Robotics*, 8, 100, 2019, 10.3390/robotics8040100.
43. Ansari, F., Erol, S., Sihn, W., Rethinking Human-Machine Learning in Industry 4.0: How Does the Paradigm Shift Treat the Role of Human Learning? *Procedia Manuf.*, 23, 117–122, 2018.
44. Fast-Berglund, A., Palmkvist, F., Nyqvist, P., Ekered, S., Åkerman, M., Evaluating Cobots for Final Assembly. *Procedia CIRP*, 44, 175–180, 2016.
45. Iyer, S., how would cobots affect the manufacturing industry?, February 9, 2022, https://techfastly.com/collaborative-robots-to-becomecommon-place-as-automation-gains-momentum-in-manufacturing/. 78 Lefranc, G. *et al.*, *Procedia Comput. Sci.*, 214, 71–78, 2022 8 Lefranc, G., López-Juárez, I., Osorio-Comparán, R., Peña-Cabrera, M., *Procedia Comput. Sci.*, 2019.
46. Bogue, R., Europe continues to lead the way in the collaborative robot business. *Ind. Robot: Int. J.*, 43, 1, 6–11, 2016. Djuric, *et al.*, 2016.
47. Marvel, J., Collaborative Robots: A gateway info factory automation, National Institute of Standards and Technology, 2014, Accessed 27.01.2020 at https://tsapps.nist.gov/publication/get_pdf.cfm?pub _id=916485.

2

Cobots as an Enabling Technique for Industry 5.0: A Conceptual Framework

Swati Lipsa* **and Ranjan Kumar Dash**

Department of IT, Odisha University of Technology and Research, Bhubaneswar, Odisha, India

Abstract

Industry 5.0 provides human interventions in critical decision-making processes in different industries like manufacturing, healthcare, and supply chain management. The use of automated robots is a key component of Industry 4.0, which faces difficulty embedding human intellect whenever required. In contrast, cobots are collaborative robots with the inherent ability to integrate artificial intelligence and human intelligence for different decision-making processes. This redirection of human ingenuity into environments in which trained individuals engage with robots is important because market demand and customer needs are abstaining from large-scale production, and consumers are expecting significantly more customization and personalization of the goods they purchase. The involvement of humans facilitates the work of robots in various disruptive activities across several industries. The digital twin, edge computing, fog computing, the Internet-of-Everything, and cobots are the main enabling techniques for implementing Industry 5.0 in reality. Industry 5.0 and collaborative robots herald the dawn of a new era in robotics and industrial production. The success of Industry 5.0 is largely dependent on how well its enabling techniques, like cobots, have been implemented to solve real-time problems. The work carried out in this article is primarily concerned with the role of cobots in the successful implementation of Industry 5.0. The potential applications of cobots in various industries are explored. The industries of interest are the manufacturing industry, the aircraft industry, the healthcare sector, and the banking sector, to name a few. A comparison of the workings of robots and cobots is presented, highlighting the key differences between them. A conceptual cobot-based cyber-physical system is presented. The explicit use of human intelligence in different domains of application

**Corresponding author*: slipsait@outr.ac.in

V. Ramasamy, S. Balamurugan and Sheng-Lung Peng (eds.) Intelligent Robots and Cobots: Industry 5.0 Applications, (19–42) © 2025 Scrivener Publishing LLC

is explored. Finally, the risks and security threats to cobots are discussed, along with some mitigation techniques.

Keywords: Industry 5.0, cobots, industry 4.0, robots

2.1 Introduction

The Sustainable Development Goals have an overarching objective to foster long-term economic development by creating skilled jobs through technological advancement. Most developing and growing nations view high-technology manufacturing as a key to meeting these goals and combating the current wave of industrial change. Despite the fast-paced technological breakthroughs that have enabled high performance in the Fourth Industrial Revolution (Industry 4.0) [1, 2], the next industrial revolution, called Industry 5.0, will bring about balanced interactions between social and technological systems to provide mass customization of services and goods with a focus on individual preferences. Industry 5.0 necessitates a socio-technical development of human involvement in industrial systems, where all operational tasks of the supply chain will be handled using intelligent techniques and supported by cutting-edge communication technology [3].

One attempt to meet Industry 5.0 [4] with smarter production processes is the widespread adoption of collaborating robots (known as cobots). Worker productivity in manufacturing can be increased by utilizing collaborative robots and relying on human potential for higher-level activities that robots are unable to perform due to their lack of flexibility. Cobots' capacity to operate in tandem with humans in a high-technology environment is a significant benefit that could lead to their increased usage in the production sector. In light of the fact that cobots are meant to assist workers rather than replace them, workplace acceptance of cobots naturally becomes a pressing concern. Simoes *et al.* [5] carried out a preliminary study to determine the characteristics that influence managers' intentions to employ cobots in industrial companies, but they did not proceed to explore the real implementation stage. Furthermore, few studies have looked at the goals and reasons for success when introducing cobots into production processes.

For optimal efficiency in a smart workplace where cobots are used in collaboration with humans, it is essential to have a firm grasp on their potential interactions with one another. It is also essential to create a shared workspace for humans and robots, replacing the traditional model of a separate human and robot section of the workplace. The layout of a shared workspace will have an impact on the company's work procedures, making the allocation of tasks between humans and cobots a crucial issue.

This study intends to outline a comprehensive understanding of the interaction between humans and cobots in the Industry 5.0 scenario by applying a socio-technical approach. Because cobotic systems are still a novel idea, many questions remain unanswered, both in terms of technical expertise and practical implementation, as well as in terms of administrative oversight. Firstly, this article outlines the many different industries that have started implementing cobots. The idea put forth provides a way forward for the development of conceptual cobot-based cyber-physical systems in a production system through the utilization of a sequence of activities. In spite of the variety of cobots available, the processes outlined here are universally applicable because of the management-oriented foundations upon which they are built. Additionally, the chapter also covers the safety and security concerns associated with cobots as well as ways to alleviate those concerns.

2.2 Industry 5.0 at a Glance

In today's world, it is difficult to envisage life without technology and industry. The emergence of new industries was a crucial factor in human progress. Much like everything else in this universe, industries are also constantly evolving to become more efficient. This continuous industrial revolution made them productive and beneficial for us (as shown in Figure 2.1). The rise of the industrial sector made mass manufacturing possible, boosting both the economy and people's lifestyles. The current era of Industry 4.0 is steadily and rapidly evolving into Industry 5.0.

The term "Industry 5.0," also known as the "Fifth Industrial Revolution," refers to a new era of industrialization in which humans collaborate with cutting-edge machinery and artificial intelligence (AI)–driven robots to boost the productivity of a workspace [6]. This comes together with a stronger emphasis on sustainability, enhanced resilience, and a reorientation toward putting people first. This new aspect, which extends far beyond just manufacturing, is a continuation of the fourth industrial revolution (Industry 4.0) and is mainly driven by advances in information technology (IT) such as big data analytics, AI, the internet of things (IoT), robotics, machine learning, cyber-physical systems, virtualization, and so on. It represents a paradigm shift from materialistic goals to the value and happiness of a community as a whole. Although this approach has been discussed before, the concept of prioritizing people and the environment ahead of profits is now establishing a new focal point for a business.

Industry 5.0's primary benefit is its potential to generate higher-paying jobs by allowing businesses to offer more customized services to their clients and giving employees more leeway in the design of their products. As more of the production process is automated, human workers can devote their time and energy to providing customers with higher-quality, more personalized services and goods. Industry 4.0 marked the onset of this trend, but Industry 5.0 accelerated it by increasing automation and interaction to develop a service-based paradigm in which humans can concentrate on rendering better performance to end-users. It is also possible that the environmental benefits of the Fifth Industrial Revolution will outweigh the costs as businesses adopt systems that consume less energy and reduce or eliminate waste.

Eventually, the advancements of Industry 5.0 will be the realization of what the designers of Industry 4.0 could have only dreamed of at the beginning of the millennium (the 2010s). As AI gets better and factory robots acquire more human-like skills, the relationship between robots, computers, and employees will deepen and enrich everyone involved. Meanwhile, organizations have become quite dynamic and flexible as a result of a growing emphasis on sustainability and resilience, ultimately benefiting society in more ways than just minimizing any negative repercussions.

Thus, Industry 5.0 is both necessary and timely, as it presents promising potential to the following:

- **Enhance Human Skills:** Through seamless, real-time connections between humans and machines, Industry 5.0 will now be able to overcome workers' limited accessibility to remote locations. As mentioned in the work [7], Industry 5.0 contradicts the idea that connected things, such as workers, must communicate within the same building but rather promotes the idea of connectedness that extends far beyond the walls of the production environment.
- **Strengthen Customer Involvement:** Industry 5.0 will revolutionize how businesses interact with their customers by allowing them to bring their customers into their showrooms and factories in a truly immersive manner using virtual reality. Moreover, using digital twin technology, consumers can also keep a close eye on the whole manufacturing process, from initial product configuration and customization to delivery.

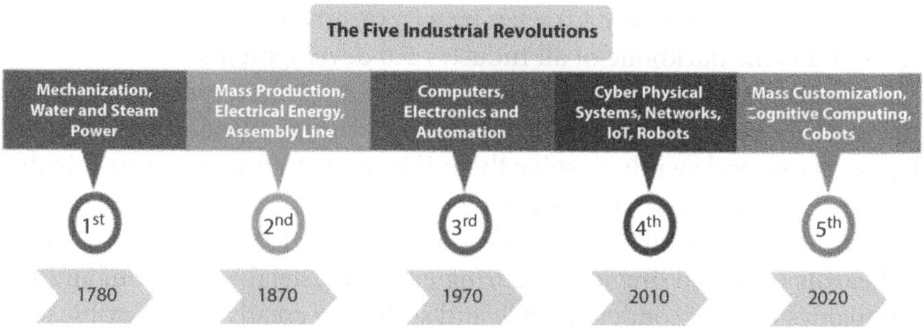

Figure 2.1 The industrial revolution over the years.

- **Increase Customization:** Industry 4.0's highly automated processes have the benefit of reducing errors, but they also limit the ability to adapt to changing consumer preferences. Workers in Industry 5.0 will take advantage of the existing machine-automated processes to complete mundane but potentially risky activities and then use the freed-up time to develop more unique and tailored offerings for their clients.
- **Move Toward Sustainable Solutions:** The cooperation of humans and machines can increase both current and past understanding of the production environment, resulting in more efficient energy management. Also, customizing products would help businesses make better use of their resources, cutting down on waste and excess output.
- **Reduce Downtime:** The Fifth Industrial Revolution will combine human knowledge with machine learning and digital twins to minimize production delays caused by malfunctioning industrial equipment. By implementing human–machine communications, a technical team can repair or replace the broken equipment by accessing the manufacturing unit remotely. This cuts down on costly downtime.

2.3 Industry 4.0 vs. Industry 5.0

The steam engine's brute force, the assembly line's efficiency, and the computer's lightning-fast processing capability each fueled a different phase of the industrial revolution. These eras of industrial development were termed "revolutions" owing to the revolutionary nature of the technology that drove them. Industry 4.0 relies heavily on advanced technological systems and is the backbone of all Industry 5.0 developments.

Unprecedented events over the previous decade have posed difficulties for the functioning of industry. In the midst of challenges, the industry needs to respond swiftly to solve them if it wants to preserve a sustainable approach to business. The industry observed a phenomenal shift in consumer interest toward customized goods and services around 2010. Such upheavals occurred previously during each industrial revolution, and the companies that responded quickly and shifted to creative and flexible procedures were the ones that ultimately succeeded.

Industry 4.0 recommends adapting to ever-evolving needs. The world is currently witnessing the beginning of Industry 5.0, the new industrial revolution. The collaborative nature of Industry 5.0 extends beyond that of its predecessor, Industry 4.0, to include increased human–robot teamwork, as presented in Table 2.1.

While the focus of Industry 4.0 is on facilitating communication between machines through cyber-physical systems, the focus of Industry 5.0 is on enhancing the human–machine interface (HMI) through the use of robots and collaborative robots. Furthermore, Industry 5.0 combines human ingenuity with robotic accuracy to produce a one-of-a-kind answer to the problems that will dominate the market in the coming decade. Thus, Industry 4.0 and 5.0 together provide guidelines for industries to succeed.

Table 2.1 Industry 4.0 versus Industry 5.0 from the perspective of robots and cobots.

	Industry 4.0	**Industry 5.0**
Objective	• Smart manufacturing • Increasing the efficiency of systems and resources	• Human centeredness • Social benefit • Sustainability
Human factors	• Repetitive actions because of automated systems • No human interference with robots • Complete virtual environment • Focuses on human-computer interaction	• More supervised operations by human intervention • Human centric decision-making system • Transitions back to an actual environment • Focuses on human–robot interaction

2.4 Key Differences Between Robots and Cobots

As robotics and other forms of factory automation become ubiquitous, it is not unusual to find such high-quality machines working on production lines. The two main categories of robot models, each with its own unique

Table 2.2 Key differences between robots and cobots.

Characteristics	Robots	Cobots
Adaptability and mobility	Robots are designed for a single, consistent task because they are bulky, difficult to maneuvre, and involve extensive programming.	Cobots are simple to program, sleek in design, and can be easily deployed, so they can rapidly reconfigure to accommodate line alterations and diverse part shapes.
Safety	Due to the fact that robots typically operate in confined work cells, these sensing traits are rarely present.	They offer advanced sensing capabilities to detect, avert, and abstain from collisions.
Automation	Robotic operations are entirely automated because there are no human interactions involved.	Unlike robots, cobots are unable to provide all the perks of complete automation.
Speed	Even with larger payloads, robots are able to operate far more quickly and precisely than humans.	Compared to robots, cobots are constrained in their speed, range, precision, and payload.
Accessibility	The acquisition cost of robots is high, and the procedures required for their integration are time-consuming and tedious.	Cobots are well-known for their cost and availability, especially for small industrial businesses.
Sensing Capabilities	Robots seldom have any other types of perception beyond vision.	Cobot can withstand a vast range of top-notch vision end-effectors and focus on a number of sensing technologies, such as energy and movement generated by workers.

set of advanced features, are traditional industrial robots and collaborative robots (Cobots). Although cobots share certain similarities with traditional robots, they are unique enough to earn their own classification. Cobots allow industries to carry out the same activities as robots, but with a lower initial investment, greater operational versatility, and fewer constraints on setup and operational range. This means that, generally, cobots cannot compete with conventional industrial robots in terms of reach, payload, and pace, but they have the advantage of being more compact and flexible in their applications. As a result, many smaller- and medium-sized organizations are turning to cobots as a less complicated and more affordable means of incorporating automation into their processes.

While both robots and cobots perform comparable tasks, the way in which they interact with humans is what sets them apart from each other. The primary differences between cobots and conventional industrial robots are typically found in these areas: cost, adaptability, and ease of operation. In order to comprehend the fundamentals and leverage the technological benefits of both collaborative robots and their industrial equivalents, it is necessary to be familiar with all of the characteristics that set them apart. The table mentioned above (Table 2.2) looks more closely at the distinctions between cobots and robots with respect to different characteristics.

2.5 Cobots as an Enabling Technique for Industry 5.0

Collaborative robots (cobots) are industrial robots with additional sensors. They are designed to put people at the center of the process, relieving them of monotonous, risky, or tedious activities. Cobots do not completely replace humans in manufacturing; rather, they function as a helpful addition to the human operators [8, 21]. Collaborative robots are designed with rounded corners and many sophisticated collision sensors to detect the presence of human workers and make operations more safe and convenient for everyone. These make use of collaborative robot arms that are equipped with force restrictions to ensure that accidental collisions do not cause any damage. These collaborative robots have a hand-guided device that lets a human control the robot's movement even when it is in automatic mode.

Cobots are becoming increasingly common in many industries, both conventional and unconventional, but they also provide some unique challenges. Cobots are similar to conventional robots; however, the success of the system will be determined during its initial implementation. Due to their unique design, which includes force-sensing technology, cobots occupy a special niche among robots.

Cobots are tuneable and can be programmed by their end users [22]. These cobots can be given the responsibility of doing tiresome, repetitive work so that humans can focus on more intellectually stimulating endeavours. The versatility and adaptability of robot and human collaboration are the real power behind industrial cobots.

Figure 2.2 depicts a few of the main advantages that cobots offer:

The past few years have witnessed the overwhelming expansion of cobots in many industries, with a growth rate of 45% [9]. The key motivation to use cobots in many industrial sectors is to enhance the capability of mass production with greater rates of flexibility and usability [10]. In many hazardous environments, the deployment of cobots can undoubtedly reduce the risk and harm that could happen to the human operator. Further, cobots can provide a sustainable working environment due to features like flexibility and mobility [11]. It is intuitive to believe that collaborative robots can work autonomously under the direct supervision of humans. In addition to this, the workspace can also be shared by cobots and human operators. The important factors pointed out in [12] provide a basis for the adoption of cobots in different industrial sectors like manufacturing and healthcare.

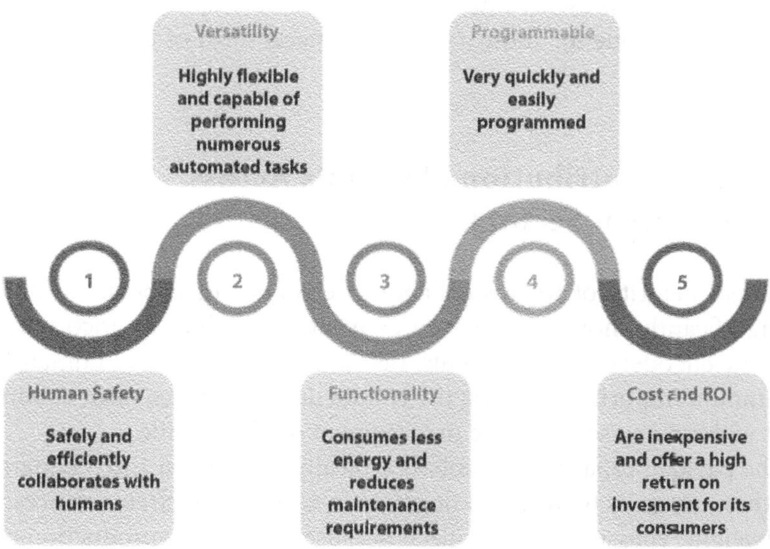

Figure 2.2 Primary benefits offered by cobots.

Table 2.3 Cobot benefits mapped to industry 5.0 traits.

Industry 5.0 cobot	Enhance human skills	Strengthen customer involvement	Increase customization	Move toward sustainable solutions	Reduce downtime
Versatility	Yes	Yes	No	Yes	Yes
Programmable	Yes	Yes	No	Yes	Yes
Human safety	Yes	Yes	No	Yes	Yes
Functionality	Yes	Yes	Yes	Yes	Yes
Consistency	Yes	Yes	No	Yes	Yes
Cost and ROI	No	No	Yes	No	No

Humans are responsible for taking key decisions at different levels in the working hierarchy, while cobots act as per the decisions taken by humans [13]. The integrated sensors of cobots make it possible to work more efficiently than traditional robots with a lower number of sensors. The adopted safety standards further open a working platform where cobots and human operators can interact with each other to accomplish some decisive tasks [14, 15]. In this regard, the prime challenge is the acceptance of cobots by human operators.

A one-to-one mapping among the features of Industry 5.0 and the attributes of cobot is presented in Table 2.3. This table upholds cobot as a key enabling technique for Industry 5.0 as it meets most of the requirements of Industry 5.0.

2.6 The Contribution of Cobots Across Different Sectors

The vast majority of today's collaborative robots, or cobots, are built to be significantly more precise than conventional robots by incorporating absolute encoders at critical joints. Because the protection of humans is of paramount importance during the cobot's design process, most cobots are built without any sort of cable management on the outside.

The payload capacities for cobots typically range from 1 to 5 kilograms, but they can go as high as 40 kilograms in exceptional circumstances. The lightweight types are widely utilized because of their low production costs, portability, and adaptability in a variety of sectors.

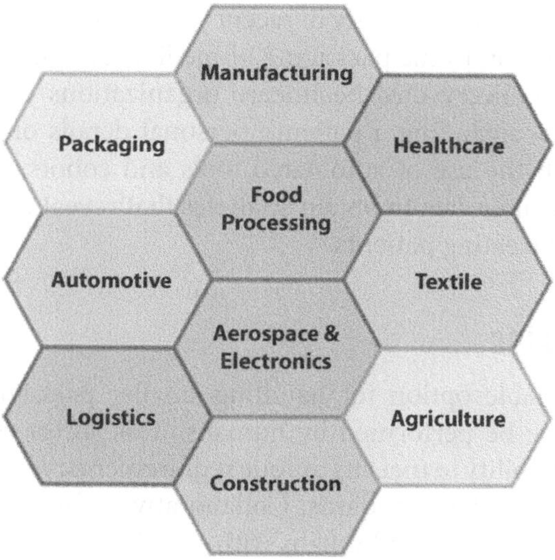

Figure 2.3 Application of cobots across different sectors.

While cobots have seen the most widespread use in the industrial sector [20] to increase efficiency and production, they are now being explored in other domains and in client-facing roles, including customer interaction, advertisement, and customer support. Some of its applications across different sectors are depicted in Figure 2.3.

A detailed discussion of the application of cobot in different industrial sectors follows next.

2.6.1 Manufacturing

The manufacturing sector is at the frontline of the collaborative robotic movement because it may have the greatest number of viable cobot applications. It is clear why this sector invests the most in collaborative robots. Cobots are widely used in the manufacturing industry to enhance output in response to increased demand, such as lifting heavy loads, handling fragile circuit boards and equipment parts, and improving inventory and supply-chain management. Furthermore, as cobots are more affordable than industrial robots, production costs for firms have greatly decreased.

2.6.2 Healthcare

As a result of their ability to integrate human interaction, cobots have found a useful application in the medical field. Its usage in the healthcare sector has advanced significantly in recent years, encompassing everything from lab automation to the packaging of medical devices to neurosurgery or even beyond. Every day, healthcare organizations collect enormous volumes of data, including a patient's personal details or their course of treatment. With the use of automated tools and cobots, medical centers may collect and alter data to obtain analytics that reveal useful insights for diagnosing and treating patients.

2.6.3 Packaging

Cobots are a viable option for handling smaller packaging chores that would otherwise be performed by humans. This sector relies on cobots because of their ability to meet its unique requirements, which include strict adherence to sanitation standards. Collaborative robots are employed to maximize output. They help humans with repetitive, delicate jobs in situations with temperatures ranging from 0°C to 50°C and low oxygen levels. Because of the ease with which cobots can be programmed, businesses all along the food supply chain can respond quickly to shifts in demand and increase the speed with which they deliver products. By helping medium and small businesses boost productivity and compete in a cutthroat market, they level the playing field and create a far more interesting workplace for their collaborators.

2.6.4 Aerospace and Electronics

Due to their necessity for high-precision automation, these sectors are strongly drawn toward collaborative robots. Cobots are becoming increasingly popular in this sector because of the benefits they provide in terms of speed, accuracy, and adaptability, as well as the minimization of ergonomic difficulties.

2.6.5 Textile

Cobots are useful in the textile industry for the same reasons that they are useful in other manufacturing sectors; their high level of accuracy makes them particularly well-suited to jobs such as cutting materials or shifting fabric components without folding or creasing them. They have progressed to the point that they can function safely alongside people, and their contributions to lowering the rate of human error have had a noticeable impact on productivity.

2.6.6 Agriculture

There are primarily two ways in which cobots excel over conventional robots in the agriculture sector. As a first benefit, their ability to control force is useful for agricultural tasks like harvesting. Because of the risk of spoilage, traditional robots are rarely employed in harvesting, but cobots can execute the work with ease because of their gentler touch. Second, they are far superior to robots in terms of safety and compatibility with humans and animals in the workplace.

2.6.7 Construction

One in five workplace fatalities happens in the construction business, making it one of the riskiest professions. Collaborative robots may come in handy and transform the construction sector by reducing risks to workers by performing dangerous operations such as pouring cement

2.6.8 Logistics

Warehouse workers, delivery personnel, and those involved in logistics are always rushing to meet tight deadlines. Cobots are widely used by logistics firms to automate administrative tasks like shipment scheduling and monitoring, invoice production, and customer support. The primary return on investment targeted here is the minimization of operational costs, a significant reduction in transit time in warehouses, and an improvement in productivity and efficiency.

2.6.9 Automotive

Due to the inevitable growing need for customization in automobiles, traditional industrial robots, which are often confined to single tasks and housed in cages, are being replaced with collaborative robots whose functions can be easily adapted to meet new demands.

2.6.10 Food Processing

Cobots are also finding their way into the kitchens of forward-thinking eateries, where they can be of great help in the preparation of food. Cooking meats and fries at just the right time is just one example of the many culinary tasks that may be handled by a well-designed cobot.

2.7 A Conceptual Cobot-Based Cyber-Physical System

A cobot-based physical system is designed as per the activities mentioned in [16] as a sequence of activities that involves a cyclic progression in the gathered knowledge of the specified problem as well as of the feasible solution. The first step of this cycle is the well-defined problem formulation, followed by a sequence of activities whose outcome is the definite design. The sequence of activities involved is presented in Figure 2.4.

The steps involved in the design of a cobot with reference to Zachman's solution strategy to solve an engineering problem are discussed below.

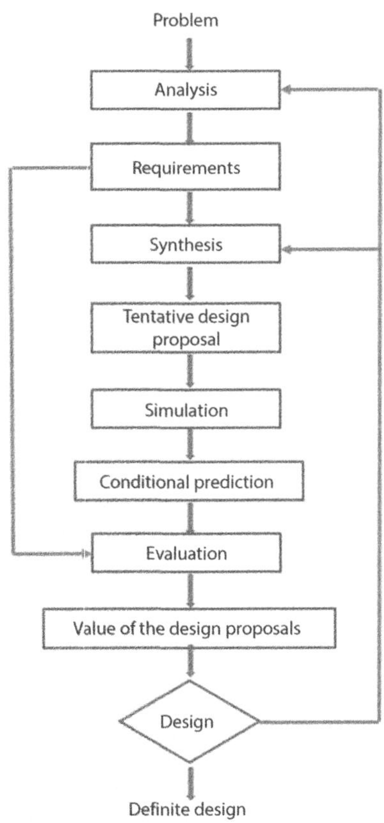

Figure 2.4 Workflow strategy for designing a cobot.

2.7.1 Need for Cobot: Problem Formulation and its Analysis

The domain-specific problem is first formulated, and, then, it is analyzed in terms of the risks, time, and budget constraints, as well as quality assurance. This involves the fundamental steps of the production system, which state that a product must be designed within a specific time and cost range without compromising its quality or degree of flexibility. The following are some of the attributes that must be analyzed:

- **Level of risks:** The risk factor must be judged from the initial design to the final deployment of the cobot at the production site. While the risk associated with its design is much less important than its deployment, it is mainly designed to assist human beings when the site of operations is hazardous and working here may be life-threatening for humans. At this time, cobots can accomplish many jobs without creating any physical barriers.
- **Time constraint:** The time from planning a product to its successful completion can be measured. It is an important factor that determines how a cobot can be helpful in meeting the time constraints, if any, imposed in the problem statement. Also, the degree of automation can be estimated here by considering time as one key ingredient. With no time restrictions, one may choose a non-automated production system unless it is infeasible in terms of cost.
- **Budget constraint:** The overall cost involved in designing a cobot necessitates the inclusion of hardware components and software. Its design will be beneficial if its utilization exceeds its cost. Also, the volume of production is another factor that cannot be avoided when calculating the overall cost. If the production persists for a long time with a satisfactory number of products, then the cobots are preferred, even if initially they seem to incur more cost.
- **Quality assurance:** Both the products' physical appearance and operational quality matter most when judging them. Accordingly, the cobot must be designed with greater precision and accuracy so that it can meet the quality in question.

2.7.2 Synthesizing Cobot: Characteristics of Design

The following characteristics of the product are considered while designing the cobot:

- Dimension
- Shape
- Material
- Mass production

2.7.3 Cobot Selection

The selection of a cobot is multiple-criteria decision-making problem. A series of evaluations is a prerequisite to the selection of a cobot. The different parameters for evaluation and selection of a cobot that best suits an application depend mainly on the following:

- the type of function that cobot will perform
- the level of human interaction
- the degree of its flexibility in the working environment
- the economical perspective of its design

2.7.4 Selection of the Gripper

The selection of grippers bears paramount importance in the design of the cobot, as they are the main functional part of the cobot while operating over physical objects. The different available grippers are listed below (Table 2.4).

2.7.5 Tentative Design Proposal, Simulation, Conditional Prediction, and Evaluation

The tentative design of a cobot is developed by considering the factors discussed earlier. It specifies the simulation environment and the proposed function that the cobot can perform. This is refined again and again to make it more quality-specific. The tentative design is simulated by a suitable programming language or software to make a digital model of the proposed cobot. It is tested and validated to check its functionality, accuracy, precision, and other factors confined to the manufacturing of the product. The basic idea behind the digital model of cobots is to ease remodelling or refining its functionality unless it performs as desired.

Table 2.4 Cobotic grippers and their application.

Types of grippers	Application	Example
Grippers with two fingers	Many industrial products	On-Robot and Robotiq
Grippers with three fingers	Delicate objects with strength and precision	Robotiq for UR Kinova Robo Cylinder from IAI
Grippers with flexible fingers	More fit to pick up different objects	Soft Robotics Gripper Robotic Gecko Gripper
Grain-filled flexible ball	To pick up delicate objects	Jamming Gripper
Grippers for palletizing	To pick up a heavy load	Parallel Gripper Vacuum Gripper
Semi-professional 3D printed robot grippers	Instantly manufactured by the user to meet his requirements	Flexy Hand Robohand

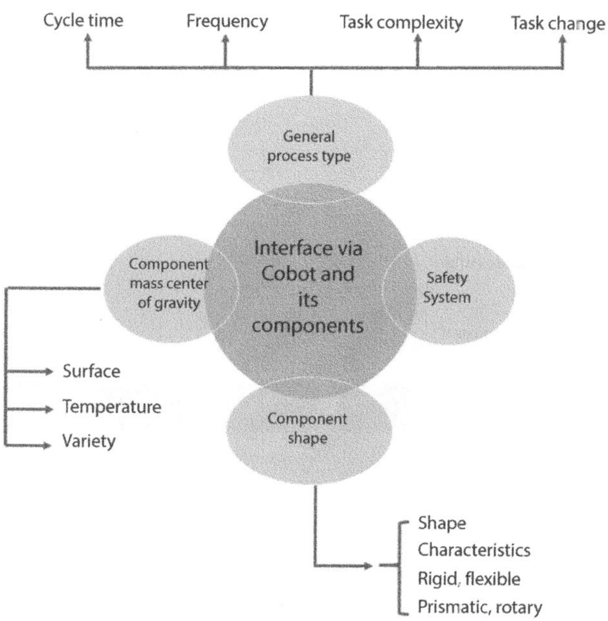

Figure 2.5 The key attributes of a cobot for HMI.

The design of the cobot as an engineering problem results in a list of some key attributes for the HMI (Figure 2.5). Further, it also drives the design of cobot as a multi-perspective system view point.

2.7.6 Design of Cobot as a Multi-Perspective System Viewpoint

Cobot provides an automation system with human interaction for high-level decision-making capabilities [23–26]. Thus, a multi-perspective systems viewpoint should be adopted for the representation of requirements and mappings. A modified framework of Zachman [16] and [17] is shown in Figure 2.6.

The matrix-like structure of the conceptual framework for the design of cobot represents multiple perspectives vertically, while various elements of this framework are classified horizontally. The multi-perspective system viewpoint consists of the following:

- **System viewpoint:** It consists of contextual information followed by conceptual information. The basic information associated with this viewpoint is the background or situational information extracted from the detailed analysis of

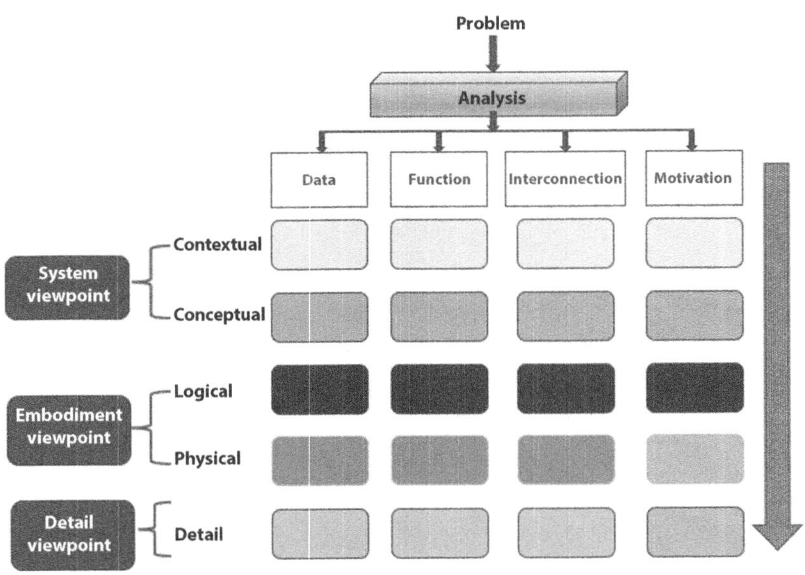

Figure 2.6 A layered-wise conceptual framework for the design of a cobot.

the problem statement. The basic motivation, the desired level of interaction, i.e., the HMI, and the working environment of the cobot are defined here. Physical design involves the design of simulation models, testing, validation, and prediction before their actual real-time deployment.

- **Embodied viewpoint:** The logical design and the physical design are the main components of this viewpoint. The functional requirements and their feasible simulation strategies are defined from this viewpoint.
- **Detail viewpoint:** The detail viewpoint is associated with a detailed specification of the cobot along with the risk and system requirements.

The selection and setup of machines are aided by performance metrics such as cycle length, frequency, the desired amount of flexibility, task complexity, and so on. This, in turn, affects the size of the workforce, the nature of workers' physical labor, and the types of decisions they take.

2.8 The Risk and Security Issues with Respect to Cobots and Their Mitigations

The HMI provides a workspace for the human operator to operate the cobots [18]. Hence, the risk associated with this workspace must adhere to safety standards like ISO TS 15066: Robots and robotic devices, collaborative robots, which provide guidelines to design and implement a collaborative robot workspace with reduced risks [19]. The key idea of this standard states that "if contact between robots and humans is allowed and incidental contact does occur, then that contact shall not result in pain or injury." This standard defines primarily two types of contact between humans and the cobot: transient and quasi-static. In transient contact, the human has the option to move away from the collaborative robot workspace, while, in quasi-static contact, the human gets trapped in the collaborative robot workspace, yielding a high level of risk that may be fatal.

The force of contact in a collaborative robot workspace must be limited by the level of human pain. Accordingly, the cobots must be designed so that any contact, irrespective of its type, between them and the human will be much less than the pain threshold. Because quasi-contact results in a high level of risk, the threshold of pain for this type of contact is normally lower than that of transient contact.

The risk assessment of a collaborative robot workspace is measured by the pilz hazard rating (PHR) value, which is defined as follows:

$$PHR = DPH \times PO \times PA \times FE$$

where DPH is the degree of possible harm, PO is the possibility of occurrence, PA is the possibility of avoidance, and FE is the frequency of exposure.

The different PHR ranges, along with the associated risks [19], are presented in Figure 2.7.

In order to mitigate the risks associated with the workspace and the resultant harm to humans, the following four mitigation techniques are adopted by different industries.

2.8.1 Safety-Rated Monitored Stop

Under this type of interaction, the cobots operate autonomously in the workspace. The human operator initiates a safety-rated, monitored stop before entering the workspace. When the human operator is in the workspace, the cobot remains powered but stationary. The cobot resumes its work when the human operator exits the workspace. A precautionary safety measure is provided with the system when a human operator enters the workspace without initiating a safety-rated monitored stop, such as a protective step that will shut down the whole power system.

PHR	Risk	Comment
1-10	Negligible Risk	Presents practically no risk to health and safety, no further risk reduction measures are required.
11-20	Very Low Risk	Presents very little risk to health and safety, no significant risk reduction measures are required, may necessitate the use of personal protective equipment and/or training.
21-45	Low Risk	Risk to health and safety is present, but low. Risk reduction measures must be considered.
46-160	Significant Risk	The risk associated with the hazard is substantial enough to require risk reduction measures. These measures should be implemented at the next suitable opportunity.
161-500	High Risk	Potentially dangerous hazard, which requires risk reduction measures to be implemented urgently.
501+	Very high Risk	Risk reduction measures should be implemented immediately, corporate management should be notified.

Figure 2.7 Cobotic application risk evaluation using the Pilz Hazard Rating.

2.8.2 Hand-Guiding

In a hand-guiding mechanism, the operator first initiates a safety-rated, monitored stop, thereby making the cobot stationary. Then, the hand of the cobot is repositioned as necessary for the next operation.

2.8.3 Speed and Separation Monitoring

In this scenario, the operator and the cobots are allowed to work together in the workspace by maintaining a minimum protective separation distance at all times.

2.8.4 Power and Force Limitation

This approach is suitable when the interaction between the human operator and the cobot is very high.

2.9 Conclusion

The rapid pace of modern technological advancement necessitates that manufacturers constantly adjust their production strategies to incorporate cutting-edge tools in order to keep up with consumer demands. Industry 5.0 is highly accessible to organizations, so they must be prepared to move in that direction. The advent of Industry 5.0 promises to bring about significant changes in the way factories operate, ultimately leading to increased efficiency and output. Due to this, factories need to adopt and employ autonomous workforces to ensure the efficient operation of their factories. Human workers are still valued in the workplace, and they are expected to keep pace with rapid advancements through creativity and collaboration. Therefore, there are many long-term benefits to incorporating cobots into industrial systems, as these autonomous workforces allow for more innovative and cost-effective systems by enhancing the capabilities of workers.

This chapter offers a comprehensive view of the interaction between humans and cobots in the industrial environment, bringing a socio-technical approach to Industry 5.0 that helps address the aforementioned issues. The idea underlying cobots is that a more accurate metaphor for teamwork is shared control over motion rather than an increase in human strength. As a result, developments in production, ergonomic design, and the integration of computers and information systems with the wide variety of tasks that employ human abilities can be anticipated.

This study adds to the existing literature by presenting a conceptual framework for cobot acceptability from an organizational perspective. The focus of this work is on the potential impact of behavioral intention on the real applications of cobots and, more broadly, on perceived work performance enhancement within the business world.

Furthermore, in the future, cobots can enhance a wide range of workflows across different industrial sectors and supply important data insights, such as predictive analytics and advanced recommendations, by connecting to other gadgets, machines, network databases, etc. Using cobots in manufacturing can give organizations a real competitive advantage in today's fast-paced, high-stakes market.

As the process of industrial transformation continues, it is likely that cobots will be used alongside a more competent, human-led workforce. The expected outcome is a significant boost in productivity and the retention of important personnel through the provision of avenues for professional growth. Additional future studies may explore approaches to enhance the deployment of cobots through: (i) making the most of concurrent advancements in AI, mobile technology, and IoT; (ii) fostering more effective human–robot interaction; and (iii) incorporating dynamic workflow planning with cobot operation.

References

1. Stock, T., Obenaus, M., Kunz, S., Kohl, H., Industry 4.0 as enabler for a sustainable development: A qualitative assessment of its ecological and social potential. *Process Saf. Environ. Prot.*, 118, 254–267, 2018.
2. German, F.A., Dalenogare, L.S., Ayala, N.F., Industry 4.0 technologies: Implementation patterns in manufacturing companies. *Int. J. Prod. Econ.*, 210, 15–26, 2019.
3. Bednar, P.M. and Welch, C., Socio-Technical Perspectives on Smart Working: Creating Meaningful and Sustainable Systems. *Inf. Syst. Front.*, 22, 2, 281–298, 2019.
4. Nahavandi, S., Industry 5 . 0 — A Human-Centric Solution. *Sustainability*, 11, 4371, 2019.
5. Simões, A.C., Soares, A.L., Barros, A.C., Factors influencing the intention of managers to adopt collaborative robots (cobots) in manufacturing organizations. *J. Eng. Tech. Manage.*, 57, 101574, 2020.
6. Prassida, G.F. and Asfari, U., A conceptual model for the acceptance of collaborative robots in industry 5.0. *Procedia Comput. Sci.*, *197*, 61–67, 2022.
7. Adel, A., Future of industry 5.0 in society: human-centric solutions, challenges and prospective research areas. *J. Cloud Comput.*, *11*, 1, 1–15, 2022.

8. Tiwari, S., Bahuguna, P.C., Walker, J., Industry 5.0: A Macroperspective Approach, in: *Handbook of Research on Innovative Management Using AI in Industry 5.0*, IGI Global Publishing Tomorrow's Research Today, pp. 59–73, IGI Global, 2022.
9. Bi, Z.M., Luo, M., Miao, Z., Zhang, B., Zhang, W.J., Wang, L., Safety assurance mechanisms of collaborative robotic systems in manufacturing. *Rob. Comput. Integr. Manuf.*, 67, 102022, 2021.
10. Calitz, A.P., Poisat, P., Cullen, M., The future African workplace: The use of collaborative robots in manufacturing. *SA J. Hum. Resour. Manag.*, 1, 2, 1–11, 2017.
11. Kopp, T., Baumgartner, M., Kinkel, S., Success factors for introducing industrial human-robot interaction in practice: an empirically driven framework. *Int. J. Adv. Manuf. Technol.*, 112, 685–704, 2021.
12. Sadok, M. and Welch, C.E., Achieving Sustainable Business Systems Through Socio-Technical Perspectives, in: *Proceedings of the 27th European Conference on Information Systems (ECIS)*, Stockholm & Uppsala, Sweden, 2019.
13. Aaltonen, I. and Salmi, T., Experiences and expectations of collaborative robots in industry and academia: Barriers and development needs. *Procedia Manuf.*, 38, 1151–1158, 2019.
14. Kildal, J., Tellaeche, A., Fernández, I., Maurtua, I., Potential users' key concerns and expectations for the adoption of cobots. *Procedia CIRP*, 72, 21–26, 2018.
15. Kardush, S.K. and Wong, E., A Techno-Economic Study of Industry 5.0 Enterprise Deployments for Human-to-Machine Communications. *IEEE Commun. Mag.*, 60, 12, 74–80, December 2022, doi: 10.1109/MCOM.001.2101068.
16. Zachman, J., Enterprise Architecture, Zachman International©, url: http://www.zachmaninternational.com/default.htm, accessed 30 Jan. 2016, 2002.
17. Djuric, A.M., Urbanic, R.J., Rickli, J.L., A framework for collaborative robot (CoBot) integration in advanced manufacturing systems. *SAE Int. J. Mater. Manuf.*, 9, 2, 457–464, 2016.
18. Bragança, S., Costa, E., Castellucci, I., Arezes, P.M., A brief overview of the use of collaborative robots in industry 4.0: Human role and safety, in: *Occupational and Environmental Safety and Health*, P. Arezes, *et al.* (Eds.), Springer, Cham, 2019.
19. https://manplusmachines.com/cobot-risk-assessment/
20. Borboni, A., Reddy, K.V.V., Elamvazuthi, I., AL-Quraishi, M.S., Natarajan, E., Azhar Ali, S.S., The Expanding Role of Artificial Intelligence in Collaborative Robots for Industrial Applications: A Systematic Review of Recent Works. *Machines*, *11*, 1, 111, 2023.
21. Michaelis, J.E., Siebert-Evenstone, A., Shaffer, D.W., Mutlu, B., Collaborative or simply uncaged? understanding human-cobot interactions in automation, in: *Proceedings of the 2020 CHI Conference on Human Factors in Computing Systems*, 2020, April, pp. 1–12.

22. Galin, R., Meshcheryakov, R., Kamesheva, S., Samoshina, A., Cobots and the benefits of their implementation in intelligent manufacturing. *IOP Conf. Ser.: Mater. Sci. Eng.*, 862, 3, 032075, 2020, May, IOP Publishing.
23. Faccio, M., Granata, I., Menini, A., Milanese, M., Rossato, C., Bottin, M., Rosati, G., Human factors in cobot era: a review of modern production systems features. *J. Intell. Manuf.*, *34*, 1, 85–106, 2023.
24. Paliga, M., Human–cobot interaction fluency and cobot operators' job performance. The mediating role of work engagement: A survey. *Rob. Auton. Syst.*, *155*, 104191, 2022.
25. Chromjakova, F., Trentesaux, D., Kwarteng, M.A., Human and cobot cooperation ethics: The process management concept of the production workplace. *J. Compet.*, 13, 21–38, 2021.
26. Schmidbauer, C., Hader, B., Schlund, S., Evaluation of a digital worker assistance system to enable adaptive task sharing between humans and cobots in manufacturing. *Procedia CIRP*, *104*, 38–43, 2021.

3

Role of Cobots and Industrial Robots in Industry 5.0

M. Makeshkumar[1]*, M. Sasi Kumar[2], J. Anburaj[3], S. Ramesh Babu[1], E. Vembarasan[1] and R. Sanjiv[1]

[1]*Department of Mechanical Engineering, KPR Institute of Engineering and Technology, Coimbatore, Tamil Nadu, India*
[2]*Department of Aeronautical Engineering, KIT-Kalaignarkarunanidhi Institute of Technology, Coimbatore, Tamil Nadu, India*
[3]*Department of Metallurgical Engineering, PSG College of Technology, Coimbatore, Tamil Nadu, India*

Abstract

The collaborative robot (cobot) takes the concept of combining robot endurance with human flexibility and problem-solving ability to a new level. Robots designed to share a workspace with humans are called cobots. Cobots are robots intended for direct human–robot interaction in shared spaces or where humans and robots are nearby. Cobot applications contrast with traditional industrial robotic applications, where the robot is isolated from human contact. Collaborative robots offer the greatest value in situations where humans need to be near the robot. Direct interaction between humans and robots may be both the greatest advantage and the greatest limitation of collaborative systems, depending on how it affects human factors such as ergonomics and psychological stress. Recent case studies have already demonstrated the successful use of collaborative robotics in various industries; for example, metal and machining; machine control, quality control, material handling, and waste science research; automotive quality control; and material handling and machine control. Finally, we outline future directions for cobots in industrial collaboration scenarios, including potential points for expansion and improvement.

Keywords: Cobots, industry 5.0, human-robot integration, flexible automation, AI integration robots, industrial robots

**Corresponding author*: makeshmech@gmail.com

V. Ramasamy, S. Balamurugan and Sheng-Lung Peng (eds.) Intelligent Robots and Cobots: Industry 5.0 Applications, (43–64) © 2025 Scrivener Publishing LLC

3.1 Introduction

The emergence of Industry 4.0 brought about a significant change in the way that manufacturing was approached. Automation, the Internet of Things (IoT), and artificial intelligence (AI) helped manufacturers streamline their operations, reduce costs, and increase efficiency [1]. However, as the industry progressed, it became apparent that human workers were still needed in many areas, such as complex assembly tasks, quality control, and problem-solving. This led to the development of collaborative robots, or cobots, that work alongside humans in a shared workspace. Cobots are robots designed to interact with humans safely and productively. They are equipped with sensors and other safety features that allow them to operate without fencing or other barriers that traditional robots require. This makes them ideal for use in small- or medium-sized businesses, where they can help workers with repetitive or dangerous tasks, freeing up human workers to focus on more complex or creative work [1–3]. Cobots have become increasingly popular in recent years, as manufacturers have recognized the benefits they can provide. They are highly versatile and can be programmed to perform a wide variety of tasks, including assembly, pick-and-place, welding, and painting. They can also be integrated with other equipment, such as conveyors or vision systems, to create a fully automated production line.

Industry 5.0 takes this a step further, bringing the focus back to human workers and their skills. In Industry 5.0, humans and cobots work together in a symbiotic relationship, with each complementing the other's strengths [2]. Cobots can help humans with physical tasks, such as lifting and carrying, whereas humans can provide the cognitive skills needed to solve complex problems and make decisions. One of the main benefits of Industry 5.0 is the increased flexibility that it provides. Cobots can be easily reprogrammed to perform different tasks, allowing manufacturers to quickly adapt to changing market conditions. They can also be used in a wide variety of industries, from automotive to food processing, and can work in almost any environment. Another benefit of Industry 5.0 is the increased safety that it provides. Cobots are designed to work alongside humans, and they are equipped with sensors and other safety features that prevent them from causing harm. This reduces the risk of accidents and injuries, making the workplace safer for all workers. Industry 5.0 also has the potential to increase productivity and efficiency [4, 5]. Cobots can work 24/7, without breaks or rest, allowing manufacturers to increase output without adding additional staff. They can also work faster and more accurately than humans, reducing the time and cost of production.

3.2 Role of Cobots

The goal of Industry 5.0 is to develop more intelligent industrial systems. By utilizing cobots, manufacturing operations can be enhanced to eliminate low-value-added jobs and utilize worker potential for more difficult tasks. Cobots are intended to put the individual at the center of the process and free up workers from dangerous activities [6]. Humans are still engaged in the production process because cobots support operators rather than replace them. Robots and people can work together in the same area. It is meant to remain at the spot that was picked out for it during the first installation. The mechanical arm of a cobot, like that of a standard industrial robot, is programmed to carry out a variety of activities in a manufacturing environment, including material handling, assembly, process duties, quality inspection, and packing which is shown in Figure 3.1. The production process is sped up by an industrial robot in part because it works non-stop. Breaks and shift changes are required for a cobot [5]. Collaborative robots are used by manufacturers of lighting, mobile phones, computers, speakers, and other products manufacturers. Collaborative robots are favored for quick implementation because they can be programmed by nonexperts and are simpler to operate than traditional robots. Cobots can increase productivity and efficiency, can upgrade the quality of the product, can increase cost savings, can maintain advanced data analytics, and can reduce waste.

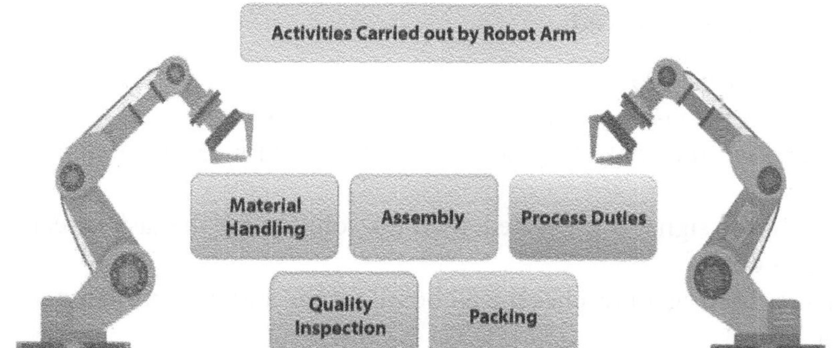

Figure 3.1 Activities carried out by cobot arm.

3.3 Programming Flowchart

- Start the cobot and establish a connection to the programming software.
- Initialize the cobot's position and set any necessary variables or parameters.
- Move the cobot to its starting position.
- Begin the main loop of the program.
- Check for any inputs or triggers that may require the cobot to perform a specific action.
- If an input or trigger is detected, then execute the corresponding action.
- If no input or trigger is detected, then continue to the next step.
- Check for any errors or faults in the cobot's sensors or hardware.
- If an error or fault is detected, then halt the program and notify the operator.
- If no errors or faults are detected, then continue to the next step.
- Check the cobot's position and orientation.
- If the cobot has not yet completed its task, then calculate the next move based on the desired trajectory and execute it.
- If the cobot has completed its task, then move it to its end position and halt the program.
- End the main loop and disconnect the cobot from the programming software.

3.3.1 Steps Involved

Step 1: Start the program.
Step 2: Get the task to be done by the robot as input.
Step 3: Then the control signals for the desired task are generated by the in board.
Step 4: If the signal is generated, then the solution for the task is searched and gets done.
Step 4.1: Then again reaches the NI board for the authentication of the task done.
Step 4.1.1: If the task is done right, then no new signals are generated.

Step 4.1.2: If the task is done partially are not completed, then new signals are generated for the pursuit of the task and again step 4 is checked.
Step 5: Display the result and stop the program.

The overall programming flowchart of Cobot is represented in Figure 3.2.

3.4 Objectives of Research in Cobots

Cobots intend to enhance human–robot collaboration. Instead of changing a person, cobots may collaborate with humans, resulting in better manufacturing and performance. Cobots are a perfect productivity device for nearly any manufacturer because they assist all organization members to achieve performance goals [7]. While running with robots,

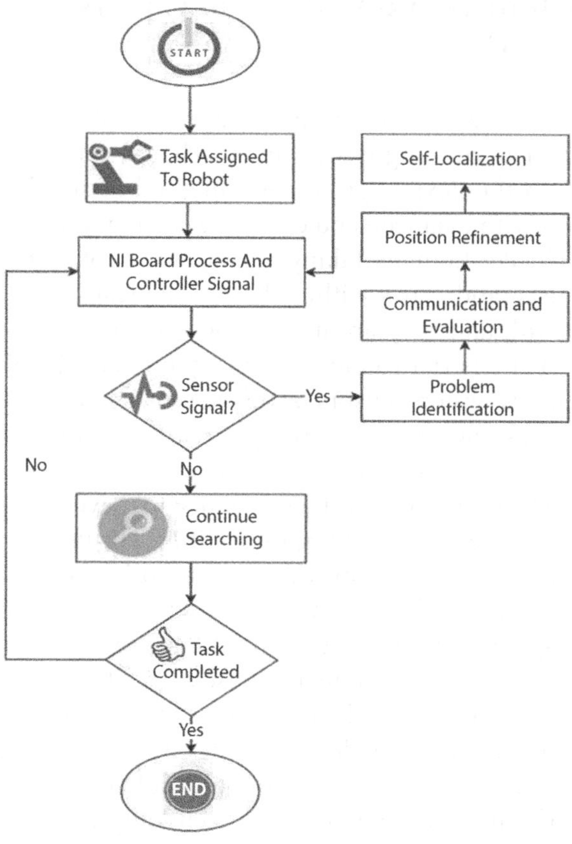

Figure 3.2 Programming flowchart.

generally, the operator loads elements to a fixture, conveyor, or turntable. Next, the operator leaves the space of the painting, and the robotic does its work. With collaborative robots, the operator can supply the component without delay to the robotic and by no means has to depart the workspace, lowering time, price, and ground area. With these days' collaborative technology, a massive number of applications see a return on investment in less than 365 days when they invest in cobots [7]. Due to the fact that those robots can be co-placed to paintings directly next to their human counterpart, there are fewer boundaries to work around and much less lost time. The flexibility and re-programmability of cobots cause them to be more investment-worth because they can be programmed to have numerous applications in one kind of sector, ranging from production to healthcare.

3.5 Capabilities and Features of Cobots for Industrial Applications

Industrial robots can be programmed again in step with the changing needs of their running environment. Hence, they can play a mass function within the flexible production method of small batches and a couple of sorts with balanced and excessive efficiency. Cobots have components much like humans taking walks, waist rotation, large palms, forearms, wrists, and claws within the mechanical structure as shown in Figure 3.3 [8], and they are used to be controlled using computers. Further, clever industrial robots have many human-like biological sensors, load sensors, vision sensors, acoustic sensors, language, and many other features as shown in Figure 3.4. Sensors improve the adaptive ability of commercial robots to the environment. Similar to particularity designed committed business robots to have higher versatility while appearing one-of-a-kind responsibilities. For instance, converting the give-up manipulators of an industrial robotic can perform distinct responsibilities. The mechanics, microelectronics, and mechatronics eras, particularly laptop technology combine, comprise the business machine age, encompassing a wide number of disciplines [9]. The advancement of the robotics generation will inevitably support the growth of other technologies because of their close ties to utility; additionally, the level of development and application of robotics technology can serve as a gauge for the state of commercial and national technology.

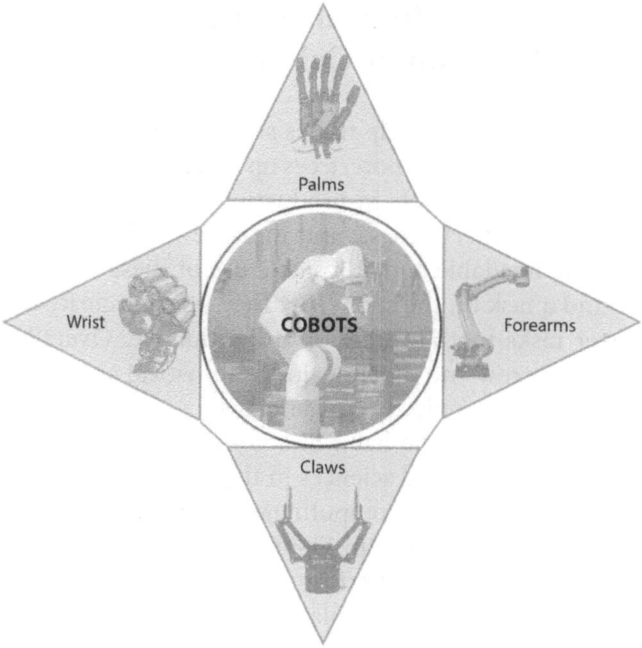

Figure 3.3 Components of a cobot.

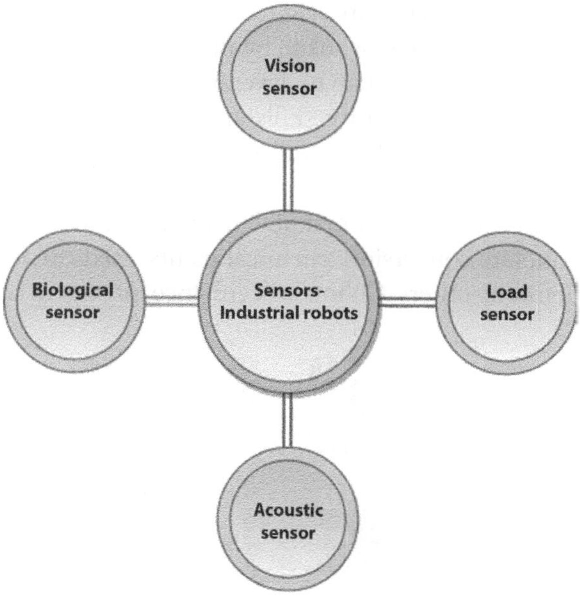

Figure 3.4 Sensors in robots.

3.6 Industrial Developments and Different Degrees of Collaboration by Cobots

Collaborative machines can help trades scale up and mechanize differing parts of result, which, furthermore, frees up more scope for detached work. The bottom line is that cobots in production can enhance product quality control, increase effectiveness, and increase results. Furthermore, the AI admits to the bendable and adjustable use of these machines, so they can forever find a task on the spot [10]. They boost security for operators by taking over tasks that are inferior and seductive. Overall, cobots are ideal for SMEs (narrow and medium-judge activities) as they can improve human work and are completely cost-effective.

Industry 4.0 concreted the habit of automation and smart production stimulated by data-compelled sciences in the way that IoT, cyber-material wholes, wearables, AR, cloud computing, machine intelligence, intelligent estimating, etc., and it is shown in Figure 3.5. Industry 4.0, which is only focused on enhancing process performance through tangible and statistical integration, erroneously downplays the significance of active human value creation [11]. By fusing the roles of mechanical components and human laborers in manufacturing, Industry 5.0 re-shifts the primary focus to human advantage. The very organization of the upcoming technological revolution, or Industry 5.0, is combatted by this shape. Cobots, which have its headquarters in Denmark, are said to be the foundation of Industry 5.0. Cobots make the same done or made by machine facilities, thereby being a part of a private tool that may be leveraged by some appendage of the workforce to request imaginative skills and create more profit. Different degrees include physical guidance, full autonomy shared control, and joint control (Figure 3.6). Cobots can be utilized as a plug-and-play solution for a variety of mechanical and manufacturing processes, such as automotive production, fare conversion, chemical plants, medical devices, and kit manufacture, among others. Cobots will enhance intelligent accountability,

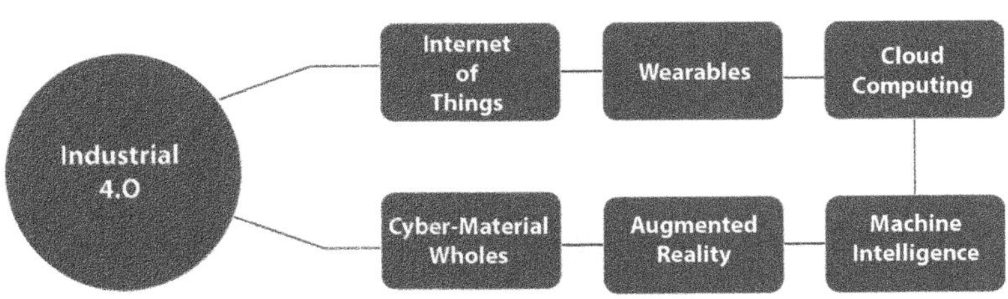

Figure 3.5 Industrial 4.0.

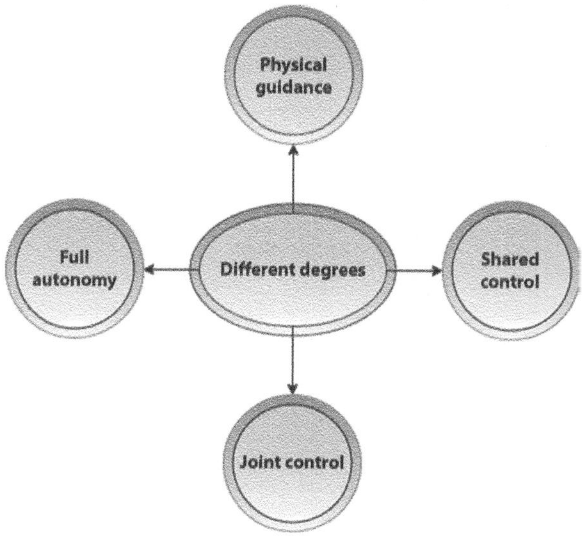

Figure 3.6 Different degrees of cobots.

drive superior products to the market, permit mass customization and personalization, develop production costs, create new job parts, and boost in essence education to consume cooperative machine intelligence because they collaborate well with humans in a trustworthy environment.

Production businesses must create use cases that demonstrate the reliability of human–machine interaction as electronics mature. Proof of concept and use cases that illustrate how or in what ways cobots might help the industry are developed to help [12]. The experts may create your cobot management operating system or mobile app for or modify an existing spreadsheet to meet changing retail demands. Fingent can operate with cobot hardware makers to create a sound program written in music that will be able to guide the development of cobots.

Universal robots and cobots are second-hand by manufacturers of ignition, movable phones, speakers, computers, and more. Because cooperative androids are smoother to program than usual modern machines, they can be registered by non-masters and are chosen for fast exercise.

3.7 Cobot Applications

Collaborative robots, or "cobots," are becoming increasingly popular in a wide range of industries due to their flexibility, safety, and efficiency [9].

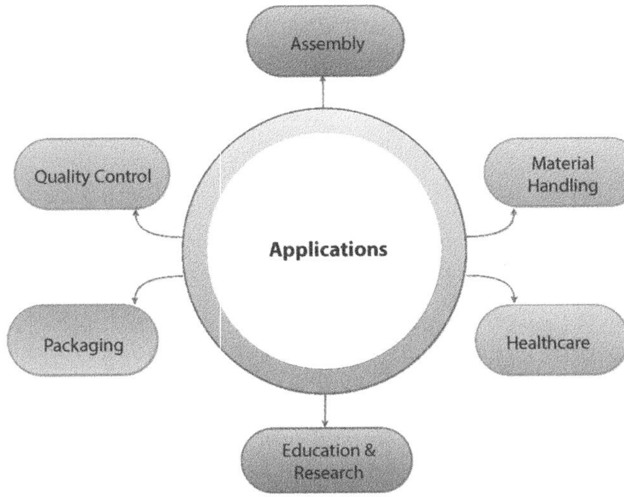

Figure 3.7 Applications.

Here are some examples of how robots are being used in different applications. Figure 3.7 shows the different applications of cobots.

3.7.1 Assembly

Cobots can be used for a wide range of assembly tasks, from inserting components to assembling complex products. They can work alongside human workers to increase productivity, accuracy, and quality. Cobots can also be programmed to perform tasks that are repetitive or require a high degree of precision, freeing up human workers to focus on more complex tasks.

3.7.2 Pick and Place

In many industries, pick-and-place operations involve repetitive, physically demanding work that can lead to injuries and decreased productivity. Cobots can handle these tasks, freeing up human workers to focus on more important aspects of the job. Cobots can also be equipped with vision systems that allow them to identify and pick up objects of different shapes and sizes.

3.7.3 Packaging and Palletizing

Cobots can also be used for packaging and palletizing tasks. They can handle a wide range of packaging materials, including boxes, bags, and pouches. By automating these tasks, manufacturers can increase efficiency and reduce the risk of injuries.

3.7.4 Quality Control

Cobots can be used for quality control tasks, such as inspection and testing. They can be equipped with sensors and vision systems that allow them to detect defects and inconsistencies. This can help manufacturers identify and address quality issues early in the production process, reducing waste and increasing efficiency.

3.7.5 Welding

Cobots can also be used for welding tasks, such as spot welding and arc welding. They can work in a wide range of environments, including hazardous or difficult-to-reach areas. By automating these tasks, manufacturers can increase efficiency and reduce the risk of injuries associated with manual welding.

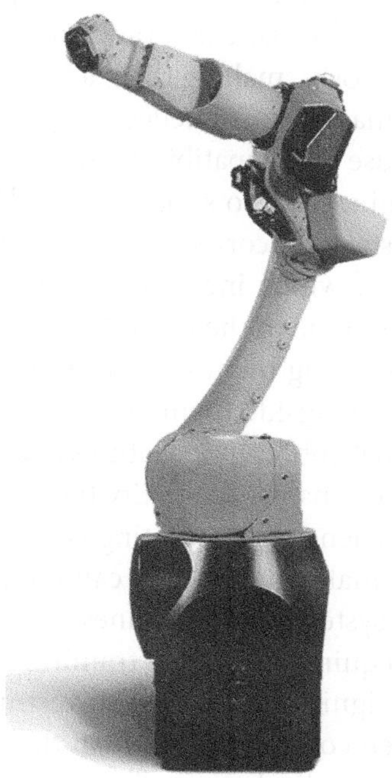

Figure 3.8 FANUC Cr15ia.

Cobots are forcibly controlled sensors to improve their safety physiognomy, yet the sensors also produce a governing class that is perfect for automating finishing demands [13]. Cobots may be used in finishing operations that include automatic polishing, sanding, and abrasive.

Automated assembly is another application that is appropriate for cobot industrialization. cobots can correctly group items, stopping mistakes and cutting down on processing times.

The FANUC CR15iA (Figure 3.8) is a collaborative robot or "cobot" designed by FANUC, a Japanese robotics company. It has a payload capacity of 15 kg and a reach of 1,441 mm, making it suitable for a variety of industrial tasks such as material handling, assembly, and inspection. The robot is equipped with sensors and software that enable it to safely work alongside human workers without the need for physical barriers. It also has a compact design and can be mounted on the floor or a ceiling, allowing for flexible deployment in different production environments.

3.8 Challenges Faced by Cobots

Cobots are used in a variety of industries, but there are several risks involved that could prevent effective implementation and deployment. Cybersecurity is one of the main dangers because cobots are interconnected with other systems, networks, and networks, making them a target for hackers and cyberattacks. Sensitive information and intellectual property may be in danger as a consequence. Because of compatibility problems with modern production methods and machinery, cobots may not be able to interact effectively and work with older systems. Incorrect programming and human error are also possibilities, both of which increase the likelihood of accidents and employee injury [14, 15]. One of the biggest challenges in using collaborative robots (cobots) is keeping them safe for prospective human operators who might be working alongside them. Thorough development testing is required to make sure that cobots have features like emergency stops, power and force limitation, and sensors to identify the presence of people in their workspace. Another challenge is integrating cobots into current manufacturing procedures and making sure they can communicate and function efficiently with current systems and machines.

Cobots might also require specialized training and technical assistance, which might require a significant financial investment from some companies. Furthermore, when cobots proliferate in the workplace, there can be worries about employment displacement. Despite these challenges, cobots have several benefits over traditional methods of production, including

Figure 3.9 Challenges of cobots.

increased flexibility, cost savings, and increased efficiency. Figure 3.9 illustrates the difficulties encountered by cobots [16, 17]. Cobots may not have capacity restrictions, but they may be unable to perform some tasks that require human ability and caution. Despite these risks, cobots can potentially transform various sectors and offer a wealth of benefits to companies that can overcome these challenges.

3.9 Economic Impact of Cobots

The use of cobots can have significant benefits as well as drawbacks. Cobots can help increase productivity and efficiency by automating repetitive chores and reducing the need for physical labour, which is a benefit [18]. Businesses may benefit from lower costs, better quality, and a shorter time to market. Businesses can become more receptive to shifts in demand by using cobots, which can be quickly reprogrammed to perform new tasks. Additionally, cobots can help companies stay competitive in the increasingly automated global economy. The effect of cobots on the economy is shown in Figure 3.10.

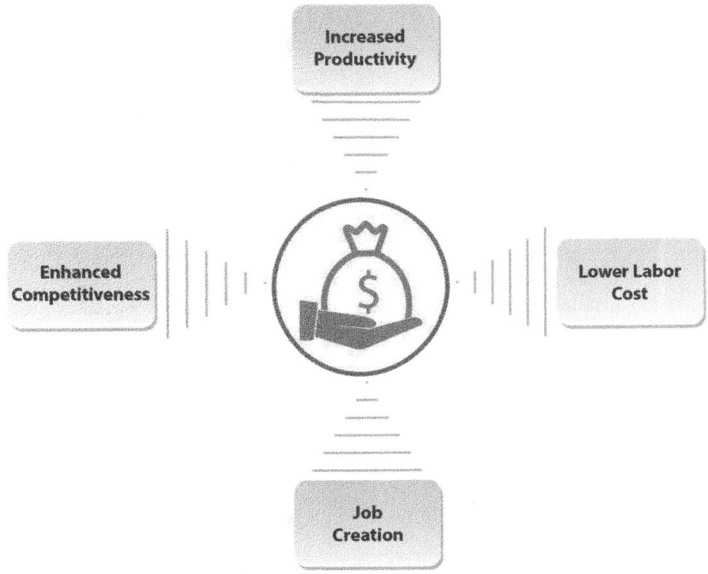

Figure 3.10 Economic impact.

On the downside, employees need to have specialized skills and technological knowledge. Jobs could be lost as cobots take over duties that had previously been handled by human employees [19, 20]. If companies are unable to successfully integrate cobots into current production processes, then they run the risk of not completely realizing the benefits of cobots. Cobots are expected to have a positive total economic effect over the long term as they increase productivity and competitiveness, even after these possible negative effects.

3.10 Components Required

Cobots, or collaborative robots, require several components to function properly. Figure 3.11 shows the components required for the cobots.

3.10.1 Robot Arm

This is the physical part of the cobot that performs the desired actions, such as picking up objects or assembling parts.

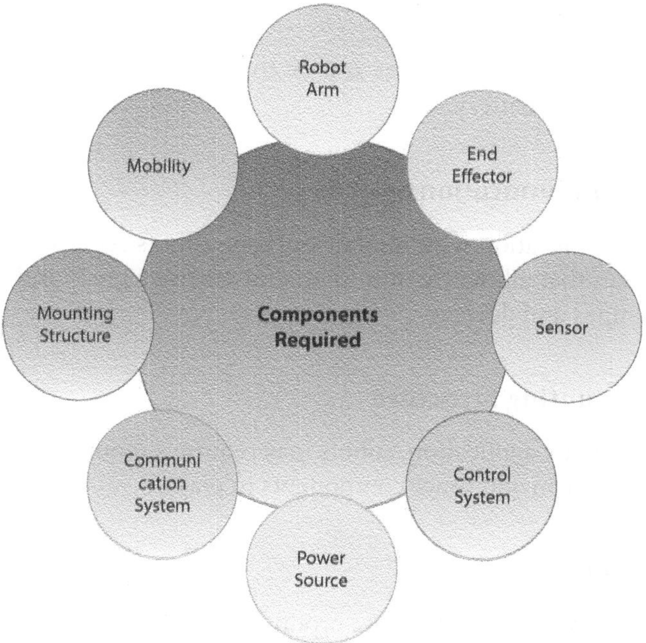

Figure 3.11 Components required.

3.10.2 End Effector

The end effector is the tool or device attached to the end of the robot arm that performs a specific task, such as a gripper or a welding torch.

3.10.3 Sensors

Cobots typically require a range of sensors to monitor their surroundings and ensure safety. These may include cameras, force sensors, proximity sensors, and tactile sensors.

3.10.4 Control System

The control system is responsible for operating and coordinating the various components of the cobot. This may include a central processing unit (CPU) labware and user interfaces for programming and monitoring the cobot.

3.10.5 Power Source

Cobots require a reliable power source to operate. This may include an electrical outlet or battery.

3.10.6 Communication System

To enable collaboration with human workers, cobots require a communication system that allows them to send and receive signals and data from other machines and devices.

3.10.7 Mounting Structure

Depending on the application, cobots may require a mounting structure to hold the robot arm and other components in place.

3.10.8 Mobility

Some cobots are designed to move around in the environment to perform tasks. This can include mobile robots that move on wheels or tracks or cobots mounted on a rail system.

Overall, cobots are complex machines that require careful design and integration of various components to function safely and effectively in a collaborative environment.

3.11 Integration of Cobots with Other Technologies

Cobots usually have a wide range of sensors, including force sensors, proximity sensors, vision sensors, and others. These sensors give robots the ability to notice and react to their surroundings, and they can be combined with other sensors and technologies to enhance their functionality. Algorithms for machine learning can be used to train robots to carry out duties more accurately and quickly. Cobots' decision-making ability can be carried out through AI [21]. Cloud computing platforms can be accessed by cobots to access more processing capacity and data storage. Since cobots can store more data and carry out more complicated duties than they could by using their internal computing capabilities. Technologies that enhance cobots' interaction with human employees, like speech recognition, can be integrated. Augmented reality is used in cobots to enhance their capabilities of cobots by providing additional information and visual guidance.

Integrating cobots with other technologies can help them perform more complicated tasks with greater accuracy, efficiency, and performance. This could increase the use of cobots across a variety of sectors and professions, boost workforce efficiency generally, and enhance worker safety [22]. Additionally, sophisticated sensors like LiDAR are being integrated into cobots, enabling them to sense their surroundings more precisely and consistently. Cobots can now carry out more difficult jobs and store more data, which is essential for many applications. The advancement of cobots and their adoption in various industries and fields depend heavily on the integration of cobots with other technologies, which is a crucial area of study and development.

3.12 Discussion

To protect humans, traditional robots carry out pre-programmed activities in an enclosed space. In place of robots, cobots are secure. To avoid hurting people, they do not move on their own and operate at slower rates than standard industrial robots. Compared to industrial robots, cobots are less expensive and more versatile. When compared to standard industrial robots, which are too large and require machinery to move, humans can transport them to various locations with ease. A collaborative robot is not meant to replace human workers, rather it is meant to enhance their capabilities. The cobot frequently assumes the shape of an arm, giving the worker access to an additional pair of hands. They frequently move around and employ collision detection to protect their human co-workers and other cobots from harm.

One of the main benefits of using cobots in Industry 5.0 is that they can help increase productivity while reducing costs. Because cobots can work alongside human workers, they can assist with tasks that may be repetitive or physically demanding, allowing workers to focus on more complex or higher-level tasks. Additionally, because cobots can be programmed to work in precise and consistent ways, they can help to improve the quality of products being manufactured. Another benefit of cobots in Industry 5.0 is that they can help to promote a safer and healthier work environment. Cobots can handle dangerous tasks, such as working with chemicals or heavy machinery, reducing the risk of injury or illness to human workers. Additionally, because cobots can work alongside humans, they can help to reduce the risk of ergonomic injuries or strain caused by repetitive motions or lifting heavy objects. However, there are also some challenges to implementing cobots in Industry 5.0. For example, there may be resistance from human workers who fear that the use of cobots will lead to job loss. It is important for companies to communicate the benefits of using cobots and to provide training and support to workers who will be working alongside these machines.

Overall, cobots have the potential to play a significant role in the future of manufacturing and industry, particularly in the context of Industry 5.0. By working collaboratively with human workers, cobots can help to increase productivity, improve product quality, and promote a safer and healthier work environment.

3.13 Future Scope

It is long existed pretended that robots are troublesome to program. But cobots destitute the mold accompanying easy-to-use educate pendant. Many want this simplification in the second place. Simple smartphone apps concede the possibility allow machine controllers to find complex programs. Robot manufacturers hope to promptly have a cobot that may be prioritized by someone accompanying little preparation or happening [23]. With their always-on links to Big Data and machine intelligence, cooperative robots can evolve into capable to perform free shifts they have figured out on their own. This progress will further boost their being in less organized manufacturing surroundings. Human–android cooperation is enabling machines to communicate and merge effectively alongside persons, freeing the ruling class from tasks that may be excessively hazardous, difficult, or tedious while reconstructing conditions and output [24]. However, many operational security challenges bear be solved for machines to obtain extensive adoption. Industry 4.0 depends on industrialization and smart production empowered by dossier-compelled electronics like Al, cognitive estimating, IoT, high-tech-material systems, wearables, and cloud calculating [25]. Industry 4.0 focuses on reconstructing process effectiveness. On the other hand, Industry 5.0 involves human advantage active optimization. That is the reason why cobots appear expected the next big wave of the modern strike. Cobots can balance robotic capacities as an individual form that produce value and applies artistic abilities. Manufacturers can use cobots as a plug-and-play resolution in different modern and production movements.

3.14 Conclusion

Because the situation of electronics is changing so quickly in the new cycle, production guests must gradually adapt their result strategy to go along with the application of cutting-edge technology to satisfy the expectations of the stock market. Manufacturing visitors must get ready to adapt to Industry 5.0's extensive bureaucracy. To increase output and efficacy, Industry 5.0 will fundamentally alter production systems. To ensure the efficient operation of their production processes, manufacturing parties must choose and employ independent workforces. Human workers are expected to keep pace with mechanical advancements by adapting to adjustments and collaborating in a cooperative environment. So, using cobots in manufacturing orders eventually has numerous advantages as these autonomous workforces enable more productive and expert result structures by assisting human employees with carrying out distinct jobs. This study contributes a comprehensive view of the synergistic connection center from two points: human laborers and cobots in the production of everything environment, causing the socio-mechanics viewpoint of Industry 5.0 to address the earlier issues. In conclusion, our study suggests an all-encompassing abstract model that emphasizes the tight collaboration between human workers and cobots. This study contributes to the body of literature by illuminating the cobot agreement from an administrative-met perspective. By anticipating the full impact of the concerned with the method of behaving goal on the actual utilization of cobots and further on perceived work achievement improvement in administrative situations, this study provides earlier drama. Future iterations of this study could be continued by refining the measurement criteria for all constructs identified in this location paper and then compiling data from the relevant population of production relationships. The abstract model that is emerging from this research may then be subjected to a practical investigation to be validated. These future developments will aid in comprehending research efforts to support cobots' expanded use in the manufacturing sector.

References

1. Stock, T., Obenaus, M., Kunz, S., Kohl, H., Industry 4.0 as an enabler for a sustainable development: A qualitative assessment of its ecological and social potential. *Process Saf. Environ. Prot.*, 118, 254–267, 2018.
2. Bednar, P.M. and Welch, C., Socio-Technical Perspectives on Smart Working: Creating Meaningful and Sustainable Systems. *Inf. Syst. Front.*, 22, 2, 281–298, 2019.
3. Longo, F., Padovano, A., Umbrello, S., Value-oriented and ethical technology engineering in industry 5.0: A human-centric perspective for the design of the factory of the future. *Appl. Sci.*, 10, 12, 1–25, 2020.

4. Sachsenmeier, P., Industry 5.0—The Relevance and Implications of Bionics and Synthetic Biology. *Engineering*, 2, 225–229, 2016.
5. European Commission, Innovating for Sustainable Growth - A Bioeconomy for Europe, 2012, Retrieved from https://ec.europa.eu/research/bioeconomy/pdf/official-strategy_en.pdf.
6. Raguseo, E., Gastaldi, L., Neirotti, P., Smart work: Supporting employees' flexibility through ICT, HR practices and office layout. *Evid.-based HRM*, 4, 240–256, 2016.
7. Franklin, C.S., Dominguez, E.G., Fryman, J.D., Lewandowski, M.L., Collaborative robotics: New era of human–robot cooperation in the workplace. *J. Saf. Res.*, 74, 153–160, Sep. 2020, doi: 10.1016/j.jsr.2020.06.013.
8. Djuric, A. and Urbanic, R.J., Using collaborative robots to assist with travel path development for material deposition based additive manufacturing processes. *Comput. Aided Des. Appl.*, 15, 4, 542–555, 2018.
9. Farel, R., Kchir, S., Lamy, X., Grossard, M., Challenges in sustainable manufacturing with industrial and collaborative robots: A case study. *International Design Engineering Technical Conferences and Computers and Information in Engineering Conference*, vol. 51791, American Society of Mechanical Engineers, 2018V004T05A040.
10. Wang, L. *et al.*, Symbiotic human-robot collaborative assembly. *CIRP Ann.*, 68, 2, 701–726, Jan. 2019, doi: 10.1016/j.cirp.2019.05.002.
11. Blomström, M., Kokko, A., Zejan, M., Host Country Compt., Labor Skills, & Tech. Transfer by Mls. *Rev. World Econ.*, 130, 3, 521–533, 1994, [Online] DOI: https://doi.org/10.1007/BF02707611.
12. Rossato, C., Pluchino, P., Cellini, N., Jacucci, G., Spagnolli, A., Gamberini, L., Facing with Collaborative Robots: The Subjective Experience in Senior and YoungerWorkers. *Cyberpsychol. Behav. Social Netw.*, 24, 5, 349–356, 2021.
13. Tirmizi, A., De Cat, B., Janssen, K., Pane, Y., Leconte, P., Witters, M., User-friendly programming of flexible assembly applications with collaborative robots, in: *2019 20th International Conference on Research and Education in Mechatronics (REM)*, IEEE, 2019.
14. Chemweno, P., Pintelon, L., Decre, W., Orienting safety assurance with outcomes of hazard analysis and risk assessment: A review of the ISO 15066 standard for collaborative robot systems. *Saf. Sci.*, 129, February, 104832, Sep. 2020, doi: 10.1016/j.ssci.2020.104832.
15. Banaś, W. and Olender, M., The evolution of the robotised workcell using the concept of cobot, in: *International Conference on Intelligent Systems in Production Engineering and Maintenance*, pp. 401–407, 2018.
16. Bragança, S., Costa, E., Castellucci, I., Arezes, P.M., A brief overview of the use of collaborative robots in industry 4.0: Human role and safety, in: *Occupational and Environmental Safety and Health*, P. Arezes (Eds.), Springer, Cham, 2019.

17. Vitolo, F., Rega, A., Di Marino, C., Pasquariello, A., Zanella, A., Patalano, S., Mobile Robots and Cobots Integration: A Preliminary Design of a Mechatronic Interface by Using MBSE Approach. *Appl. Sci.*, 12, 1, 419, 2022.
18. Maier, A. and Student, D., Industrie 4.0 – Der Große Selbstbetrug, in: *Manager Magazin*, 2015, February 13, Retrieved from http //www.managermagazin.de/magazin/artikel/digitale-revolutionindustrie-4-0-ueberfordert-deutschen-mittelstand-a-1015724.html.
19. Gheorghe, G., Concepts and mechatronics and cyber-mixmechatronics constructions, integrated in COBOT type technology platform for intelligent industry(4.0), in: *International Conference of Mechatronics and Cyber-Mixmechatronics*, pp. 281–300, 2019.
20. Foderaro, E., Cesta, A., Umbrico, A., Orlandini, A., Simplifying the AI Planning modeling for Human-Robot Collaboration, in: *2021 30th IEEE International Conference on Robot & Human Interactive Communication (RO-MAN)*, IEEE, 2021.
21. Rizopoulos, A.S. and Greene, C.M., Analytically Hierarchy Process for Augmented Reality Device Selection in Cobotics, in: *IIE Annual Conference. Proceedings*, Institute of Industrial and Systems Engineers, IISE, pp. 1–6, 2020.
22. Stone, R.T., Pujari, S., Mumani, A., Fales, C., Ameen, M., Cobot And Robot Risk Assessment (CARRA) method: an Automation Level-Based Safety Assessment Tool to Improve Fluency in Safe Human Cobot/Robot Interaction, in: *Proceedings of the Human Factors and Ergonomics Society Annual Meeting*, vol. 65, CA: SAGE Publications, Los Angeles, pp. 737–741, 2021.
23. Nogueira, R., Reis, J., Pinto, R., Gonçalves, G., Self-adaptive cobots in cyber-physical production systems, in: *24th IEEE International Conference on Emerging Technologies and Factory Automation (ETFA)*, IEEE, pp. 521–528, 2019.
24. Faccio, M., Bottin, M., Rosati, G., Collaborative and traditional robotic assembly: a comparison model. *Int. J. Adv. Manuf. Technol.*, 102, 5-8, 1355–1372, 2019, doi: DOI:10.1007/s00170-018-03247-z.
25. Elprama, B.V.S.A., El Makrini, I., Jacobs, A., Acceptance of Collaborative Robots by Factory Workers: A Pilot Study on the Importance of Social Cues of Anthropomorphic Robots, in: *International Symposium on Robot and Human Interactive Communication (RO-MAN)*, 2016.

4

The Evolution of Cobots in Intelligent Transportation Systems

Rahul Jethwa[1], Rohit Raj Pradhan[1], Sushruta Mishra[1]* and S. Balamurugan[2]

[1]Kalinga Institute of Industrial Technology, Deemed to be University, Bhubaneswar, Odisha, India
[2]Albert Einstein Engineering and Research Labs (AEER Labs), Vice Chairman-Renewable Energy Society of India (RESI), Tamil Nadu, India

Abstract

The use of robotic technology, particularly collaborative robots, or cobots, has made a profound impact on the evolution of transportation within smart cities. Leveraging the Internet of Things (IoT), these intelligent robots have revolutionized Intelligent Transportation Systems (ITSs). As cities grapple with growing populations and technological advances, robots and cobots have become increasingly essential. Connected devices have vastly expanded data collection in urban settings. Machine learning (ML) techniques are instrumental in harnessing this data flood, enhancing transportation system intelligence. However, road safety remains a paramount concern for societal well-being. The paradigm shift toward robots and cobots in smart mobility necessitates translating abundant quantitative data into insightful qualitative information. ML-powered applications imbue these technologies with intelligence, enabling them to tackle traffic management issues and enhance passenger safety. ITS encompasses various services and applications, including driverless cars, passenger information systems, and public transportation network management. These systems are integral to the blueprint for smart cities, fostering efficient transit, bolstered road safety, and reduced energy consumption and environmental impact. This study embarks on a comprehensive exploration of the dynamic interplay between robots, cobots, and smart transportation, highlighting their transformative potential. A variety of ML methods and IoT applications that support the synergy within ITS are explored in this chapter.

**Corresponding author*: sushruta.mishrafcs@kiit.ac.in

V. Ramasamy, S. Balamurugan and Sheng-Lung Peng (eds.) Intelligent Robots and Cobots: Industry 5.0 Applications, (65–84) © 2025 Scrivener Publishing LLC

***Keywords*:** Cobots, smart transportation, robots, traffic, machine learning, IoT

4.1 Introduction

Robotic fee collection, vehicle-to-vehicle communication, and vehicle-to-infrastructure interaction stand as some of the most promising initiatives on the global stage. In the realm of Intelligent Transportation System (ITS), cutting-edge wireless, electronic, and automated technologies find application. These innovations possess the potential to seamlessly unite users, infrastructure, and a diverse array of vehicles—ranging from cargo trucks and personal automobiles to public transit systems—on the roadways. The spectrum of automated and in-vehicle technologies extends to encompass automatic guideways, accident avoidance systems, precise bus docking, and an array of robotic advancements. Within this landscape, robots and cobots take center stage, reshaping the concept of ITS. Leveraging their capabilities, these mechanical entities hold the power to enhance trip optimization, reduce unnecessary travel distances, and catalyze the utilization of alternative transportation modes. By mitigating traffic congestion and optimizing route planning, robots and cobots contribute to a reduction in time spent on the road, diminish dependency on foreign oil, and foster improved air quality. Moreover, the integration of these technologies extends to broader system management, impacting fuel consumption and vehicle design in multifaceted ways:

Effective Route Planning and Timing: Robots and cobots make it possible to choose the best route and time it for increased effectiveness.
Traffic Alleviation: Through dynamic traffic management, these technologies contribute to reduced congestion.
Demand and Pricing Management: Robots and cobots facilitate adaptive demand-based pricing strategies.
Adaptive Vehicle Functionality: It leads to vehicle adjustments tailored to varying weather and terrains.
Precision Formation Movement: These technologies enable tightly coordinated movement of compact vehicle units.

The integration of robots and cobots is still in its infancy, especially in developing countries like India, even if the idea of ITS has made significant progress in developed countries. Regardless of a country's level of development, the path to a surface transportation system that is safer, more financially feasible, and environmentally conscientious depends on the thoughtful integration of robots and cobots. The trajectory ahead underlines a common worldwide commitment to use cutting-edge technologies in the quest of excellence in transportation [1].

The purpose of this study is to conduct a thorough, self-contained investigation into how ML methodologies and Internet of Things (IoT) applications can be synergistically integrated within the framework of ITS, with a focus on the function of robots and cobots. The goal of the study is to get a thorough understanding of the rapid developments in these sectors and to identify any prospective areas that require more research. Remarkably, the current body of research emphasizes a discernible gap in ML coverage, particularly concerning the domain of smart parking and intelligent lighting systems within the context of robotic and cobotic integration. The research will focus on the unique contributions that robots and cobots make to these applications and their potential for having a transformational effect, even if ITS applications like parking management, accident detection, and route optimization are well-studied topics. The study intends to provide light on the crucial interactions between robotic and cobotic technologies and many aspects of smart urban mobility, taking into account commonplace situations like school commutes, everyday errands, business trips, and doctor's visits. The study also recognizes the crucial role that intelligent transport systems play in the broader smarter cities movement, which is in line with the principles of improved urban life and environmentally friendly transport initiatives. The research aims to glean practical insights for the integration of robotic and cobotic ITS technology by taking inspiration from the pro-active strategy of the New Zealand Transport Agency (NZTA). The research aims to contribute to a comprehensive blueprint for the integration of robots and cobots within the framework of contemporary urban transportation by synthesizing standards and specifications that govern the development, acquisition, and seamless integration of these technologies into the national road system. In conclusion, this study's ambitious goal is to map out the uncharted region at the nexus of ML, IoT, and robots in ITS. The project aims to pave the path for a more effective, sustainable, and technologically sophisticated urban transportation ecosystem by identifying untapped niches, enhancing the contributions of robots and cobots, and interacting with the changing landscape of smart mobility.

4.2 Uncovering Challenges in Intelligent Transportation System

The issues are recognized and categorized logically based on the known literature, into three groups:

- lack of a traffic control system;
- vehicle and homeland security system operation; and
- manufacturing and logistics industries.

• **Problem 1. Lack of Traffic Control Systems:** When there are many vehicles, the complexity of the systems rises, and they are unable to manage the crowd, which reduces mobility, reduces fuel consumption, prolongs travel times, and increases pollution. Traffic management systems are designed to efficiently handle large masses of traffic.

• **Problem 2. Vehicle and Homeland Security System Operation:** This term refers to the safety and oversight of the transportation network and vehicles. It aids in keeping track of the automobile's travel and in the quick identification of the car and the driver. The problem has been highlighted as the absence of such effectiveness.

• **Problem 3. Manufacturing and logistics industries:** Since the advent of e-commerce, there has always been a strong need for workers in the logistics industry. In the current lockdown situation caused by the COVID-19 epidemic, this demand is only increasing, while assembling a sizable number of personnel in factories might be hazardous to their health. Depending on their workload, the employees must manage the operation in accordance with the clock. When consumer demand is this great, the economy finds it difficult to keep up.

• **Solution 1. Dynamic Traffic Control System:** An innovative concept that makes use of cobots and robots to manage traffic effectively in order to address the issue of insufficient traffic control systems building a dynamic traffic control system that is capable of real-time evaluation and adaptation by strategically placing these intelligent creatures at crucial locations. Robots, which are outfitted with sensors and cameras, can instantly optimize signal designs and lane directions based on traffic patterns and density. This flexible strategy lessens traffic jams, uses less fuel, and improves mobility in general. A network of robots works together to control traffic in metropolitan areas through centralized coordination, guaranteeing a smooth flow and promoting cleaner, more sustainable transportation.

- **Solution 2. Robotic Enhancement for Security:** A robot- and cobot-powered solution is recommended for addressing the challenges of vehicle and homeland security. This solution enables efficient monitoring and rapid identification by establishing a wireless communication network between vehicles and infrastructure, incorporating distinct digital IDs for each vehicle. At checkpoints, strategically placed robots and cobots improve security by using cutting-edge sensors for quick identification. This integrated system also records the route histories of the vehicles, facilitating investigations. This groundbreaking method improves safety, simplifies operations, and fortifies security for the transportation network.
- **Solution 3. Cobots in Manufacturing and Logistics:** The use of robots to automate routine chores has been one of the largest advances in the Industry 4.0 revolution. This was a significant improvement for many firms because it increased production by saving both time and human work. Cobots are robots that have been created expressly to work alongside and safely cooperate with human workers in the same environment. Artificial intelligence (AI) models are given data from sensors attached to the cobots in order for them to be independent. As the precise position and speed of the operators can be ascertained in real time, this helps with collision avoidance and subsequent navigation. The cobot automatically shuts down in the event of a system failure. In any warehouse, the bots may deploy sensors, transport supplies between workstations, and point human employees in the right direction. These devices can complete the remaining job, allowing workers to concentrate on picking as much as possible.

4.3 The Role and Application of Cobots in Manufacturing and Logistics

Cobots are robots that have been created expressly to work alongside and safely collaborate with human workers in the same environment. AI models are fed data from sensors attached to the cobots in order for them to be independent. As the precise position and speed of the operators can be ascertained in real time, this helps with collision avoidance and subsequent navigation. The cobot automatically shuts down in the event of a system failure. By identifying trends in both recent and historical data, AI can even forecast these circumstances for improved performance. Cobots primarily serve the manufacturing and logistics sectors. Since the emergence of e-commerce, there has been a strong need for labor in the logistics industry, and this demand has continued to increase as a result of lockdown restrictions during the Covid-19 outbreak. Cobots save the day when there are health issues associated with assembling a big number of workers in industries. Even many small and medium-sized businesses in India and abroad have implemented them for their business continuity due to their accessibility and adaptability. Consider the US-based Locus Robotics as an illustration. CEVA Logistics and Geodis are just a couple of the many worldwide logistics companies who receive its cobots. In any warehouse, the bots may deploy sensors, transport supplies between workstations, and point human employees in the right direction. The remaining work can be done by these machines, allowing workers to concentrate on picking as much as possible.

Cobots are in high demand worldwide, but especially in India and the automotive industry. Over 150 cobots are currently present on Bajaj Auto's assembly lines, where they are used for tasks including machine tending. This presents a challenge for many semi-skilled or unskilled workers who get daily pay for these repetitive chores in a society where unemployment is already a concern. Before being given the updated job requirements that deal with preventative maintenance, validations, and the like, these employees need to be upskilled [2].

In Figure 4.1, the cooperative MiR500 or mobile industrial robot safely avoids humans and other physical barriers. The robot automatically navigates using state-of-the-art hardware and cutting-edge algorithms to choose the fastest route to its goal. To prevent expensive delays in the logistics process, it will autonomously reroute when it encounters barriers. The MiR500 has the most recent laser scanning technology, which provides a 360° sight for maximum safety. The detection range of pallets using 3D cameras is between 30 mm and 2000 mm above the floor. The MiR500 picks up, moves, and unloads pallets on its own with the help of the MiR500 Pallet Lift, freeing up staff members for more useful work.

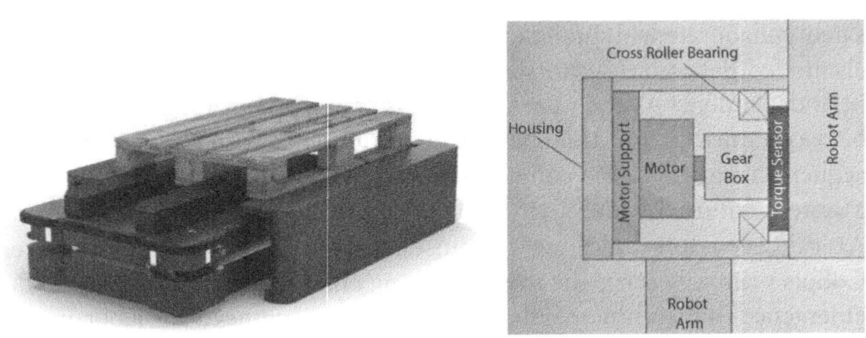

Figure 4.1 MIR500 autonomous mobile robot with obstacle avoidance and navigation.

When using the MiR500, the function of wires, magnets, or QR codes is replaced by smart navigation software, so there is no need to alter the facility layout. To ensure a quick and affordable implementation, just download CAD (Computer-Aided Design) files of the facility to the robot or use its laser scanners to produce a map.

4.4 Advancing Technologies Facilitating Robot and Cobot Operations in Intelligent Transportation Systems

The proliferation of robotics and automation has seen a remarkable surge in various sectors, particularly in manufacturing. This section categorizes and delves into the integral technologies that facilitate robot and cobot operations. Table 4.1 highlights some vital cobot-based technologies related to intelligent transport models.

Table 4.1 Enabling cobot-driven technologies of ITS.

Enabling technologies	**Infrastructures resources**	**Vehicle resources**
Locating referencing	• Warehouse mapping systems • Object recognition databases	• Onboard sensors • Vision systems • Lidar systems
Data acquisition	• Environmental sensors • Assembly line feedback systems	• Tactile sensors • Motion sensors • Force sensors
Data processing	• Central processing units	• Onboard microprocessors • Robot OS
Communications	• Intra-factory communication	• Bluetooth modules • NFC modules
Information distributions	• Centralized dashboards • Factory-wide alert systems	• Mobile devices • In-vehicle units
Information utilization	• Job scheduling algorithms • Inventory management systems	• Obstacle avoidance

Many professionals in the manufacturing and automation sector are acquainted with established technologies such as Wi-Fi communication and vision systems. However, innovations continually emerge, leading to the development of more advanced AI algorithms tailored for robot-specific functions. Despite the rapid technological advancements, it is pivotal to understand that these technologies are not only exclusive to specialized applications but are also readily available in the open market [3].

4.5 Redefining Smart Transportation: The Synergy of Robotics, Cobots, and Predictive Analytics in ITS

This section examines the convergence of smart transportation with robotic automation. While IoT and ML remain instrumental in addressing transportation issues, the integration of robots and cobots offers another layer of innovation. Six primary smart transportation challenges are highlighted below, along with the respective ML and IoT solutions, further enhanced by robotic capabilities.

4.5.1 Enhancing Urban Mobility with Robotic-Enabled Route Optimization

Urban congestion has always been a major concern, intensified with the exponential growth in vehicles. Route optimization addresses this by providing efficient routes, thus mitigating congestion and emissions. In this space, ML and IoT work in tandem, but the incorporation of robots and cobots can further aid in real-time data gathering and immediate intervention during traffic challenges. For instance, a model has been devised which employs the Markov decision process algorithm combined with vehicle-to-vehicle communication. In this system, autonomous robotic vehicles can actively communicate with each other, making real-time adjustments to their routes, reducing overall congestion. A combination of k-means and DBN has been proposed for traffic network architecture optimization. Furthermore, in exploring short-term traffic forecasting, various models like BN-SARIMA (seasonal autoregressive integrated moving average), FF-NN (feedforward neural network), and NARX (nonlinear autoregressive network with exogenous inputs) have been tested. Cobots can be deployed at strategic urban intersections to aid in immediate traffic redirection based on data predictions. For estimating journey times and deviations, an FF-NN trained with a backpropagation is deployed, complemented by cobots equipped with sensors to actively detect traffic anomalies and react accordingly. In one comprehensive approach, a logistic regression was paired with SVM (support vector machine), a decision tree, random forest, and a multi-layer perceptron to predict potential congestion. A parallel research employed a stacked autoencoder to predict traffic flow using extensive data. Scalable enhanced road side units are equipped with tools like a 4G network to monitor traffic. These units are akin to cobots: they house features like wireless modems, radio-frequency modules, weather sensors, and cameras. The Master Control Center interprets this data to adjust road speeds and conditions. A breakthrough was the introduction of vehicular ad-hoc network (VANET) with mobile navigation apps. Here, vehicles, functioning like mobile robots, communicate using IEEE 802.11p, popularly known as Wi-Fi to update road conditions. In scenarios where VANET data is sparse, these vehicles leverage Google Maps. This dual-source system was further investigated to discern correlations between congestion, emissions, and Google's "popular times" statistics, emphasizing the fusion of traditional data sources with robotic intelligence [4].

4.5.2 Revolutionizing Parking Efficiency with Robotic and Cobot Assistance

The challenge of streamlining parking spaces is not just about availability but also efficiency and ease. The integration of robots and cobots can revolutionize how parking applications are perceived and utilized. The primary objective remains to provide parking spots for vehicles promptly. With the incorporation of IoT devices, it has become feasible to recognize occupied parking slots in real time and relay this data to a centralized system for spot availability analysis. Robotic solutions can take this a step further. Consider mobile robots that can guide vehicles to the nearest available spot, minimizing the time taken to search for parking. In an innovative approach, researchers employed Lifecycle Modeling Language to propose a lifecycle methodology tailored for IoT-enhanced smart city service systems. Using the example of a city retail center's smart parking, they demonstrated how this methodology can be brought to life. Their proposed scenario uses robots to actively designate parking spaces to vehicles as they approach open mall entrances, ensuring efficient space utilization. Further, for spot identification, instead of just relying on cameras, robotic agents equipped with sensors can patrol parking areas. In a particular study, convolutional neural networks (CNNs) were applied to data captured from parking lot cameras but imagine the accuracy boost when these are combined with on-ground cobots, equipped with advanced sensors that can relay real-time data. Moreover, traditional ML algorithms are enhanced by the application of robots in the field. For example, while the SVM and MRF (Markov random field) algorithms have been amalgamated to discern open slots from camera imagery, robotic aids can provide a physical, on-site check to validate these findings. In one approach, the researchers integrated ultrasound sensors to detect the availability of parking spaces. Now, these sensors can be fitted on cobots, which can move around and provide dynamic, real-time data, which is then transmitted using Wi-Fi modules to cloud servers. This direct data allows for increased accuracy when users check for spot availability through smartphone applications. For a more integrated solution, combining both hardware and software is key. Magnetic sensors, which detect parked vehicles, can be embedded in robot assistants that patrol parking lots. As they identify occupied spaces, gateway devices strategically positioned transmit this data to remote servers. With the use of specialized algorithms, the nearest available parking spot can then be recommended to incoming vehicles. Imagine a scenario where a robot or cobot can physically lead a car to its parking spot. Furthermore, genetic algorithms optimized for robotic implementations can then find the most suitable parking spot for an incoming vehicle, ensuring efficiency, ease, and time-saving. The essence here is to bring robots and cobots into the traditional parking narrative, highlighting their potential roles in enhancing parking efficiency and user experience.

4.5.3 Enhancing Street Lighting with Robotic and Cobot Integration

In the age of smart cities and interconnected devices, a harmonious merger of robotic technology with lighting infrastructure presents a new horizon for urban development. The integration of robots and cobots has the potential to bring about a transformative shift in the perception and utilization of street lighting. The significance of managing and optimizing lighting in modern city frameworks cannot be overstated. Smart street lights using IoT devices are already gaining traction but imagine the additional capabilities brought about by integrating robots into the mix. To truly revolutionize the current street lights, it is essential to equip them with advanced sensory apparatus and this is where robots and cobots can play a pivotal role. For starters, robots equipped with light sensors, IR (infrared) sensors, and GPS (Global Positioning System) can actively patrol streets, ensuring that the lights are functioning properly, offering maintenance checks, and even replacing bulbs as needed. The GPS tech in these robots not only pinpoints the location of faulty lights but can also provide real-time data on ambient light conditions, facilitating dynamic adjustment of lighting. A particularly innovative method, suggested by some researchers, involves equipping street lights with IR sensors and IR LEDs. With robots having integrated IR sensors, they can detect sunrise and sunset, making dynamic decisions about switching the lights on or off. Cobots, on the other hand, can hover around busy streets or pedestrian zones, using these sensors to determine if additional light is required or if the lights can be dimmed, depending on the vehicular and foot traffic. Moreover, visualizing a city where each lamppost is not only a source of light but also a hub of connectivity brings us to a scenario where lampposts double up as Wi-Fi hotspots. Robots patrolling the streets can connect to these hotspots, upload collected data to central servers, and even download updates or new instructions. Furthermore, cobots equipped with cameras can serve a dual function; they can provide real-time feedback on the effectiveness of the lighting while also acting as security agents. By constantly monitoring surroundings, they ensure a safer environment, enhancing the utility of an ordinary lamppost. In essence, as the world drives toward the era of smart cities, the combination of robots and cobots with smart lighting solutions promises an integrated, efficient, and secure urban environment. The core idea here is to envision a scenario where robots and cobots become an integral part of the street lighting ecosystem, ensuring optimal lighting conditions, energy efficiency, and enhanced security [5].

4.5.4 Robotic Intervention in Accident Detection and Prevention

Safety on the roads is paramount, especially in bustling urban landscapes. While preventive measures remain crucial, the introduction of robots and cobots is set to revolutionize accident detection and mitigation. A forward-thinking approach to enhancing road safety is by incorporating robots equipped with advanced sensors and AI-driven algorithms to monitor and interpret traffic conditions in real time. These robots can be strategically placed at accident-prone zones or busy intersections. Through their sophisticated sensors, they can detect erratic driving patterns, predict potential collisions, and even measure vehicle speeds. On the other hand, cobots, primarily installed within vehicles, work collaboratively with drivers. They process vast amounts of data from internal and external vehicle sensors, alerting drivers of impending dangers like blind spots or sudden obstacles. These cobots can also interface with other vehicles on the road, fostering a collaborative driving environment where each vehicle is aware of the others' intentions, effectively reducing the chances of accidents. Furthermore, to address the aftereffects of accidents, robotic drones equipped with cameras and other sensors can be dispatched to accident sites, ensuring immediate assessment and relaying critical information to emergency responders. This rapid response can help in efficient rescue operations and traffic management. One groundbreaking study presents an integrated approach where robots and cobots collaborate within an IoT cloud ecosystem. In this framework, devices installed in selected vehicles continuously gather data, such as GPS coordinates, and relay it to a centralized cloud server. These data, stored and analyzed in a hybrid database system, can be accessed in real time by robots stationed across the city and by cobots within vehicles, creating a comprehensive safety net for drivers.

4.5.5 Robotic Solutions for Road Anomalies Detection

The enhancement of road conditions through robotic technology is vital for modern smart transportation. Robots, equipped with a variety of advanced sensors, play an instrumental role in detecting and reporting road anomalies such as potholes, cracks, and other infrastructure degradations. These irregularities not only affect the commute experience but can also lead to accidents, thereby emphasizing the importance of timely detection and rectification. A landmark study delved into the potential of robots integrated with deep learning methodologies, specifically focusing on the capabilities of CNNs and deep CNNs for pavement distress diagnosis. These robots, moving along roads, utilize their on-board CNNs to scan, detect, and report irregularities in real-time. In comparison to manual surveillance, this robotic approach promises greater accuracy and efficiency. Moreover, cobots can be integrated within maintenance vehicles, providing assistance to human operators by highlighting detected anomalies and suggesting potential repair methodologies. This collaborative effort ensures timely and effective road repairs, leading to safer commuting experiences.

On the other hand, cobots, installed within transit vehicles, interact with passengers, ensuring that they have a valid ticket, providing route guidance, and even offering details about nearby attractions, eateries, and shops. Their functionalities include the following:

- Authenticating fare transactions swiftly.
- Promoting a user-centric smart card system that simplifies the travel experience.

- ➢ Offering real-time GPS-based route guidance to optimize travel time.
- ➢ Advertising nearby points of interest through digital interfaces.
- ➢ Ensuring reliable Wi-Fi connectivity for passengers, enriching their transit experience.
- ➢ Streamlined operations with enhanced passenger satisfaction and safety.

4.5.6 Advanced Vehicle Tracking or Transportation Monitoring

Many businesses now require vehicle tracking or transportation monitoring systems in order to effectively manage their fleets and supply chain operations. With the use of GPS trackers, transportation companies may readily access the vehicle's position, information, and statistics in real time. This enables transportation firms to monitor crucial assets in real time. In addition to tracking a driver's position, IoT devices may monitor their conduct and give details about their driving patterns and downtime. IoT has reduced operational and fuel costs as well as maintenance costs in fleet management systems. Real-time tracking has made it much simpler to put wise decisions into practice in terms of transportation monitoring, allowing drivers to spot problems with their vehicles right away and take preventative measures as needed [6].

4.6 A Comparative Analysis of Cobot and Predictive Protocols in Enhancing Safety and Sustainability in ITS

It has been observed that an increase in the number of vehicles on the road is closely related to an increase in the overall number of accidents. According to data already available, it is projected that more than 50,000 people died in Asia in 2013 as a result of avoidable or thoughtless incidents. Road accidents can happen for a variety of reasons, including distracted driving, natural disasters, dangerous lane changes, collisions caused by traffic, inappropriate turning, and animal or pedestrian crossing. Figure 4.2 denotes the generic functionalities of a cobot.

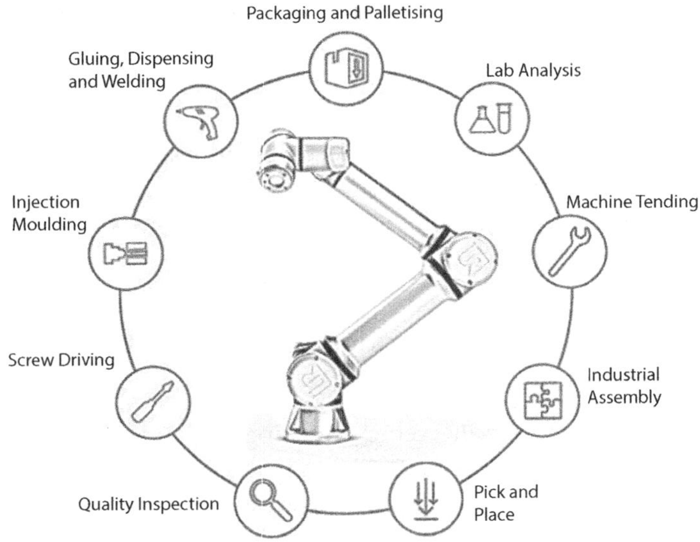

Figure 4.2 Functional diagram of an operational cobot.

4.6.1 Advancing Eco-Friendly Transportation Through Robotic and Cobot Integration

With the escalating number of vehicles, there is a consequential rise in fossil fuel consumption. As vehicles predominantly rely on these fuels, the resultant carbon emissions adversely affect the environment. In the wake of these concerns, robotic and cobot interventions offer sustainable alternatives to counteract the environmental degradation. Mehar and team showcased an avant-garde blueprint. Their strategy accentuates the deployment of electric vehicles equipped with robotic systems and cobot assistants. These automated entities are embedded with a plethora of sensors, constantly monitoring diverse vehicular parameters ranging from brake efficiency to passenger safety. Concurrently, cobots assist in gathering and processing data on energy consumption, bestowing users with insights to circumvent the inherent limitations of electric vehicles and promote eco-friendly transit [7]. Because each cobot is concentrated on a single point of decision, it can be quickly created and deployed and is simple to upgrade. The group of cobots supports a full business or a portion, such a function or a business unit, and functions as an enterprise-class system because they are loosely connected. Cobot systems provide the benefits of quick modular flexibility, security, and large-scale coordination. Cobots are developed using technology characterized by speed, scalability, reliability, expansiveness, and maintainability. Figure 4.3 presents a schematic representation of a cobot for clarity.

4.6.2 Robotic and Cobot Enhanced Collision Avoidance in Traffic

Traffic mishaps, often a byproduct of myriad causes, have been of great concern. One of the most prevalent and avoidable causes is needless collisions between vehicles. The integration of robots and cobots in recent years has been a groundbreaking approach toward preventing such unwanted occurrences. In 2014, Mbachu and Onuora, highlighting the grim statistics of road accidents, introduced a robotic technique for the timely detection and prevention of potential vehicle collisions. The methodology relies on two distinct arrival factors: the time delay of the reflected signal and the angle at which the signal is received [8]. In a later study, Hamid and his team in 2016 delved into a comprehensive evaluation of cobot-integrated collision mitigation solutions. These solutions utilized real-time environmental data procured by sensors in mobile robots and cobots. Not only did they help in automated parking, reducing driver strain, but they also played a crucial role in collision prediction and avoidance. In the absence of corrective driver action, the cobot takes autonomous control, steering the vehicle back to safety [9].

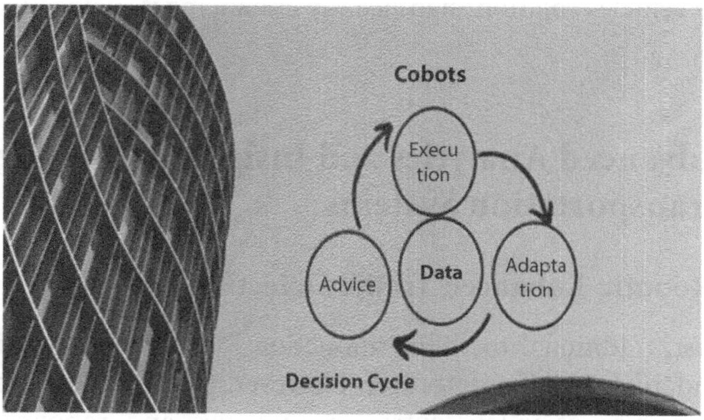

Figure 4.3 Schematic representation of cobot.

4.6.3 Revolutionizing Transportation: Robotic-Driven Autonomous Vehicles

The realm of autonomous transportation, spotlighting self-driving vehicles, marks a revolutionary advancement poised to supersede traditional human-operated vehicles. Anchoring this transition is the pervasive integration of robots and cobots, striving to sculpt a cohesive smart community landscape. This community vision is the foundation stone for evolving smart cities backed by sustainable transport ecosystems. Li *et al.*, in their groundbreaking 2016 research, conceptualized a robotic-driven model for autonomous vehicles. This model was meticulously crafted by analyzing a plethora of driver attributes including age, gender, and even emotional states. The cobot-integrated driver model showcased significant reductions in collision incidents, harmonized with driver preferences, and offered a heightened safety shield, predominantly by mitigating human-induced errors. Within this model, robotic mechanisms seamlessly manage vehicle dynamics such as speed modulation, direction adjustments, lane transitions, and traffic light responsiveness. Furthermore, cobots were deployed to simulate and incorporate human behavioral dynamics, ensuring that the autonomous vehicles bore an intricate understanding of human driving nuances [10]. Further expanding the scope of robotic integration in autonomous transportation, Dr. Kamran and his team, in 2016, addressed the intricacies of autonomous vehicle navigation in expansive outdoor environments. They innovatively brought cobots to the forefront by unveiling a state-of-the-art digital visual gyroscope. This mechanism leveraged imagery obtained from a monocular camera embedded within the self-driving vehicle. The cobots subsequently processed this visual data to predict and adjust the vehicle's rotation, significantly enhancing its navigation prowess [11].

4.7 Advanced Analytics and Insights in Intelligent Transportation Systems

4.7.1 Robotic-Enhanced Traffic Detection

In the past, traditional tools like induction loops and magnetic sensors were standard for traffic detection. However, as the 2020s commenced, the convergence of video-centric technologies and robotics emerged as the cornerstone of ITS. With the integration of computer vision and robotic algorithms, ITS has seen significant strides, notably in real-time multi-object detection systems. The power of ML, reinforced by robotics—employing methods like clustering, neural networks, and advanced visual algorithms—has become the gold standard in this realm [12].

4.7.2 Advanced Road/Lane Detection with Robot

Robotics is now indispensable for road/lane detection. Their capability to identify and separate road or lane markings from miscellaneous environmental elements is pivotal for the autonomous vehicles of today. Augmented by mathematical algorithms, ML, and neural networks, robots offer unparalleled insights into the behaviors and movement patterns of surrounding vehicles.

4.7.3 Elevating Precision in Navigation

Accurate horizontal positioning is a linchpin in contemporary ITS. While the virtues of AI and ML are well recognized, the addition of robots ensures instantaneous, granular-level accuracy, especially in challenging navigation scenarios like winding roads. They adeptly manage multifaceted challenges, ensuring top-notch accuracy, quick response times, and user-friendly interfaces [13].

4.7.4 Cobot-Driven Vehicle Detection

ITS leans heavily on cobots to swiftly and precisely recognize vehicles and diverse moving entities. Even as video surveillance is fundamental, cobots optimize the processing of superior images, achieving real-time detection in scenarios with dense vehicular populations. With neural networks deeply integrated, these cobots excel in surveillance and recognition tasks [14].

4.7.5 Robotics and 5G Routing for Transportation

With the paradigm shift toward autonomous vehicles, robotics have become central in navigating the complex landscape of routing. Leveraging the capabilities of the 5G network, robotics, powered by mathematical models and advanced optimization techniques, is a reshaping approach to comprehending and addressing urban and freeway transportation [15].

4.7.6 Robotic Traffic Optimization for Efficient Commuting

Robotics has redefined traffic optimization. From intelligent route decisions to smart traffic signal management, the robotic touch ensures that transit is always a step ahead. These systems, enriched with dynamic traffic assessments and advanced algorithms, make commuting more efficient than ever before [16].

4.7.7 Robotic Traffic Flow Prediction for Safer Commutes

Robots, fortified with advanced predictive mechanisms, excel in anticipating both immediate (from minutes to an hour) and distant traffic scenarios. These foresights are crucial in preempting and addressing potential traffic challenges. Today's robots, equipped with intricate traffic models, play an integral role in ensuring seamless traffic at the granular, broader, and overarching network levels. Their data-driven insights, coupled with on-the-fly analysis, promise a traffic future that's both smooth and safe [17] .

4.7.8 Robotics and ITS Data Transformation

The new-age robotic systems have revolutionized data dynamics in ITS. From sensors to smart cards, they offer varied avenues of data collection. Furthermore, they address pressing challenges of data storage and seamlessly integrate missing data points, transforming voluminous traffic data into actionable strategies [18].

4.8 Conclusion

The combination of ML and IoT technologies has been carefully investigated in the context of smart transportation. This work has demonstrated how ML techniques may be applied to the large amount of IoT data, particularly in applications like route optimization, parking management, and accident detection. A similar narrative unfolds when shifting focus to the integration of cobots and robotics. Cobots and robots have the same disruptive potential as IoT and ML, offering to reimagine transportation systems by boosting effectiveness, safety, and user experience. In both situations, there are recurring themes of time management, money management, safety, and environmental preservation. These themes are emphasized by cobots and robots, which provide very precise route optimization, convenient parking options, and state-of-the-art accident avoidance. The paper highlights the important advancements in smart transport made possible by IoT and ML. Cobot and robot integration promises to advance technology. Exploring the capabilities of cobots and robots expands the potential for ML applications, in line with the rising demand for broader coverage. In conclusion, the introduction of cobots and robots has the potential to alter urban mobility, much as IoT and ML have transformed smart transportation. In this journey, which focuses on efficiency, sustainability, and safety, cobots and robots are positioned as key players in the development of ITS.

References

1. Robotiq, Cobots EBook, 2018. [ResearchGate].
2. Zhu, Z. and Hu, H., Robot learning from demonstration in robotic assembly: A survey. *Robotics*, 7, 17, 2018. [ResearchGate].
3. Mohammed, M., Khan, M.B., Bashier, E.B.M., *Machine Learning: Algorithms*, Press, Boca Raton, FL, USA, 2016. [RoboticBlogs].
4. Araújo, A., Kalebe, R., Girão, G., Filho, I., Gonçalves, K., Melo, A., Neto, B., IoT-Based Smart Parking for Smart Cities, in: *Proceedings of the 2017 IEEE First Summer School on Smart Cities (S3C)*, Natal, Brazil, 6–11 August 2017, pp. 31–36. [Google Scholar].
5. Robots and robotic devices — Collaborative robots [ISO].
6. Lee, J., A survey of robot learning from demonstrations for human robot collaboration, [AVS].
7. Statista, Number of Vehicles in Use Worldwide 2006–2014, Available online: [Statista].
8. Robots and robotic devices Safety requirements for industrial robots Part 1: Robots, ISO Standard [ISO].
9. Barattini, P., Morand, C., Robertson, N.M., A proposed gesture set for the control of industrial collaborative robots, in: *2012 IEEE ROMAN*, [Semantic Scholor].
10. Implementation of collaborative robot applications: A report from the industrial working group, High Speed Sustainable Manufacturing Institute, Tech. Rep., 2017, [ACM library].
11. Zhang, F., Guo, F., Huang, H., A Research on Driving Cycle for Electric Cars in Beijing, in: *Proceedings of the 28th Chinese Control and Decision Conference (2016 CCDC)*, Yinchuan, China, 28–30 May 2016, vol. 100081, pp. 4450–4455. [Google Scholar].
12. Sriwilai, A. and Pattaraprakorn, W., The Study on the Effect of Electric Car to Energy, in: *Proceedings of the 2016 13th International Conference on Electrical Engineering/Electronics, Computer, Telecommunications and Information*

Technology (ECTI-CON), Chiang Mai, Thailand, 28 June–1 July 2016. [Google Scholar].

13. Li, L., Liu, Y., Wang, J., Deng, W., Oh, H., Human dynamics based driver model for autonomous car. *IET Intell. Transp. Syst.*, *10*, 545–554, **2016**. [Google Scholar] [CrossRef].
14. Lin, Y., Li, L., Jing, H., Ran, B., Sun, D., Automated traffic incident detection with a smaller dataset based on generative adversarial networks. *Accid. Anal. Prev.*, 144, 105628, 2020. [Science Direct].
15. Hamdi, M.M., Audah, L., Rashid, S.A., Mohammed, A.H., Alani, S., Mustafa, A.S., A review of applications, characteristics and challenges in vehicular ad hoc networks (VANETs), in: *Proceedings HORA 2020-2nd International Congress on Human-Computer Interaction, Optimization and Robotic Applications*, 2020, article number 9152928. [Google Scholar].
16. Hu, X., Chen, L., Tang, B., Cao, D., He, H., Dynamic path planning for autonomous driving on various roads with avoidance of static and moving obstacles. *Mech. Syst. Signal Process.*, 100, 482–500, 2018. [Science Direct].
17. Intelligent transportation systems (ITS): A systematic review using a Natural Language Processing (NLP) approach, *Heliyon*, 7, 11, 2021. [ScienceDirect].
18. Banziger, T., Kunz, A., Wegener, K., Optimizing human-robot task allocation using a simulation tool based on standardized work descriptions. *J. Intell. Manuf.*, 31, 1–14, 2018. [IWF].

5

Low/No-Code Software Development of Cobots Using Advanced Graphical User Interface

N. L. Padma Swati[1], S. Vanshita Gupta[1], Vallidevi Krishnamurthy[1]*, Sujithra @ Kanmani[1] and Surendiran Balasubramanian[2]

[1]Vellore Institute of Technology, Chennai, Tamil Nadu, India
[2]National Institute of Technology Puducherry, Tamil Nadu, India

Abstract

Low/No-code Development Platforms established a new paradigm in application development, offering ease of use around the environment that allowed non-IT professionals to build applications with very minimal or no coding experience. This can largely enhance digital transformation and process automation due to the following reasons: it makes data analysis and manipulation easy, flexible, cost-effective, and easy to use because of drag-and-drop features. On top of that, they provide shorter development cycles and better risk management. According to Gartner, the demand for app development services may grow in the market at least fivefold compared with IT's capacity to deliver; therefore, traditional IT projects may be delayed due to long queues. Low/no-code platforms help resolve this challenge by enabling faster application development.

On the other hand, already-existing low/no-code development platforms, including Mendix, OutSystems, Appian, Zoho Corporation, and Quickbase, can be based on GUI. In each of these single cases only, this permits making a developmentally intuitive environment for users with no programming knowledge. Hence, these platforms are easily accessible to a great circle of users with no formal coding skills.

Collaborative robots, or cobots, are designed to share tasks alongside humans in industrial environments, those tasks having been specifically identified as packaging, welding, and palletizing. These robots were initially programmed using the C and C++ programming languages, but they are now being trained by instructors.

**Corresponding author*: vallidevi.k@vit.ac.in

V. Ramasamy, S. Balamurugan and Sheng-Lung Peng (eds.) Intelligent Robots and Cobots: Industry 5.0 Applications, (85–106) © 2025 Scrivener Publishing LLC

Integrating techniques that encompass machine learning—a segment of artificial intelligence—with deep learning significantly enables cobots to perform. These neural networks function much like the human brain and will learn from huge amounts of data to create a robust link between low/no-code platforms and cobot technology.

The development process for cobots with low/no-code technology includes some vital steps. First, the physical cobot will be developed, and then it is linked with the GUI. Further, the cobot will be trained with AI-based code modules from a repository. After this initial phase of training, the humans rigorously test the cobots. In case the performances are not expected, further trainings are easily administered by configuring through the GUI options. By this iterative process, the cobots are made continuously optimized to their tasks.

Overall, using low/no-code platforms in cobot development marks a sea change in democratizing higher robotics while making them more efficient. These platforms help speed up and make affordable the development process of cobots for varied industrial applications, hence improving productivity and collaboration between humans and robots. The various aspects of this are explained in detail in subsequent chapters, in order to arrive at a thorough understanding of the potential and implementation of low/no-code software development in the domain of collaborative robotics.

***Keywords*:** Low/no code, cobot, GUI, application development, process automation, drag-and-drop, machine learning, RPA

5.1 Introduction

5.1.1 Low/No Code

Low/no code is used as a method that designs and develops an application using interactive graphical features. It replaces the traditional way of coding or writing programs with less or no coding experience. It helps non-IT professionals develop an application without using a single line of code. Low code requires basic coding skills to build up complex applications, while, in the case of no code, no basic programming knowledge is required.

Low/no-code platforms have created a great impact in the world of programming. These platforms are handled with ease because of their visual workflow, drag-and-drop features, and more. By 2025, it is projected that approximately 70% of newly developed applications in organizations will rely on low-code or no-code technologies, a significant increase from the previous figure of less than 25% in 2020, as indicated by research conducted by Gartner [1].

In 2011, Forrester first regarded the low-code platform as a productive platform to build custom apps. In 2012, the first no-code platform "Bubble" was launched, which replaced traditional coding. In 2014, Forrester coined the term "low code" in order to differentiate or classify the different development platform that focuses on simplicity and ease of use. A new method by which the already existing program code could be read by a software bot that categorizes the working of every code block and fetch the appropriate functionalities that can be customised for the new software to be developed is discussed in [9]. A new method will enable a software bot to read existing program code, label the functionality of each code block, and fetch appropriate functionalities that can be customized for new software development [9].

The adoption of low-code and no-code platforms is revolutionizing software development, empowering individuals with limited coding skills to create both internal and external technology solutions. This transformative trend not only addresses the skill shortage but also provides numerous benefits to business users, as highlighted by research [1].

- Within conglomerates embracing low-code platforms, a majority of employees (56%) are actively utilizing applications built on these platforms.
- Feedback from two out of five low-code users indicates enhanced collaboration, accelerated development timelines, and reduced costs. Specifically, low-code projects have demonstrated a 53% cost reduction and are completed 56% faster.
- The growing significance of low-code solutions is evident in their integration into critical business operations. Around 33% of low-code users have leveraged the technology to develop mission-critical applications.
- Recognizing its advantages, software developers acknowledge the flexibility of low-code platforms, with 51% of their regular development tasks being potentially executable on such platforms.

The traditional way of developing an app requires hours of coding and its time consuming but Rapid Application Development (RAD) via low/no code prioritizes rapid prototype building, which enables the non-talent developers to take a stab at building apps with no hours of coding.

5.1.2 Analysis of Various Low/No-Code Platform

The analysis of various low/no-code platforms is being depicted as shown below [2].

5.1.2.1 Microsoft Power App

It is a low-code development platform for mobile and website applications. It has pre-existing templates and user interfaces and prevents us from starting from scratch, and it extends its functionality by using Azure functions and custom connectors to proprietary or on-premises systems. Supported databases are OneDrive, Salesforce, SQL server, etc.

5.1.2.2 Outsystem

Outsystem is a full stack low-code platform. This low-code platform increases the team's productivity dramatically; with a push of one click button, the teams can deploy and make changes to the app they are developing. A few of the supported databases are SQL, Azure, Oracle, IBM, etc.

5.1.2.3 Kissflow

It is a new breed cloud platform that manages all kinds of workflows. It is simple and intuitive interface makes it easy for all users to use a simple platform instead of multiple tools. The drag-and-drop features let the user automate the human-centric approval processes. The supported database is SQL.

5.1.2.4 Bubble

Bubble is a powerful and the first no-code platform that designs and deploys web applications without coding, and it creates designs and customer user experiences using drag-and-drop features. It has code-free data storage, and there is no limit on the number of users. Supported databases are PostGres, MySQL, and Microsoft SQL.

5.1.2.5 Mendix

Mendix, as the pioneering low-code platform, specializes in developing proactive and intelligent applications. The components used in this platform are reusable. Tasks that are time-consuming are made to perform in the background so that the user need not wait and work within the app. It builds apps effortlessly using visual development tools. Supported databases are MariaDB, MySQL, IBM, etc.

5.2 Cobots

A cobot is also known as a "collaborative robot," which has an in-built feature that enables the safety of humans. Cobots provide collaboration through consistency and precision. Collaborative robots have created important developments in the robotics industry; the first automation technology that allows safe operation among humans is as shown in Figure 5.1.

There are several differences between traditional robots and cobots. Cobots are smaller and easier to program than traditional robots. The cobots learn their tasks through physical teaching and servo-assist training [16].

5.2.1 Types of Cobots

1. **Power and Force limiting:** These cobots have the ability to sense force and stop operation to avoid collisions.
2. **Safety-rated monitored stop:** These cobots are traditional robots that are installed with sensors, which discontinue when a person is detected in the work envelope (robot range of movement).
3. **Hand guiding:** These cobots are directly operated by the operation. Hand-guiding cobot in automatic mode allows the operator to manually guide it. These cobots are generally used in lifting heavy objects.

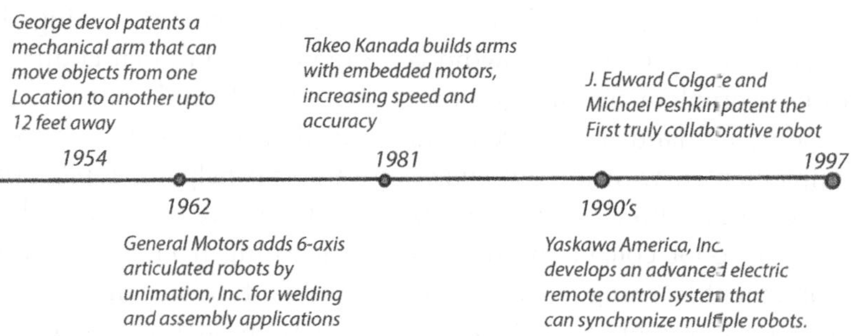

Figure 5.1 Timeline of cobots [20].

4. **The Speed and Separation:** These cobots share similarities with safety-rated monitored stop cobots, but they differentiate themselves by incorporating an additional visual system during installation.

5.2.1.1 Uses

Cobots are used to perform many tasks like machine tending, welding, material handling, packaging, and work assistance. Cobots are perfect for many industries and businesses, which involve processes (CNC machining and precision manufacturing, and metal fabrication) that require consistent, repeatable steps [18].

5.2.1.2 Advantages of Cobots

- Fast installation
- Quickly programmed
- Can be used in a different department
- Fast return on investment (ROI) and fast payback periods
- More consistent and accurate than humans
- Increased productivity and process optimization

5.3 Design of Low/No-Code–Based Cobot Development

The recent trend of low/no-code platforms is because of their drag-and-drop features and visual tools, which can be integrated with cobot programming, producing the best results [14]. With cobots, having simple programming options and low/no code requiring minimal coding skills gives us a problem-free integration of low/no code and cobots. To meet the scaling demands of different industries that use cobots as a tool can be implemented with low/no-code software that helps in rapid production and reduces hours of training and time. Extensive research in the field of human–robot collaboration, including learning from demonstration, reinforcement learning, collaborative task semantics, and more, has been focused on improving specific enabling functions such as visual perception, action recognition, intent prediction, safe on-line motion planning, and others. The primary objective of this research is to facilitate human awareness, thereby enabling cobots to exhibit flexible behavior, as opposed to the rigid action sequences found in traditional cobot programs. The recent trend of low/no-code platforms is because of their drag-and-drop features and visual tools, which can be integrated with cobot programming, producing the best results [14]. With cobots, having simple programming options and low/no code requiring minimal coding skills gives us a problem-free integration of low/no code and cobots. To meet the scaling demands of different industries that use cobots as a tool can be implemented with low/no-code software that helps in rapid production and reduces hours of training and time. Extensive research in the field of human–robot collaboration, including learning from demonstration, reinforcement learning, collaborative task semantics, and more, has been focused on improving specific enabling functions such as visual perception, action recognition, intent prediction, safe on-line motion planning, and others. The primary objective of this research is to facilitate human awareness, thereby enabling cobots to exhibit flexible behavior, as opposed to the rigid action sequences found in traditional cobot programs [15].

The recent statistics explaining the exponential rise say that, by 2025, the market for cobots is expected to be $12 billion as shown in Figure 5.2. But still when the cobots are made in the traditional way, it did not account for a large piece of the pie when compared to robots. Traditional cobot programming methods use either a teach pendant or offline software.

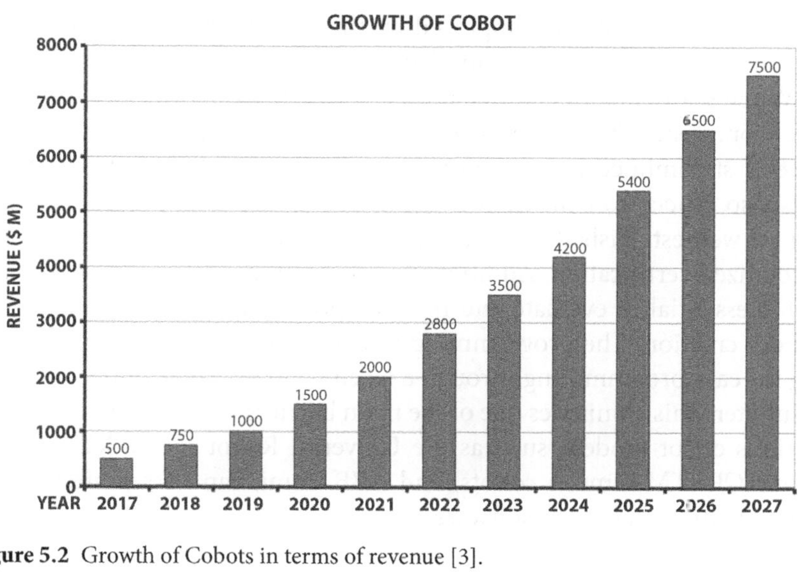

Figure 5.2 Growth of Cobots in terms of revenue [3].

This method requires some programming knowledge, which becomes a disadvantage for the growth of cobots.

Considering the advantages and functionalities of low/no code and cobots, we can make their integration efficient and balance the industrial demand.

For the creation of a low/no-code platform, the first approach is data collection. To address the shortcomings of the first approach and extend the list of supported platforms, without being limited to HTML or code that can be displayed as UI components, we can consider the editor's architecture based on the intermediate data model (Editor -> JSON -> code).

Data files can be extracted from the relevant cobot repository. It is not necessary for the data format to be in JSON because JSON files can pose uncertainty when linked with untrusted browsers or services, despite offering certain advantages.

To ensure no limitations on the UI component, developers employ an intermediate data model, allowing for seamless integration of a wide range of UI components through a comprehensive drag-and-drop functionality. The editor's primary task is to accurately convert the UI elements into the intermediate data model without any loss of fidelity.

Converting the data model from the repository, which encompasses a comprehensive database infrastructure composed of multiple databases used for data analysis, sharing, and reporting, necessitates the utilization of a converter or compiler to generate the implemented code. This process presents significant challenges, as it demands a profound expertise not only in the technologies employed by the editor but also in the specific domains where the code is compiled.

When searching for cobot brands, it is beneficial to prioritize those that offer comprehensive education and training resources. This approach facilitates a streamlined training process and reduces implementation times. It is also crucial to consider the quality of the training programs, focusing on well-established cobot training curriculums that lead to industry-recognized certification. Additionally, conducting a cobot risk assessment test is essential to evaluate and mitigate potential risk factors during the cobot's creation. The programming of cobots has been simplified, allowing for easy programming through a teach pendant, smartphone, or tablet computer. This eliminates one of the main barriers to successful upskilling. Various cobot models, such as the Universal Robot series, KUKA LBR, Fanuc CR, TM-Omron cobots, and ABB Yumi, can be seamlessly integrated with low/no-code software.

After the hardware components of cobots are manufactured, the software that is built using low/no code is integrated with the cobot. With human guidance, the cobot is prone to different sets of instructions that can be accessed through a repository designed based on low/no code. For the cobot to understand which pathway is needed for accessing the desired application and repository, cobots use neural networking [19] (deep learning) and machine learning (supervised learning) as shown in Figure 5.3.

Example:
For performing the task of drilling:

- Input: The human provides the drill location as a specification during runtime.
- Output: The cobot autonomously drills a hole precisely in the specified location, ensuring motion is constrained to the axis of the drill bit.
- Initially, the cobot utilizes neural network processing and machine learning techniques to interpret the input information.
- For signalizing the output, the decoded pathway is followed, and it reaches the data stored in the repository via the low/no code application (drag and drop button).
- Finally, the task is accomplished.

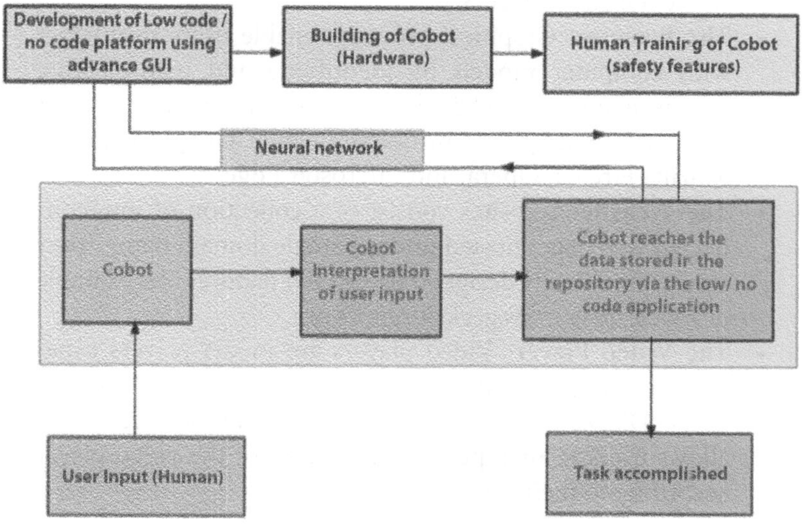

Figure 5.3 Architectural design of COBOT using low/no-code software [4].

5.4 Graphical User Interface Features

A Graphical User Interface (GUI) acts as the intermediary between users and electronic devices such as computers and smartphones. It employs visual elements like icons, menus, and graphical representations to facilitate user interactions. Unlike text-based interfaces, GUIs present information in a graphical format rather than text, enhancing the user experience. GUI descriptions are manipulated using input devices such as a mouse, trackball, stylus, or by direct touch interaction on a touch screen

The initial form of human/computer text interface relied on keyboard input, commonly referred to as a DOS (disk operating system) prompt. This interface required users to interact with the computer system by typing commands through the keyboard. This method needs the commands to be spelled in an exact way. This feature makes it an inefficient interface. The introduction of GUI has made computers and technologies more accessible to average and non-IT professionals. GUI is not fundamentally machine-centered; they are standardized in digital applications because the design is centralized by the user.

The four types of GUI are as follows [5]:

- **The Operating System:** An operating system (OS) with a GUI serves as a bridge between the hardware of a system and the user. Some examples of operating systems that incorporate GUI are Microsoft Windows, Linux, Android, Apple iOS, and MacOS.
- **The Browser:** Graphical browsers enable users to interact with computer models by identifying and manipulating the visual and geometric properties of the models. Popular examples of graphical browsers include Google Chrome, Mozilla Firefox, Opera, and Microsoft Edge.
- **The Website:** Websites consist of a collection of interconnected web pages hosted under a single domain name. They can be graphically designed using GUI features. An example of a website is www.geeksforgeeks.org.
- **The Video Player:** Video players are media players capable of playing video data from various sources such as local disks, DVDs, VCDs, CDs, and streaming services. GUI allows users to manipulate the features of the video player according to their preferences.

Low-code and no-code approaches empower professional developers to rapidly create applications without the need for manual coding. These approaches also enable non-developers such as business analysts, office administrators, and small-business owners to build and test applications. Low-code and no-code approaches empower professional developers to rapidly create applications without the need for manual coding. These approaches also enable non-developers such as business analysts, office administrators, and small-business owners to build and test applications [17]. By leveraging GUI-based features, individuals with little to no knowledge of traditional programming languages or machine code can design applications using drag-and-drop functionality.

In both cases, the GUI allows users to easily integrate components and third-party Application Programming Interfaces (APIs), rearrange modules, and conduct iterative testing until the desired app functionality is achieved. A well-designed GUI also aids in proper documentation of the application's components, facilitating easy access for cobots when retrieving desired applications from the repository.

The building blocks for low/no-code applications encompass various components with distinct functionalities:

- **Window:** The window component serves as the visual element that displays information on the screen. Manipulating a window is simple, as it can be opened or closed with a single click and easily moved around by dragging it. In multitasking environments, multiple windows can be open simultaneously, each performing different tasks. GUI offers various types of windows, including container windows, browser windows, text terminal windows, child windows, message windows, and more.
- **Menu:** Menus consist of a list of choices that allow users to select an option. A menu bar is typically displayed horizontally across the screen, offering a pull-down menu when clicked. Context menus, on the other hand, appear in response to specific user actions, such as right-clicking the mouse. They provide relevant options based on the context of the action.
- **Icons:** Icons are graphical representations used to represent files, programs, web pages, and other entities within a GUI. They offer a convenient and quick way to access documents, launch programs, or perform other actions by simply clicking on them.

- **Controls:** Controls, also known as widgets, enable users to interact with and manipulate information in an application. They can display lists of items, facilitate navigation through links and tabs, and allow data manipulation through check-boxes, radio boxes, and other elements.
- **Tabs:** Tabs are associated with view panes and often feature text labels or graphical icons. They facilitate the switching between different sections or widgets within an application. Tabs are commonly used in web browsers, such as Internet Explorer, Firefox, Opera, Safari, and others, allowing users to open multiple web pages simultaneously and switch between them seamlessly.

These components collectively contribute to the user-friendly and visually appealing interface of low/no-code applications, enabling users to create and interact with software solutions without extensive programming knowledge.

The flowchart of GUI working with low/no code is given below in Figure 5.4.

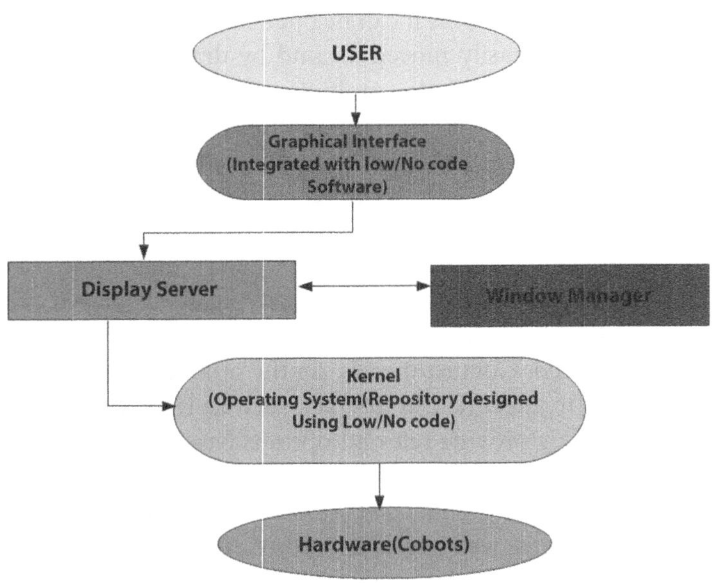

Figure 5.4 Working of GUI [6].

GUI design principles in low/no-code platforms align with the model-view-controller software architecture. This architecture effectively separates the internal data representations from the user-facing presentation, empowering users to explore and leverage various functionalities without relying on manual coding. Through visually interactive elements, users can interact with data, and these elements are designed to respond appropriately based on the type of data they hold, facilitating seamless completion of user tasks. The low/no-code platform acts as a repository, storing all the data using interactive features, enabling cobots to efficiently extract information, and performing the necessary tasks.

5.5 RPA vs. Low Code No Code in Cobot Development: "Low Code or RPA? Who Wins?"

Robotic Process Automation (RPA) helps to automate each and every task that can be performed manually. Virtual bots, through programming and training, are capable of executing tasks that were traditionally performed by humans. These tasks include activities like information filtering, data migration, and form completion. The primary objective of RPA is to offload repetitive and mundane work to the bots, allowing human workers to dedicate their time and efforts to more high-value and meaningful tasks. By automating routine processes, RPA enhances productivity and enables human workers to focus on tasks that require critical thinking and decision-making abilities.

5.5.1 Working of RPA [7]

RPA encompasses various technologies, but the fundamental concept involves programming both bots and intelligent bots. By adhering to predefined rules established by programmers, intelligent bots, leveraging artificial intelligence (AI), are equipped to perform specific tasks. These bots learn from user interactions, including clicks and mouse movements, and once they accumulate sufficient data, they can autonomously execute tasks without human intervention.

RPA finds extensive application in diverse business tasks:

- Scraping and extracting data from websites or other systems.
- Extracting relevant data and content from emails, PDFs, and various document formats.

- Facilitating data, file, and folder migrations.
- Transferring information seamlessly between disparate applications.
- Automating form-filling processes.
- Copying and pasting information across different systems.

By leveraging RPA, organizations can automate repetitive processes, significantly enhancing efficiency and productivity. This empowers human workers to focus on higher-value activities, driving innovation and strategic decision-making.

5.5.2 Cobots and RPA [8]

RPA is used with cobots to automate processes in factories and warehouses with frequent tool changes. As with software bots, using RPA with cobots can eliminate time-consuming, monotonous, and process steps and improve productivity. RPA can enable the use of a single cobot across a range of processes, increasing the cost-effectiveness of having cobots, especially for small to medium-sized businesses.

In addition, cobots are often small and lightweight and easy to move around the facility. RPA makes them easy to reprogram and redeploy. Examples of cobot tasks with RPA include the following:

- **Pick and Place** can cover various activities from loading machines such as automated welders to packaging in warehouse and logistics operations. Cobots also can provide more precise movements and handle larger weights when compared with human operators.
- **Process Finishing** activities tend to be repetitive tasks. Cobots can do the same process again and again without taking a lot of time by making it judicious. Many finishing processes require large amounts of force that cobots can consistently supply without fatigue. Cobots also can provide delicate motions when needed.
- **Quality Control and Inspection** can be implemented with cobots using high-resolution cameras. The images can be compared with reference images or computer-generated models of ideal parts. With RPA, these cobots can be reprogrammed easily and often, ensuring high levels of quality resolution from processes.

Cobot programming has two ways of using RPA. Some cobot makers have developed built-in RPA support and no-code development interfaces. In addition, specialized companies are actively creating comprehensive sets of development environments that enable no-code and low-code application development, specifically tailored for integration with cobots from various robot manufacturers. These development suites provide users with user-friendly tools and interfaces, empowering them to build applications without extensive coding knowledge while seamlessly incorporating cobots into their workflows. In addition, an open-source RPA community has emerged. The bot building is linked with the cobot software and uses neural network and machine learning.

Cobot programming using RPA tools is implemented in various ways:

- Physically moving the cobot arm(s) from position to position to create the program.
- Using a stylus to trace the cobot arm(s) required path and identify actions that need to take place at specific locations.
- User-friendly GUIs equipped with intuitive drag-and-drop or point-and- click functionalities.

5.5.3 The Downfall of RPA [13]

- RPA is not specifically designed to optimize complex processes on a large scale. Its primary focus is on streamlining simpler procedures, such as data entry or formatting data for reporting purposes.
- Failures in RPA implementations can occur, and, despite efforts to mitigate them, they cannot always be completely avoided. Nonetheless, RPA offers greater adaptability and long-term reliability when it comes to repetitive tasks compared to human employees.
- RPA bots primarily interact through user interfaces, making them sensitive to even minor changes in those interfaces. Unlike humans, programmed robots lack the ability to adjust their behavior accordingly, which can lead to broken processes. The same holds true for changes in the underlying processes themselves.
- Deploying numerous bots for process automation often necessitates transitioning their management and maintenance from individual departments to the IT department.

This integration into the corporate IT infrastructure introduces additional complexity and overhead, requiring careful oversight to ensure proper functionality without disrupting other IT processes or increasing risks.
- While RPA offers a tactical and relatively straightforward approach to achieve quick process efficiency gains, it can inadvertently divert attention away from more strategic projects. These projects may involve developing new systems to support disruptive business processes or replacing legacy core systems that hinder progress.
- Although the combination of RPA and AI to address complex processes is an intriguing prospect, it is still in its early stages. This can lead to fragmentation and longer setup times. While some vendors are attempting to merge AI with RPA to overcome this challenge, this approach is still nascent and often comes with a higher price tag.

5.5.4 Low/No Code's Edge Over RPA

Table 5.1 explains the difference between low/no code and RPA in terms of application use and when to use.

- **The Agile Low/No Code**
  Low/no-code applications allow to entirely create a new one that adapts quickly to the current requirement, rather than mending and tweaking old applications. Low/no code's built-in codes, workflows, and firms are able to become more agile in the long run by reducing the amount of coding they need and by automating their testing.
- **Enhanced Applications**
  RPA is limited in enhancing existing applications as it works within the confines of the current configuration. In contrast, leveraging low/no code enables quick and easy addition of features or complete redesign, offering greater speed and simplicity
- **Flexibility**
  RPA is known for its rigidity and slow adaptation to change, while low/no code provides the flexibility to easily adjust and adapt to new requirements. It streamlines the engineering process and simplifies system restructuring, empowering

Table 5.1 Comparison between low/no code and RPA [10].

+	**Low code/no code**	**Robot process automation (RPA)**
Applications	• Database managing applications • Realistically any transactional business process • Business process • Omni-channel platforms demand seamless integrations with Application Programming Interfaces (APIs) and business logic, as well as real-time charts, live data grids, and dynamic lists that offer instant data visibility and insights.	• Streamlines and automates repetitive and mundane tasks, including application login, file and folder management, data copying and pasting, form filling, structured and semi-structured data extraction from documents, and web scraping. • Typically employed in repetitive and low-value back-office tasks and activities. • Occasionally utilized as a short-term solution to address issues with malfunctioning or inadequately integrated software/systems.
Users	Mendix, Outsystem, Zoho creators, and Knack	UIPath automation, SignNow, and Pega Platform
When to use	• Automation spanning across the entire enterprise. • Developing fresh applications and websites. • Improving and optimizing current applications. • Seamlessly integrating with intricate legacy systems.	• Automating simple, repetitive, and routine tasks

businesses to experiment with new products and swiftly respond to evolving market needs.

- **Easy to Manage**
  RPA bots require ongoing management, maintenance, and security measures. In contrast, low/no-code platforms reduce maintenance efforts with ready-to-use templates and simplified development procedures. This results in fewer bugs and a less stressful development process.
- **Time Matters**
  Both low-code systems and RPA contribute to increased productivity by enabling more work to be done in less time, allowing users to focus on other challenging or engaging projects. However, RPA entails a complex setup and significant time investment in identifying and analyzing the processes to automate. On the other hand, low/no code empowers organizations to create applications at the same speed as ideation, providing long-term agility and automated testing benefits.

5.6 Conclusion

Low/no-code platforms are revolutionizing the software application development landscape, including the cobot industry [11]. In the post-pandemic era, where organizations are increasingly adopting remote solutions, these platforms play a crucial role in expediting digital transformation. While both RPA and low/no code offer distinct use cases, advantages, and limitations, the need for swift innovation poses a significant challenge for businesses relying on RPA due to its susceptibility to change. With the increasing demands of various industries and processes, it would be wise enough to choose a low/no-code platform to build a cobot that does not need any coding skills, when compared to RPA.

5.7 Pros and Cons

Pros of low/no code are shown in the below discussion [12].

- Easy to use: These platforms provide a user-friendly interface that simplifies the application development process, allowing individuals without extensive technical knowledge to quickly create digital products.

- Rapid time-to-market: With straightforward logic, users can build solutions within a few hours, ensuring faster product deployment and reducing time-to-market.
- Robust security: The advantage of vendor lock-in is that source code inaccessibility protects against accidental security errors, providing a relatively high level of security for users.
- Scalability and customization: Leveraging coding capabilities, clients can work with built-in templates and customize them to create highly scalable and personalized products.
- Empowering a wide range of users: These platforms enable employees to independently create and deploy applications, reducing reliance on the IT department and allowing them to focus on strategic projects.
- Low/no-code community support: There are communities of similar users available to help with any problems or challenges faced during development.
- Cost-effective: Whether building from scratch or utilizing low/no-code solutions, these platforms offer a more affordable and efficient development process.
- Increased productivity: Low-code tools save time by reducing repetitive manual coding, enabling developers to create more applications in less time.
- Minimal training required: These platforms are designed to be user-friendly, requiring minimal training and allowing anyone to quickly start building applications.
- Easy updates: Applications built with low/no code can be easily updated, allowing for continuous improvement and adaptation.
- Empowering internal application development: These platforms provide a safe environment for employees to develop applications that address their unique needs, reducing reliance on third-party vendors and mitigating shadow IT risks.

5.7.1 Pros of Cobots

- Lower error rates: It reduces error percentage while doing any processes.

- It has greater ease of use in its operation.
- Cobots can seamlessly operate alongside human workers in the same workspace, fostering collaboration and enhancing productivity.
- The implementation of cobots requires minimal upfront investment and avoids substantial modifications to the work environment, particularly when a dedicated space for the cobot is unnecessary.
- Simplified programming and seamless integration capabilities.
- The investment in building cobots can be recouped within a short timeframe.

All these features make low/no-code platform an excellent choice for designing a robot to perform tasks and activities.

5.7.2 Cons of Low/No Code

- Limited flexibility for custom updates: Making general updates in low/no code–based products can be time-consuming and hindered by the vendor's initiatives. Despite the perceived flexibility, managing and customizing these products may prove challenging.
- Potential integration issues: When opting for a low/no-code platform, ensuring seamless integration with existing software is crucial. However, integrating low/no-code–based solutions with legacy systems and third-party services can often be difficult.
- Costly maintenance: Effectively designing a low/no-code environment may require the expertise of specialized engineers. Despite their apparent simplicity, low/no-code platforms have complex characteristics that demand proper and regular maintenance by specialized professionals.
- Need for dedicated developers: Developing applications using low/no code approaches may still necessitate dedicated developers, contradicting the goal of reducing costs and development time.
- Vendor lock-in risks: Relying on a single low-code vendor may lead to vendor lock-in, posing challenges in the future, particularly if developers become heavily reliant on coding with that specific platform.

- Potential for shadow IT: If adequate precautions are not taken, then the use of low/no-code platforms can contribute to shadow IT, where employees utilize unauthorized hardware, software, web services, or cloud applications, leading to potential security risks.

5.7.3 Cons of Cobots

- Cobots rely on human assistance and supervision, as they are designed for collaborative work alongside humans.
- They have limitations in terms of reach, payload, speed, and accuracy compared to more powerful six-axis robots.
- Cobots may not be suitable for high-speed and high-payload applications.
- They excel in applications with lower payloads.
- Implementing safety precautions can lower productivity, resulting in reduced operating speeds and increased integration costs.
- Conducting a thorough risk assessment is necessary to ensure appropriate safety measures.
- If required, the installation of expensive fencing can incur substantial costs.

References

1. Roe, D., What's Behind the Explosion of Low-Code and No-Code Applications, accessed 31 December 2022, <https://www.reworked.co/information-management/whats-behind-the-explosion-of-low-code-and-no-code-applications/amp/>.
2. Talesra, K. and Nagaraja, G.S., Low-Code Platform for Application Development. *Int. J. Appl. Eng. Res.*, 16, 5, 346–351, 2021, ISSN 0973-4562.
3. Collaborative Robot Market to Grow by 60% in 2018, accessed 31 December 2022, <https://www.roboticstomorrow.com/story/2018/10/collaborative-robot-market-to-grow-by-60-in-2018/12661/>.
4. What are the Different Types of Collaborative Robots?, Last accessed on 31 December, 2022, https://processsolutions.com/what-are-the-different-types-of-collaborative-robots/.
5. Graphical User Interface (GUI), Last accessed on 31 December, 2022, https://www.techopedia.com/definition/5435/graphical-user-interface-gui.

6. What is GUI in Computer? How does a GUI work?, Last accessed on 31 December, 2022, https://eevibes.com/computing/introduction-to-computing/introduction-to-gui-in-compute-system/.
7. Robotics Process Automation (RPA) and Lowcode Process Automation, Use Cases Benefits and More, Last Accessed on 31 December, 2022, https://www.mendix.com/blog/rpa-low-code-process-automation/.
8. Low Code, No Code, Fast Code, or RPA. Full tutorial 2021, Last Accessed on 31 Dec, 2022, https://www.xme.digital/post/low-code-no-code-fast-code-rpa-tutorial.
9. Balwal, U. and Krishnamurthy, V., Bot Assisted Software Development. *Int. J. Innov. Technol. Explor. Eng. (IJITEE)*, 8, 6S, 42, 2019.
10. Low Code or RPA? Who Wins?, Last Accessed on 31 December, 2022, https://www.pillir.io/edgeucation-center/blog/low-code-or-rpa-who-wins.
11. Why Should You Use Low-Code Automation Instead Of RPA, Last Accessed on 31 December 2022, https://kyanon.digital/why-should-you-use-low-code-automation-instead-of-rpa/.
12. Robotic Process Automation (RPA) and Low Code: When should you use what?, Last Accessed on 31 December, 2022, https://www.outsystems.com/blog/posts/robotic-process-automation-low-code/.
13. Downfall of RPA: https://www.advsyscon.com/blog/why-rpa-fails-robotic-process-automation/#:~:text=RPA%20is%20most%20common%20outside,and%20includes%20additional%2C%20unnecessary%20steps
14. Steps to design low and no code platform : https://codeornocode.com/tutorial/how-to-build-a-low-code-platform/
15. Quixy, LCNC - Low Code No Code Platform, pp. 46, from https://quixy.com/lcnc/.
16. Man+Machines, Differences Between Robots and Cobots, 2022, January 18, Retrieved from https://manplusmachines.com/differences-robots-cobots/.
17. Zoho Corporation, Low-Code Platform, Zoho Creator, pp. 50, from https://www.zoho.com/creator/low-code-platform/.
18. WiredWorkers, Top 7 Cobot Applications in Manufacturing, 2021, May 27, Retrieved from https://wiredworkers.io/cobot-applications/.
19. Investopedia, Neural Network, From https://www.investopedia.com/terms/n/neuralnetwork.asp.
20. History of cobots, https://www.universal-robots.com/about-universal-robots/news-centre/the-history-behind-collaborative-robots-cobots/.

6

Future Workforce for Industry 5.0

Huzina Saheal and Sheikh Sajid Mohammad*

School of Business Studies, University of Kashmir, Srinagar, J&K, India

Abstract

Industry 5.0 refers to an arrangement whereby humans work alongside smart machines and robots. It involves utilizing cutting-edge technology to enable robots to assist people in working more efficiently. The transition from Industry 4.0 to Industry 5.0 will create new industrial sectors and job opportunities. The World Economic Forum 2020 has predicted that, by 2025, 85 million jobs currently performed by humans will be automated. Subsequently, about 97 million new jobs might be required for the occupations of the future, which are anticipated to be performed in collaboration with people, robots, and computers. These forecasts suggest that employment in the upcoming years will be significantly impacted by the new industrial revolution. As a result, the vast majority of people in the labor market of today will need to acquire new skills, as there will be a significant change in nearly every job. Employees should have self-management abilities and must also learn new skills or enhance existing ones in order to assess, compare, and comprehend information. In order to overcome the obstacles in the transition to Industry 5.0, personnel must be resilient, stress tolerant, and adaptable. The chapter will offer critical reflections about the benefits to workforce in Industry 5.0, challenges that workers will have to face due to transition to Fifth Industrial Revolution, and the in-demand skills that workers should possess to be prepared for Industry 5.0. The chapter also entails the issues related to integration of technology and human workforce and considerations that managers should keep in mind before integrating technology and workforce. Finally, recommendations for preparing workers for future of work by reskilling and upskilling them are also presented.

**Corresponding author*: sheikhsajid298@gmail.com; sajid.scholar@kashmiruniversity.net

V. Ramasamy, S. Balamurugan and Sheng-Lung Peng (eds.) Intelligent Robots and Cobots: Industry 5.0 Applications, (107–130) © 2025 Scrivener Publishing LLC

***Keywords*:** Industry 5.0, employee skills, cobots, reskilling, upskilling

6.1 Introduction

Industry 5.0 aims at creating a more interactive environment between humans and machines throughout plant operations. Here, human engagement would be cognitive rather than physical, necessitating significant modifications to the industrial world that prioritizes the well-being of workers during the manufacturing process. The article "Industry 5.0: From Virtual to Physical" [1] was the first source to introduce the concept of Fifth Industrial Revolution. Human contribution to Industry 5.0 involves combining critical thinking and cognitive abilities with the high speed of automation [2]. As such employees are required to upskill and reskill in order to increase their added value. The idea of humans and machines cooperating could be made practical by assigning repetitive and labor-intensive tasks to robots while leaving personalization to humans. As a result, all monotonous, repetitive tasks will be performed by robots, allowing workers to focus more on their creative abilities and to closely monitor the systems to improve overall quality of output. Two major objectives of Industry 5.0 stand out from a broader perspective. Firstly, algorithms facilitate the collaboration of humans and robots to provide customized production or services. Secondly, creation of a sustainable economy. Thus, Industry 5.0 aims to create a sustainable bio-economy on a global scale while also paving the way for a system where people, robots, and algorithms can work in cooperation [3].

Industry 5.0 would foster mass personalization and customization of products for consumers by using human intellect and computers through a process known as cognitive computing. To enhance operational efficiency and intellectual capabilities, Industry 5.0 would seamlessly incorporate advanced technologies in current workforce. The cobots would also be capable of understanding the requirements of the human operator, determining whether help is required and providing it as needed. The workforce will be able to work alongside robots in peace because they will be able to successfully communicate with and collaborate with their robotic co-workers which will lead to a process of production that is efficient and adds value [4].

Industry 5.0 is expected to generate more employment than it will take away. The development of a new breed of industrial robots, maintenance, planning, training, and programming are few of the professions that are anticipated to emerge in the disciplines of artificial intelligence (AI) and robotics.

Recent studies have shown that bringing people and technology together would not result in layoffs. According to estimates, Industry 5.0 will help increase the current 1.2 billion global manufacturing workforces by up to 4% by 2030 [5]. With a rise in R&D (Research and Development) investments, intelligent sales and marketing strategies, and improved distribution methods, supply chain management across various industries would witness enhanced creativity and adaptability. Moreover, there would be a notable decrease in the cost per unit of value delivered. The need for a competent workforce will increase as new technologies develop by the time Industry 5.0 is completely implemented. Industry 5.0 expects trained workers to offer high-value tasks in manufacturing and standardization.

Industry 5.0 has been explored in the literature since 2017 [2, 3, 6–9]. Despite the growth of knowledge, few studies have looked at the connection between Industry 5.0 and employee skills [8, 10, 11]. From this perspective, Paschek *et al.* [8] highlighted the necessity of looking at the in-demand employee skills required for Industry 5.0. Against this backdrop, this chapter aims to provide a brief explanation of the idea of Industry 5.0, the requisite skills for working in Industry 5.0, the underlying principles upon which Industry 5.0 is built, and the potential difficulties the workforce may face in adapting to Industry 5.0. The chapter also contributes in providing some ways to effectively integrate employees with technology for raising their morale and productivity.

6.2 Underlying Principles of Industry 5.0

There are three main core pillars on which Industry 5.0 is based: human centricity, sustainability, and resilience. These are discussed below.

6.2.1 Human Centricity

In human-centric approach, needs and interests of humans are assumed to be fundamental to the industrial process. Here, technology is viewed as a way to assist, serve, and adjust the production process to meet the various needs of the worker. It involves creating a safe and inclusive work environment to emphasise physical and mental well-being of individuals, ultimately safeguarding their fundamental rights, such as privacy, human dignity, and autonomy. As such, industry must recognize societal boundaries in order to avoid leaving anyone behind. This has several implications for safe and beneficial working environments, human rights, and the skill requirements for workers [12]. By placing an emphasis on people and their welfare, the human-centric paradigm promotes individual and group choice. This strategy increases sustainability, accessibility, and worker satisfaction.

6.2.2 Sustainability

The idea of "Industry 5.0" aims to improve a sectors' financial performance while defending the needs and interests of its people and safeguarding the environment. It is beneficial to business owners as well as potential investors, consumers, and clients due to the availability of more competitive items. Industries must be sustainable if they want to respect the constraints of the planet. Companies must embrace a circular economy approach in order to reduce energy consumption and CO_2 emissions as well as waste management of by products to avoid endangering future generations [13]. Industry 5.0 calls for the implementation of AI and additive manufacturing to improve resource management, effectiveness, and management of waste and overproduction.

6.2.3 Increased Resilience

Resilience is the ability to deal with vulnerabilities that might emerge at several levels, such as those at the factory floor, the supply network, and the industrial system, in order to adapt to change. When social and environmental conditions are steady, industry innovation frequently focuses on boosting the efficacy of production lines and supply chains. But these efficiency gains frequently come at the expense of decreased resilience. Research on industrial resilience can help in comprehending the local, regional, and technical dangers that industries are increasingly exposed to [4]. It can create and implement mitigation measures that could serve as the pillars of an industry that is both optimal and resilient in the future.

With the help of cutting-edge technology along with adaptable production lines, innovative materials, remotely managed factories, and risk tracking and management in real time, industry can attain the resilience it needs. Digital technology will play a special role in this. Despite the fact that digital interconnection will make it possible for a variety of resilient technologies (such as data collection, automated risk analysis, and mitigation measures), an increased reliance on digital technologies exposes industry to technical disruptions caused by malfunctions as well as cyberattacks [14]. A crucial part of creating the cybersecurity needed for the resilient industry of the future will be played by research and innovation. In order to adapt to the new normal, resilience is required to assure continuous operation and enable speedy recovery from unforeseen crises or breakdowns.

6.3 Benefits for Workers in Industry 5.0

One of the most crucial and fundamental transformations characterizing Industry 5.0 is the shift from technology driven growth to a fully human-centric approach. To ensure inclusivity and prevent any individual from being overlooked the industry must take societal limits into account. This has a number of ramifications, including fewer human-factor failures, a safe and supportive work environment, the creation of jobs and a greater role for human workers, as well as workforce empowerment, which are explored below.

6.3.1 Reduction of Human-Factor Failures

The two primary approaches used in manufacturing to recognize and address process errors brought on by the human element are proactive and reactive strategies. While the reactive strategy focuses on failures that have already occurred, the proactive approach takes prospective outcomes of employees' behavior into consideration [15]. Human error–related incidents can be avoided with the aid of Industry 5.0 technologies by having workers just carry out monitoring duties or refraining from taking part in the production process altogether [7]. Industry 5.0 can even monitor and evaluate what it accomplishes on its own thus preventing failures as an automated factory operated by machines.

6.3.2 Safe and Inclusive Work Environment

The two major concerns related to the use of new technologies are the potential for accidents from hazardous employment and the loss of jobs. But with proper application, new technologies could enhance workforce welfare and job satisfaction by making workplaces more inclusive and secure [10]. By replacing humans in repetitive dangerous and physically demanding tasks, robots have the potential to reduce workplace accidents, thus resulting in a safer working environment. Robotics technology still has a lot of potential, especially when combined with AI. Establishing smart workplaces requires careful consideration of a number of factors, including the potential biases that must be avoided when developing AI-driven technology to be used in manufacturing processes. When closely collaborating with industry workers, intelligent robots must avoid undermining their respect for all workers' dignity. Virtual and augmented reality tools, as well as AI-based technology, can be used to assist the worker to perform more specialized tasks that would otherwise require specialized knowledge and training [16]. As a result, there may be greater opportunities to integrate individuals with cognitive impairments into the workforce.

Similar to this, mobile robots and exoskeletons may minimize the physical demands of specific occupations, thus enabling women to do tasks that were previously exclusively performed by men. A wide range of new opportunities will emerge as the workforce becomes ever more digital. The digitization of industrial processes makes remote labor feasible, enabling individuals to find work even in remote locations and enhancing the manufacturing process' resilience [17]. The recent COVID-19 crisis, in which the ability of several organizations to operate was hindered by geographic limitations, has adequately demonstrated the potential of digitalized remote operation. Safeguarding and promoting employees' physical health at work is only one aspect of their safety and well-being. Mental health and well-being must be equally prioritized while designing digital workplaces. Businesses and other organizations should promote mental health and a culture of well-being as an integral part of their corporate culture with the use of cutting-edge technologies and digital solutions. In addition, productivity gains and the reduction of chronic disease and absenteeism are anticipated to result in financial gains and savings. According to the WHO (World Health Organization), productivity losses caused by mental health conditions including depression and anxiety are predicted to cost the global economy $1 trillion annually. Additionally, it is predicted that spending $1 on the treatment of common mental illnesses would result in enhanced productivity and health, worth $4 [18].

6.3.3 Job Creation and Better Roles for Human Workers

According to a recent study on skills gaps by Deloitte, there would be 3.4 million open positions in the coming 10 years but only 1.4 million competent candidates to fill them [19]. For many of these roles, robots are ideal, and cobots in particular can collaborate with humans. Cobots are advantageous because they may replace risky, monotonous, and menial professions while allowing human workers to move into positions of higher value. In contrast to other industrial robots, cobots can encourage enhanced skill sets and increase productivity, which can lead to more job prospects within firms. These machines can assist people in manufacturing and processing jobs rather than substituting for them. Additionally, cobots can assist manufacturers in expanding their workforce as they try to raise outputs and create new product lines and move workers from mundane jobs into positions with higher job satisfaction and pay [20]. Certainly, there are many chances to increase industrial productivity and innovation with the help of a networked and collaborative workforce. Additionally, technology offers the potential to increase work satisfaction and safety while also giving human workers access to more exciting tasks and promoting employment progress.

6.3.4 Empowering Workers

The job description and the narrative that surrounds the industrial worker change significantly in Industry 5.0. The employee should be viewed as an "investment" position for the business rather than as a "cost," allowing both the business and the employee to grow. It suggests that, in order to achieve their goals, the business is interested in investing in the abilities, skills, and general well-being of its employees. Such a strategy places a higher value on and appreciation for human capital, than just balancing worker costs with financial earnings [10]. Technology serving people, rather than the other way around, is a necessary condition for Industry 5.0. In an industrial setting, this means that production technology is tailored to the diverse demands of industry workers rather than requiring constant adaptation on the part of the workers. The workplace is more inclusive and the employee has greater control. To accomplish this, workers must play a significant role in the creation and use of new industrial technologies, such as robotics and AI. This process keeps people at the center of the production while maximizing the advantages of technology for both the employer and the employee.

6.4 Challenges for Workers in Industry 5.0

Since workers need to learn how to interact with advanced robots and acquire knowledge on how to operate together with intelligent machines, it is important to identify any potential challenges faced by workers to facilitate a smooth transition to the new industry. The workforce in Industry 5.0 needs to be aware of how their engagements with intelligent equipment and robots will work. Given the highly automated production processes, developing workers' skills can be a challenging task. For example, educating workers to adopt cutting-edge technology and causing behavioral change for working with machines can be tricky. Industry 5.0 demands AI, collaborative robotics, internet of things, customized software-connected factories, and real-time data [21]. To be able to work with sophisticated robots, employees must, therefore, have both technical and competency skills in addition to soft skills. The workforce may find it challenging and time-consuming to adapt to these cutting-edge technologies and manage the transition into new jobs. Another challenge posed to workers by Industry 5.0 is work polarization wherein semi-skilled employment will decline and the workforce could possibly split into two groups: extremely trained and qualified; and low-paid and unqualified employees [22]. Consequently, this may create the skilled and unskilled divide in the society. The heightened connectivity and widespread adoption of universal communications protocols in Industry 5.0 expose vital industrial systems and production lines to potential cyber security risks. Because authentication is required in the sector to communicate with various devices, there is a need to upskill the workers in ICT (Information and Communications Technology). Businesses will be able to avoid security issues owing to ICT-skilled personnel by deploying IoT (Internet of Things) nodes and quantum computing [23].

6.5 Industry 5.0 and Employee Skills

Industry 5.0 is expected to see an increase in automation as well as the introduction of new industrial sectors and job opportunities. In this area, it is estimated that 85 million jobs would have been automated (transferred to robots) by 2025. At the same time, up to 97 million new roles could be required for future jobs to be performed in collaboration with robots and computers [24]. These predictions suggest that staff skill sets should change to meet Industry 5.0. Cyber-physical systems are expected to develop into more complex systems; therefore, people must focus more on exercising strategic judgement when dealing with more complex issues [11]. Employees will work with cobots and algorithms in the future to delegate routine and dangerous tasks [25]. By showcasing their creativity, employees might handle the operations more skilfully and offer a better added value. It is a must for obtaining the skills that are in demand across Industry 5.0 in order to realise this potential.

Skill refers to the personal, technical, or interpersonal qualities necessary for a worker to perform required job [26]. Besides, the "skills gap" is described as the inability to complete the work [27]. Industry 5.0 is projected to transform how business is conducted, which will cause changes in the skills needed to accomplish the work, resulting in the emergence of skills gaps. Businesses who take pre-emptive measures to fill these skills gaps will have a strategic edge in the times to come. According to the Future of Jobs Report (2020), the in-demand skills needed by industry can be divided into five groups. These are discussed in detail below.

6.5.1 Problem Solving

Problem solving encompasses a wide range of skills, such as complex problem-solving; creative problem-solving; originality and initiative; as well as analytical thinking and ideation. With the progression of automation in Industry 5.0, there is an opportunity for employees to contribute greater value by tackling more complex issues [28]. In such a scenario, employees will have the opportunity to engage in critical thinking. Thus, it becomes essential for staff members to apply their problem solving skills not only in meeting the customized demands of consumers but also in interacting with robots and algorithms.

6.5.2 Working with People

The digital transformation brought about by Industry 5.0 is expected to have an impact on employee communication. It will be necessary for employees to effectively collaborate not only with intelligent machines but also with individuals both inside and outside the organization. Moreover, leaders must demonstrate a leadership style that aligns with the dynamic nature of the business and the competitive environment, ensuring that employees remain adaptable and motivated. Transformational leadership and collaborative leadership styles are regarded as useful tools because of Industry 5.0's emphasis on workers. Increased involvement, empowerment, and participation in relevant strategies are aims of Industry 5.0 [29–31]. Workers can therefore adjust to the change and contribute more to achieve organizational goals. Social influence has the potential to alter the beliefs, values, attitudes, and actions of others. The social influence of employees and leaders will increase if they exhibit leadership behavior in line with Industry 5.0 within the organization [31]. Influencing individuals both within and outside of the organization will likely be one of the skills that will be in demand in the Industry 5.0 era.

6.5.3 Use of Technology and its Development

The application of technology, its advancement, monitoring, control, design, and programming abilities are all included in this category.

In the Industry 5.0, having technical capabilities is essential. The need for sophisticated technological abilities has already started to emerge for well-known professions like data analyst, big data professional, expert in AI and machine learning, business automation, and digital marketing and strategy [24].

Utilizing technology effectively, especially through digitization, creates a self-perpetuating economic advantage [32]. As a result, workers with expertise in both the use and development of technology have the opportunity to provide something unique to organizations. Furthermore, these workers are not just found in IT (Information Technology) areas. The methods of carrying out some specific duties may also alter in other sectors including HR (Human Resource), marketing, production, and sales. For instance, job interviews conducted by utilizing AI may make it easier for HR departments to hire AI specialists [33]. Data analysts may soon manage promotion tasks in marketing departments in a similar manner.

6.5.4 Self-Management

Employees are required to have the management abilities necessary to successfully adapt to Industry 5.0. These competencies can be categorized into two types which are referred to as "self-management skills" including (a) active learning techniques, and (b) flexibility, stress tolerance, and resilience. These soft skills can be categorized as power skills because these aid employees in ensuring their survival.

The application of active learning and learning approaches is one of the competencies under self-management. For the workforce to adapt to Industry 5.0, they must understand the transformation and develop and acquire learning skills. By 2025, when Industry 5.0 is anticipated to begin, 85 million jobs are expected to be lost due to a shift in the labor-machine divide [9, 34]. Employees must reskill or even upskill to manage these gaps. Active learning, which does not just concentrate on receiving in the information being offered, is one of the fundamental prerequisites for reskilling and upskilling. Questioning, debating, and reasoning are all aspects of active learning [35]. Employees that are proficient at active learning might acquire new skills and, as a result, find their work meaningful. According to the Business Characteristics Model, motivation, job satisfaction, and performance will rise when the person finds significance in their work. As a result, the worker may influence the structure and well-being of society.

Stress tolerance, resilience, and flexibility make up the remaining self-management skills. According to Wu *et al.* [36], the ability to successfully adjust in the face of stress and adversity is known as resilience. Dündar *et al.* [37] classified flexibilities into four categories: income flexibility, functional flexibility, geographical flexibility (remote working), and time flexibility (flexible working hours). Functional flexibility is one of the different types of flexibility that can be viewed as an employee skill. Functionally flexible workers are able to maintain several positions within an organization and quickly adjust to shifting technology environments and varying workloads. Functional flexibility is anticipated to be one of the abilities that will be needed in the near future, given the growing roles that employees will undertake.

6.5.5 Cross-Disciplinary Skills

Both general and specialized skills are out of synchronization in today's world. Organizations are aware of the abilities they are lacking because they are familiar with the technology that will be needed in the future. As a result, organizations are in a better position to properly educate and train employees. Employee participation in the creation of training programs should be encouraged in order to match their demands. An employee must possess a strong entrepreneurial attitude, which includes proactivity and the capacity to think creatively, in the field of cross-disciplinary abilities. A more diverse industrial workforce will present new issues; therefore, they should also have an accommodating mindset to deal with them. Employees must develop transformational abilities that continually challenge the status quo and lead knowledge transfer from other disciplines, but, most importantly, they must maintain an open mind to change.

6.6 Issues Related to Integration of Robots into Organizations

Following are some of the organizational issues emerging from human-robot co-working.

6.6.1 Learning to Work with Robots

Peoples' knowledge of robots comes from books and films. Nonetheless, majority of employees will not be familiar with human–robot co-working. Although it is believed that interacting with robots will be simple and clear, reality may not match expectations. There is need to learn how to interact with the robot in order to get what is actually desired. Human connection often involves nonverbal communication. Robots might or might not be able to understand these nonverbal clues. People will need some time to get used to and understand how to work with robots. There will also be many kinds of robots. Some will possess more sophistication than others. Humans would not be able to discern between different-skilled robots with ease. Robot manufacturers must work on how to design machines which are able to explain their capabilities precisely. To work with various robot types, people will need to master new skills.

6.6.2 Issues Relating to Laws and Regulations

Robots are commonly defined as machines that possess the capacity to perceive, analyze and perform actions [38]. However, this definition is both straightforward and flawed. The Oxford Dictionary defines a robot as a machine capable of autonomously executing various intricate tasks particularly those programmable by a computer. Similarly, the Merriam Webster dictionary defines robot as a humanoid device capable of mimicking human functions such as walking and talking. Nevertheless, these definitions also exhibit flaws with one relying on a vague notion of automatically executing a series of complex actions. The opposing viewpoint presumes that a robot resembles a human person. These imprecise definitions show that there is still a need to come up with a more precise one. Co-working with robots would not be possible without the proper legislation, which will lead to numerous issues. A legal definition of a robot must be first established. Although the term "robot" has scientific definitions, only the legal definition of the term "robot" is enforceable by businesses and organizations. Automated machines come in a wide variety. A robot is also an automated device. Machine automation levels are specified [39]. A legislation that regulates the collaboration between humans and robots should clarify the definition of what constitutes a robot versus an automated machine. This legislation, along with other relevant regulations, should specify, at the very least, the types of robots that are allowed in workplaces, their roles and responsibilities, the decisions that robots are authorized to make, the types of malfunctions that robots may experience and who would be held accountable, the allocation of responsibilities in the event of robot malfunctions, whether robotic software is required to follow rigid pre-programmed rules or has the ability to make independent decisions, the necessary certifications for robots, the organizations responsible for certifying them, and their respective duties.

6.6.3 Personal Preferences for Utilizing Robots at Work

Every person has a different personal choice for utilization of a given technology. There will be people who are excited about working with robots and some who are against this idea. Organizations interested in utilizing human–robot collaboration in the workplace should be aware of these individual inclinations. It would not be simple or even practicable for firms to convert to human–robot co-working settings when majority of workers do not possess a favorable attitude toward robots. As human–robot co-working is not yet a reality, media portrayals of working with robots rather than actual experiences have a greater influence on people's preferences.

6.6.4 Negative Attitude Toward Robots Due to Shrinking Human Workforce

A survey found that people are most concerned about robots on a personal and societal level because they fear losing their jobs [40]. An unfavorable attitude toward robots is likely to develop as they begin to replace humans in certain occupations. This viewpoint will be refuted by the argument that more jobs will be generated. History demonstrates the soundness of this counterargument. Humans will concentrate on occupations requiring creativity, artistry, research, and development as robots take over routine tasks. Nonetheless, this viewpoint presents a conundrum. It is unlikely that those losing their jobs will have the knowledge, experience, and abilities necessary to take on the responsibility of positions requiring innovation. Others may assert that the employment of robots will result in higher unemployment. If robots lead to unemployment, then the worker unions will be fiercely opposed to them. Constituents that have this mindset will contact their legislators, either by supporting them or passing robotic legislation and regulations. Because of this, humans may initially have a bad opinion of robots because they lead to job loss, at least in the early stages of the robotic age.

6.6.5 Humans Competing with Robots or Robots Complementing Humans

Professionals and scientists will disagree on the positions and functions that robots play. An easier solution will be to use robots to supplement human labor. The majority of people might even favor submissive robot assistants. Instead of overseeing a human assistant, these robots will help humans do tasks. Also, when robots take over laborious and tedious work, humans will have more time for jobs demanding greater levels of intelligence and creativity. The easiest, simplest, and least troublesome way to create human–robot co-working settings is to have robots that complement humans. On the contrary, some individuals might contend that jobs currently held by people should be filled by robots. They will contend that doing this will raise workers' productivity. Humans naturally compete with one another for jobs inside companies. The introduction of robots will make management and organizational behavior more challenging [41].

6.6.6 Psychological Consequences of Human–Robot Co-Working

Modern technologies have a psychological impact on people. For instance, the addiction to video games is being recognized as a significant psychological problem. Some people are very reliant on their smartphones. A genuine phobia is nomophobia, or the dread of not having a mobile device. It is described as a stressful condition brought on by being unable to use or having restricted access to a mobile phone. At this point, it is unclear if there will be a "robophobia" or "norobophobia." Yet it is highly possible that the introduction of robots will have a lot of psychological effects on a large number of employees that have never been observed before. The psychological issues connected to deploying robots at work and their effects on organizational culture will be a research agenda in organizational robotics.

6.6.7 Societal Consequences of Human–Robot Collaboration

Being social beings, humans interact in all facets of their lives, including job and engaging with people. It is normal practice in many firms to hold social activities at work to boost employee performance. When more robots are used in human workplaces, there may be a drop in the number of people employed. As a result, there may be limitations on how people connect socially. Even if the number of human employees stays the same, integrating robots into workplaces may have hitherto unimagined effects on the social dynamics there. Some people might favor socially intelligent robots over others. Some individuals might think that robots' social behavior is artificial and that it is just computer programming that causes them to act in such a way. Different workers will have different perspectives on how to interact socially with robots. For example, children treat robots the same way they treat their living pets [42]. The future of human–robot connection may be comparable to how the younger generation of today interacts with cell phones and views them as an integral part of their daily lives.

6.6.8 The Shifting Functions of Human Resources Departments

One of the key duties of HR departments' is to carry out work analyses, create job descriptions, and fill open positions with qualified candidates. With the exception of those sectors that use industrial robots, the great majority of businesses now employ people. Human resource departments will encounter new difficulties as robots become an integral component of enterprises. They will need to determine the jobs that robots will take over in addition to their current duties. They will essentially decide which tasks to give to robots. The role of HR departments will grow in significance and responsibility. Eventually, HR departments will undergo changes, and they might even receive new titles. There are organizations and enterprises today that promote themselves as "green." They assert that they are environmentally conscious. Even though these activities increase prices, "green" firms and organizations aim to behave properly toward the environment. There may even be companies in the future calling themselves "human." Even though it costs more to be a "human" organization, these businesses will only hire people, even if using robots would be less expensive. These "human" organizations will assert that they treat people with social responsibility and that they provide jobs for people. Leading responsibilities of HR management departments include drafting job descriptions, undertaking work analyses, and developing business processes. The majority of the work that HR management departments have already completed will need to be redone before integrating robots into their businesses.

6.7 Considerations for Integration of Humans and Smart Machines in Industry 5.0

The collection of workplace automation considerations for firms is discussed in this section.

6.7.1 Augmenting Workforce Through Automation

Several researchers have advocated for a fundamental shift from the idea that automation would replace human jobs and toward the consideration of how machines might effectively assist and supplement the workforce [43]. Companies must evaluate how to use automation successfully in order to enhance workers' well-being and productivity by reducing unfavourable job characteristics such as wait times, health and safety risks, scheduling challenges (such as feeling rushed or working long hours), and repetitive labor. Job displacement is a serious fear, particularly in the Industry 5.0; therefore, employers must find methods to use automation in this way [44].

6.7.2 Select Tasks Carefully

It is crucial to consider how robots should be integrated into the workforce because employees prefer to work on projects that foster positive attributes and lessen negative ones. Participants consider social interaction, physical activity, autonomy, problem solving, and task diversity as the most beneficial work characteristics. Health and safety hazards, repetitive work, and feeling rushed have been the major drawbacks of the job. Robots could improve positive job characteristics and minimize negative ones in a number of ways, such as minimizing interruptions and promoting physical and mental well-being [45]. Therefore, while choosing activities, it is crucial to evaluate these complex choices and take robot reprogramming into account. In this context, participatory design methods can be useful because they assist firms in determining which activities people dislike depending on their preferred task, thus having a significant impact on the adoption and implementation of technology.

6.7.3 Retrain and Retain

Robots in the workplace provide people a chance to learn new skills, which might be a tactic to draw in and keep the next generation of workers interested in embracing new technology. However, when employees train their co-workers on the job, training can also benefit positive work characteristics like interpersonal interaction and problem solving [46].

6.7.4 Ensuring Health and Safety

By carrying out hazardous operations like welding or working with corrosive chemicals, robots can help to promote occupational health and safety. Additionally, they can be utilized to lessen the frequency of milder hazards, such as back problems from unaided lifting, repetitive stress injuries from using tools repeatedly, and prolonged chemical exposure. Robots may also promote psychological well-being by promoting social interaction, whether it be with other robots or with human employees [47].

6.8 Reskilling and Upskilling the Workforce for Industry 5.0

Employment will be significantly impacted by Industry 5.0 in the years to come. Almost all jobs will change, some rather significantly, and the substantial portion of workers will have to learn required skills. One of the most important business concerns today is being ready for the future of work, yet most businesses are not prepared to deal with it. Two distinct types of modifications will be necessary for the workforce to get ready for the future of work: upskilling, in which workers pick up new skills to do their existing jobs, and reskilling, in which workers pick up new skills to take on different or entirely new positions. The need for reskilling will be especially intense in industries with high operational demands, such as transportation, manufacturing, and retail, as well as in jobs that are closely related to operations, such as claim processing, maintenance, and order picking in warehouses [48]. Industries and professions that employ a large workforce and involve numerous predictable and repetitive operational tasks are expected to undergo more substantial transformations compared to other sectors. This is primarily due to the potential for automation or digitalization.

Employers can use a variety of strategies to close skill gaps. They can hire new candidates with required credentials form outside the company and also retrain the current employees by developing their capabilities internally so as to prepare them for new positions. Alternately, they can employ a hybrid strategy that combines hiring a skilled contract labor to meet immediate demands with internal skill development. Nearly every operation-intensive industry will require a structured solution to the problem of evolving skill requirements in the upcoming years. Planning, shaping, and shifting are the three major processes needed to design, construct, and implement such a strategy.

6.8.1 Workforce Planning

The skills required to achieve a company's strategic goals must be taken into account when determining future skill requirements. A crucial prerequisite for any "right skilling" program is the identification of potential effects of automation and digitalization within an organization and the development of a strong strategy to take advantage of such opportunities. This is because an organization's overall technological strategy influences skill demands [8]. During the planning phase, organizations should evaluate the skills required for future role and match them with existing abilities of their workforce. This analysis enables them to strategically allocate resources overtime and identify any skill gaps that need to be addressed to meet the requirements of both current and upcoming positions. It is crucial for businesses to assess the underlying factors that can either support or hinder the success of a reskilling program. A company's workforce openness to change as well as the capabilities of its current HR and training infrastructure are some of those. The next step for a corporation is to identify priorities for the skills that have the greatest impact on a large number of individuals and the roles that necessitate significant skill transition. The organization should then create appropriate content and delivery approaches to each prioritized group.

6.8.2 Skill Shaping

All initiatives aimed at developing skills should revolve around a centralized talent transition hub and corporate academy. This hub would oversee the implementation of reskilling initiatives, guide employees through their learning journeys, and strategically allocate retrained workers to the areas within the organization that have the highest demand for their newly acquired skills. The hub will also bear the responsibility for evaluating the effectiveness and impact of the program in addition to testing, changing, and enhancing curriculum and learning systems using agile approaches. This will assist reskilling system of an organization in growing and changing to meet the demands of the industry. Companies must create compelling value propositions for their workers in order to convince them of the advantages of learning new skills and to recruit external talent to fill specialized positions for which there are not enough internal applicants.

6.8.3 Shifting Skill Profile

Finally, businesses must build and deploy the infrastructure and training necessary to reskill people in order to transform the skill profiles of their whole organizations. While all employees might need to improve their knowledge of general subjects like the business value and applications of analytics and digital technology, certain workers would require far more intensive and specialized reskilling for specialized new jobs.

6.9 Conclusion

The Fifth Industrial Revolution necessitates three main elements—intelligent systems, intelligent devices, and intelligent automation—to integrate with human intelligence. This industrial revolution relates to human–machine interaction to make jobs easier and quicker. It brings the personalization idea to the next stage. Businesses that fail to embrace modern and intelligent technology will become irrelevant. The transition can only be carried out by workers who have the necessary skills. Therefore, organizations will be able to adapt to this shift by identifying the necessary skills to fill the skill gaps. In this context, it is projected that Industry 5.0 would change the functions and obligations of HR departments within organizations. In order to assist employees in picking up new skills, businesses should run education and training programs. Employees with active learning skills should be able to upskill and reskill, which are seen to be crucial in the Industry 5.0 adaption process because they may learn new skills that will make their work more meaningful. An initiative to reshape skills should be centered on a talent-transition hub and corporate academy that supervise the implementation of reskilling programs, allocate people to learning journeys, and provide reskilled staff to the parts of the business that need them the most. Organizations must create compelling value propositions for their workers in order to convince them of the advantages of learning new skills and to recruit external talent to fill specialized positions for which there are not enough internal applicants. Finally, businesses must build and deploy the infrastructure and training necessary to reskill people in order to transform the skill profiles of their whole organizations. While all employees might need to improve their knowledge of general subjects like the business value and applications of analytics and digital technology, certain workers would require far more intensive and specialized reskilling for specialized new jobs. Conversely, the challenge of skill development for the workforce is enormous because of increasingly automated production systems. Virtual training is crucial for developing a trained workforce without jeopardizing an ongoing process' productivity or threatening an employee. It is particularly crucial for vocations and tasks that need repeated actions or awkward postures because they involve some level of danger. For example, virtual training can considerably help a variety of the workforce by providing them with adequate, economical training without exposing them to potentially hazardous training scenarios. This is especially true when human posture analysis is included. In general, Industry 5.0 is projected to revolutionize the manufacturing processes and systems by facilitating increased human–robot collaboration to deliver more individualized products to clients.

This chapter offers insights into the benefits workforce will have in Industry 5.0, potential challenges workers will face due to transition from Industry 4.0 to Industry 5.0, and the skills necessary for workers to sustain in Industry 5.0. Future studies should focus on how the skills required for Industry 5.0 will differ according to industry segments. An empirical analysis of the adaptation level of employees to Industry 5.0 should also be conducted. Also, an assessment should be made of how robots will intersect with work attributes of employees. There is scant research that specifically addresses the organizational challenges arising from the collaboration between humans and robots in the workforce. Future work should identify such issues for successful implementation of Industry 5.0.

References

1. Rada, M., INDUSTRY 5.0-from virtual to physical, Linkedin, 2015, [mrežno] Dostupno na: https://www. linkedin. com/pulse/industry-50-from-virtual-physicalmichael-rada [Pristupljeno 20.9. 2021.].
2. Atwell, C., Yes, Industry 5.0 is Already on the Horizon, *Mach. Des.*, 2017, Available at: https://www.machinedesign.com/automation-iiot/article/21835933/yes-industry-50-is-already-on-the-horizon (Accessed: 14 July 2024).
3. Demir, K.A., Döven, G., Sezen, B., Industry 5.0 and human-robot co-working. *Procedia Comput. Sci.*, *158*, 688–695, 2019.
4. Industry 5.0: Adding the human edge to Industry 4.0: SAP insights, Retrieved May 5, 2023, from https://www.sap.com/insights/industry-5-0.html.
5. Selvaraj, A., Role of humans in Industry 5.0, Columbus, 2019, [Online] October, 10.
6. Demir, K.A. and Cicibas, H., Industry 5.0 and a Critique of Industry 4.0, in: *Proceedings of the 4th international management information systems conference*, Istanbul, Turkey, pp. 17–20, 2017, October.
7. Nahavandi, S., Industry 5.0—A human-centric solution. *Sustainability*, *11*, 16, 4371, 2019.
8. Paschek, D., Mocan, A., Draghici, A., Industry 5.0—The expected impact of next industrial revolution, in: *Thriving on future education, industry, business, and Society, Proceedings of the MakeLearn and TIIM International Conference*, Piran, Slovenia, pp. 15–17, 2019, May.
9. Yücebalkan, B., Endüstri 4.0'dan Endüstri 5.0'a Geçiş Sürecine Genel Bakiş. *Pearson J.*, *5*, 9, 241–250, 2020.
10. Breque, M., De Nul, L., Petridis, A., *Industry 5.0: towards a sustainable, human-centric and resilient European industry*, European Commission, Directorate-General for Research and Innovation, Luxembourg, LU, 2021.

11. Doyle-Kent, M., Industry 5.0 and its social impacts, Doctoral dissertation, TU Wien, Vienna, 2021, Retrieved from: https://repositum.tuwien.at/handle/20.500.12708/17416.
12. Xu, X., Lu, Y., Vogel-Heuser, B., Wang, L., Industry 4.0 and Industry 5.0—inception, conception and perception. *J. Manuf. Syst.*, *61*, 530–535, 2021, doi: 10.1016/j.jmsy.2021.10.006.
13. Adel, A., Future of Industry 5.0 in society: Human-Centric Solutions, challenges and prospective research areas. *J. Cloud Comput.*, *11*, 1, 40, 2022, doi: 10.1186/s13677-022-00314-5.
14. Jang-Jaccard, J. and Nepal, S., A survey of emerging threats in cybersecurity. *J. Comput. Syst. Sci.*, *80*, 5, 973–993, 2014, doi: 10.1016/j.jcss.2014.02.005.
15. Kunovskiy, I., Industry 5.0: Digital Transformation of manufacturing, 2022, May 11, Retrieved December 4, 2022, from https://hqsoftwarelab.com/blog/industry-5-0/.
16. Why augmented and virtual realities matter, 2019, September 10, Retrieved May 5, 2023, from https://direct.mit.edu/books/book/4550/chapter/203214/Why-Augmented-and-Virtual-Realities-Matter.
17. Ambrogio, G., Filice, L., Longo, F., Padovano, A., Workforce and supply chain disruption as a digital and technological innovation opportunity for Resilient Manufacturing Systems in the COVID-19 pandemic. *Comput. Ind. Eng.*, *169*, 108158, 2022, doi: 10.1016/j.cie.2022.108158.
18. Kovacevic, R., Mental health: Lessons learned in 2020 for 2021 and forward, 2021, Retrieved December 4, 2022, from https://blogs.worldbank.org/health/mental-health-lessons-learned-2020-2021-and-forward.
19. Pajula, S., 2018 manufacturing skills gap study, 2020, Retrieved December 4, 2022, from https://www2.deloitte.com/us/en/pages/manufacturing/articles/future-of-manufacturing-skills-gap-study.html.
20. The robotics industry creates jobs - collaborative robots, 2016, Retrieved December 4, 2022, from https://www.universal-robots.com/blog/robots-create-jobs/.
21. Kent, M.D. and Kopacek, P., Do we need synchronization of the human and robotics to make Industry 5.0 a success story? *The International Symposium for Production Research*, pp. 302–311, Springer, 2020.
22. Rashidi, M., Mohammadi, M., Sadeghlou Kivi, S., Abdolvand, M.M., Truong-Hong, L., Samali, B., A decade of modern bridge monitoring using Terrestrial Laser Scanning: Review and Future Directions. *Remote Sens.*, *12*, 22, 3796, 2020, doi: 10.3390/rs12223796.
23. Ghobakhloo, M., Fathi, M., Iranmanesh, M., Maroufkhani, P., Morales, M.E., Industry 4.0 ten years on: A Bibliometric and systematic review of concepts, sustainability value drivers, and success determinants. *J. Cleaner Prod.*, *302*, 127052, 2021, doi: 10.1016/j.jclepro.2021.127052.
24. The Future of Jobs Report 2020, 2020, Retrieved December 4, 2022, from https://www.weforum.org/reports/the-future-of-jobs-report-2020.

25. Gotfredsen, S., Bringing back the human touch: Industry 5.0 concept creating factories of the future, Manufacturers Monthly", dostęp, 11, 2018, 2016.
26. Turan, N., Çalışma yaşamında yetenek, beceri, yetkinlik, yeterlilik, Nobel, Ankara, 2015.
27. Erikli, S., Sinop ilinde işgücü piyasasının temel sorunu: Beceri uyumsuzluğu. *Çalışma Dünyası Dergisi, 2*, 7, 18–33, 2015.
28. Güğerçin, S., How employees survive in the Industry 5.0 era: In-demand skills of the near future. *Int. J. Discip. Econ. Adm. Sci. Stud. (IDEAstudies), 7*, 31, 524–533, 2021, doi: 10.26728/ideas.452.
29. Little, B. and Little, P., Employee engagement: Conceptual issues. *J. Organ. Cult. Commun. Conflict, 10*, 1, 111–120, 2006.
30. Eraslan, L., Liderlikte post-modern bir paradigma: dönüşümcü liderlik. *J. Hum. Sci., 8*, 1, 1–32, 2011.
31. Aprilisa, E., Realizing Society 5.0 to Face the Industrial Revolution 4.0 and Teacher Education Curriculum Readiness in Indonesia, in: *Proceeding International Conference on Science and Engineering*, vol. 3, pp. 543–548, 2020, April.
32. Knudsen, E.S., Lien, L.B., Timmermans, B., Belik, I., Pandey, S., Stability in turbulent times? The effect of digitalization on the sustainability of competitive advantage. *J. Bus. Res., 128*, 360–369, 2021.
33. Suen, H.Y., Chen, M.Y.C., Lu, S.H., Does the use of synchrony and artificial intelligence in video interviews affect interview ratings and applicant attitudes? *Comput. Hum. Behav., 98*, 93–101, 2019.
34. Skobelev, P.O. and Borovik, S.Y., On the way from Industry 4.0 to Industry 5.0: From digital manufacturing to digital society. *Industry 4.0, 2*, 6, 307–311, 2017.
35. Petress, K., What is meant by" active learning?". *Education*, 128, 4, 566, 2008.
36. Wu, G., Feder, A., Cohen, H., Kim, J.J., Calderon, S., Charney, D.S., Mathé, A.A., Understanding resilience. *Front. Behav. Neurosci., 7*, 10, 2013.
37. Dündar, GI., Tüzüner, V.L., Atay, S.E., Covid-19, from the perspectıves of human resource practıces and employees, in: *Covid-19 and new business ecosystem*, D. Hıdırlıoğlu, S.S. Aktuğ, O. Yılmaz (Eds.), pp. 153–173, Gazi Kitabevi, Ankara, 2020.
38. Bekey, G.A., *Autonomous robots: from biological inspiration to implementation and control*, MIT Press, United States, 2005.
39. Sheridan, T.B. and Verplank, W.L., *Human and computer control of undersea teleoperators*, Massachusetts Inst of Tech Cambridge, United States, Man-Machine Systems Lab, 1978.
40. Ray, C., Mondada, F., Siegwart, R., What do people expect from robots?, in: *2008 IEEE/RSJ International Conference on Intelligent Robots and Systems*, pp. 3816–3821, IEEE, 2008, September.
41. Bartlett, B., Estivill-Castro, V., Seymon, S., Dogs or robots: why do children see them as robotic pets rather than canine machines?, in: *Proceedings of*

the fifth conference on Australasian user interface, vol. 28, pp. 7–14, 2004, January.

42. Cormier, D., Newman, G., Nakane, M., Young, J.E., Durocher, S., Would you do as a robot commands? an obedience study for human-robot interaction, in: *International Conference on Human-Agent Interaction*, pp. 1–3, 2013, August.
43. Van Meerbeek, I.M., De Sa, C.M., Shepherd, R.F., Soft optoelectronic sensory foams with proprioception. *Sci. Rob.*, *3*, 24, eaau2489, 2018.
44. Autor, D.H., Why are there still so many jobs? the history and future of Workplace Automation. *J. Econ. Perspect.*, *29*, 3, 3–30, 2015, doi: 10.1257/jep.29.3.3.
45. Welfare, K.S., Hallowell, M.R., Shah, J.A., Riek, L.D., Consider the human work experience when integrating robotics in the workplace, in: *2019 14th ACM/IEEE international conference on human-robot interaction (HRI)*, pp. 75–84, IEEE, 2019, March.
46. (PDF) the importance of training and development in employee, Retrieved May 5, 2023, from https://www.researchgate.net/publication/332537797_The_Importance_of_Training_and_Development_in_Employee_Performance_and_Evaluation.
47. Sauppé, A. and Mutlu, B., The social impact of a robot co-worker in industrial settings, in: *Proceedings of the 33rd annual ACM conference on human factors in computing systems*, pp. 3613–3622, 2015, April.
48. Ellingrud, K., Gupta, R., Salguero, J., Building the vital skills for the future of work in operations, in: *What now*, Sneader, K., Singhal, S., Sternfels, B. (Eds.), pp. 66–73, 2020.

Part 2
APPLICATIONS

7

Intelligent Robots and Cobots: Concepts and Applications for Industry 5.0 Transformation

G. Ganesh Kumar[1]* and S. Kanakaprabha[2]

[1]*Department of Information Technology, Hindusthan Institute of Technology, Coimbatore, Tamil Nadu, India*
[2]*Department of Computer Science and Engineering, Rathinam Technical Campus, Anna University, Coimbatore, Tamil Nadu, India*

Abstract
In this chapter, we explore the concepts and applications of intelligent robots and cobots in the context of Industry 5.0 transformation. We begin by providing an overview of Industry 5.0 and its goals of merging human capabilities with advanced technologies. We then delve into the concepts of intelligent robots and cobots, highlighting their characteristics, functionalities, and potential benefits. Subsequently, we discuss various application areas where intelligent robots and cobots are revolutionizing industries, including manufacturing, healthcare, logistics, and more. Furthermore, we address the challenges and considerations associated with implementing intelligent robots and cobots, such as safety, ethics, and human–machine interaction. Finally, we conclude the chapter by outlining the future prospects and potential impacts of intelligent robots and cobots in driving the Industry 5.0 transformation.

Keywords: Industry 5.0, intelligent robots, cobots, automation, artificial intelligence, human–robot collaboration, industrial applications

**Corresponding author*: gkumarcse@gmail.com

V. Ramasamy, S. Balamurugan and Sheng-Lung Peng (eds.) Intelligent Robots and Cobots: Industry 5.0 Applications, (133–158) © 2025 Scrivener Publishing LLC

7.1 Introduction

In recent years, we have witnessed unprecedented advancements in technology that have transformed various industries and revolutionized the way that we live and work. From the rise of artificial intelligence (AI) and automation to the Internet of Things and virtual reality, these innovations have had a profound impact on our society. However, as we move forward, there is a growing recognition that the future lies not only in the development of advanced technologies but also in the harmonious integration of humans with these technologies. This vision is encapsulated in the concept of Industry 5.0.

Industry 5.0 represents a paradigm shift that focuses on the symbiotic relationship between humans and technology. It emphasizes the importance of collaboration, coexistence, and cooperation between human workers and intelligent machines. While Industry 4.0 introduced the concept of automation and connectivity in manufacturing, Industry 5.0 takes it a step further by emphasizing the unique capabilities and creativity that humans bring to the table.

At its core, Industry 5.0 recognizes that humans possess qualities that machines cannot replicate: empathy, intuition, creativity, and complex problem-solving skills. Rather than replacing human workers, the aim of Industry 5.0 is to augment their abilities with advanced technologies to unlock new levels of productivity and innovation.

This merging of humans and technology opens up exciting possibilities across various industries. In manufacturing, it enables the creation of flexible production lines where humans and robots work together seamlessly. In healthcare, it empowers medical professionals with AI-powered tools for diagnosis and treatment, enhancing patient care. In education, it revolutionizes the learning experience, allowing personalized and adaptive teaching methods.

However, as we embark on this new era, challenges must be addressed. Ethical considerations, such as data privacy and security, as well as the potential impact on employment and workforce dynamics, need to be carefully navigated. Additionally, ensuring equal access and opportunities for all individuals is crucial to prevent technology-driven inequalities.

In this series of discussions, we will delve into the various aspects of Industry 5.0, exploring its potential applications, benefits, and challenges. We will examine real-world examples, Figure 7.1 discuss the societal implications, and explore the ethical considerations involved. By understanding the merging of humans and technology, we can navigate this transformative era with foresight and shape a future where technology enhances the human experience rather than replaces it.

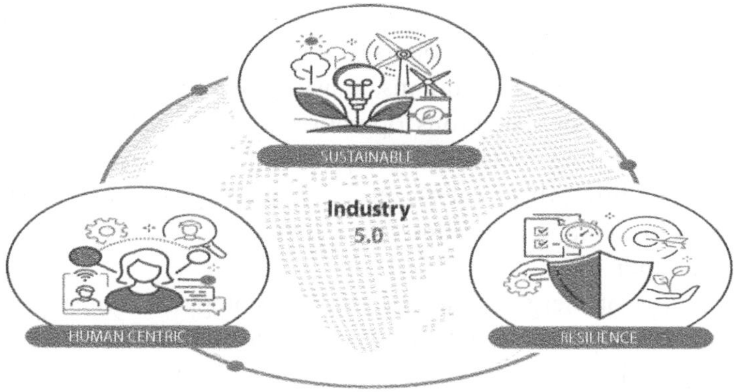

Figure 7.1 Industry 5.0 technologies.

7.1.1 Industry 5.0: Merging Humans and Technology

Throughout the chapter, real-world examples and case studies will be included to illustrate the concepts and applications of intelligent robots and cobots. The chapter aims to provide insights into the transformative potential of these technologies in shaping the future of industries and fostering human–machine partnership in the context of Industry 5.0.

Industry 5.0 denotes the next phase of manufacturing revolution, structure upon the thoughts of Industry 4.0 and aiming to merge the capabilities of humans with advanced technologies. It recognizes the importance of human intelligence, creativity, and problem-solving skills in conjunction with the power of automation, robotics, and AI. Industry 5.0 seeks to create a harmonious partnership between humans and machines, leveraging the strengths of both to drive innovation, productivity, and sustainable growth.

Unlike previous industrial revolutions that focused primarily on automation and machine-driven processes, Industry 5.0 emphasizes the integration of human-centric technologies into the production and decision-making processes. It recognizes that humans possess unique qualities such as intuition, empathy, and adaptability that are crucial in complex and non-routine tasks.

The goal of Industry 5.0 is to leverage advanced technologies to augment human capabilities and provide a more inclusive, flexible, and efficient work environment. Rather than replacing human workers, it envisions a future where humans and machines collaborate seamlessly, with each contributing their respective strengths. This collaboration can lead to enhanced productivity, increased job satisfaction, and improved quality of work.

Industry 5.0 envisions a range of technologies that enable this collaboration, including intelligent robots and cobots. These technologies are designed to work alongside humans, assisting them in physical tasks, performing repetitive or dangerous operations, and handling heavy workloads. By integrating sensors, AI algorithms, and advanced machine learning capabilities, these robots and cobots can perceive their environment, adapt to changing conditions, and interact with humans in a safe and intuitive manner.

The concept of merging humans and technology in Industry 5.0 goes beyond the factory floor. It extends to other industries such as healthcare, logistics, agriculture, and construction, where human–machine collaboration can drive significant advancements. By leveraging the expertise and judgment of human professionals and combining it with the precision and efficiency of intelligent robots and cobots, Industry 5.0 paves the way for innovative solutions, improved outcomes, and sustainable growth.

In the following sections of this chapter, we will explore the concepts and applications of intelligent robots and cobots, examining how they contribute to the Industry 5.0 transformation by facilitating human–machine collaboration. Figure 7.2 discussed the will delve into their characteristics, benefits, application areas, and the challenges associated with their implementation. By understanding the potential of these technologies and addressing the considerations involved, we can unlock the full potential of Industry 5.0 and shape a future where humans and machines work together to achieve unprecedented levels of productivity and innovation.

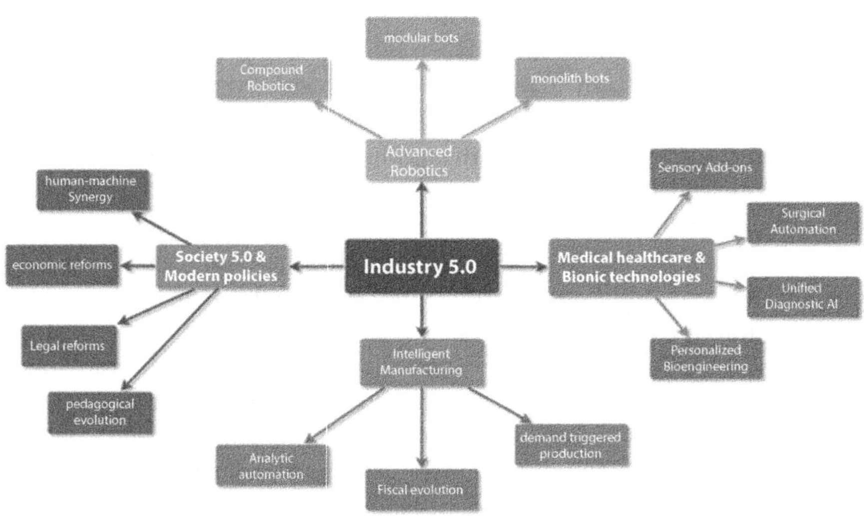

Figure 7.2 Block diagram for Industry 5.0.

7.1.2 The Role of Intelligent Robots and Cobots

In the background of Industry 5.0, intelligent robots and collaborative robots, also known as cobots, play a pivotal role in enabling the merging of humans and technology. These advanced machines possess the ability to perceive and interact with the world around them, making them valuable partners in various industries.

Intelligent robots are designed to perform complex tasks autonomously, utilizing AI, machine learning, and sensor technologies. They can analyze vast amounts of data, make decisions, and execute precise actions with speed and accuracy. In manufacturing, for example, intelligent robots can handle repetitive and physically demanding tasks with high precision, boosting productivity and efficiency.

Cobots, on the other hand, are designed to work collaboratively together with humans. Unlike old-fashioned industrial robots that function in isolation, cobots are equipped with progressive sensors and safety topographies that allow them to work safely together with human labours. They can assist with tasks that require dexterity, precision, or heavy lifting, whereas humans focus on more cognitive or supervisory roles. Cobots are highly flexible and can be easily programmed and reprogrammed, making them adaptable to changing production needs.

The integration of intelligent robots and cobots in Industry 5.0 brings several advantages. Firstly, it enhances the overall productivity and efficiency of operations. By automating repetitive and physically strenuous tasks, these machines free up human workforces to focus on more multifaceted and value-added activities. This leads to improved product quality and increased output.

Secondly, intelligent robots and cobots have the potential to enhance workplace safety. They can handle hazardous or dangerous tasks, reducing the risk of injuries for human workers. Cobots, with their built-in safety features, can detect the presence of humans and adjust their actions accordingly, minimizing the chances of accidents.

Moreover, Figure 7.3 these machines can augment human skills and capabilities. By collaborating with robots, human workers can leverage the precision and speed of machines to achieve higher levels of performance. The combination of human creativity, problem-solving abilities, and emotional intelligence with the computational power and efficiency of robots creates a powerful synergy that drives innovation and breakthroughs in various industries.

However, the integration of intelligent robots and cobots also raises concerns and challenges. There are ethical considerations surrounding

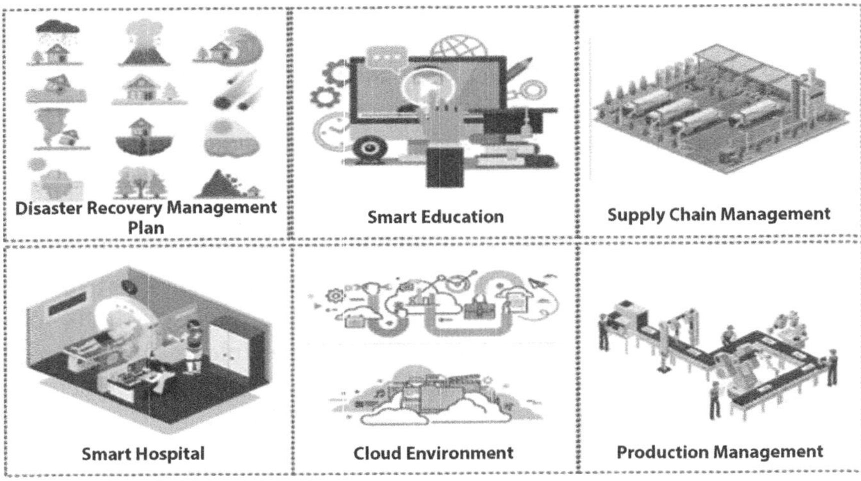

Figure 7.3 Future of Industry 5.0 in society.

the potential displacement of human workers and the need to ensure fair distribution of benefits and opportunities. Moreover, the design and deployment of these machines must prioritize safety, transparency, and accountability to prevent unintended consequences.

In conclusion, intelligent robots and cobots play an energetic character in Industry 5.0 by enabling the collaboration and integration of humans and technology. They enhance productivity, improve workplace safety, and augment human skills, leading to increased efficiency and innovation. However, cautious attention must be assumed to the ethical and societal insinuations to safeguard a balanced and inclusive future.

7.2 Systematic Review

Diseph Emma-Ikata and Mary Doyle-Kent [1] proposed the evaluation of task reconstruction in accordance with Industry 5.0 supplies took into consideration the readiness of humans, who have been working in the field of manufacturing for many years, to learn and enhance their skills. The adoption of independent robots and collective robots (cobots) posed critical issues that were uncovered during the study. The results indicate that there is currently a limited level of consciousness and willingness for Industry 5.0. Consequently, the findings emphasize the necessity for educational institutions and providers to play a pivotal role in addressing these challenges and facilitating the transition to Industry 5.0.

Dr. Ranjeet Yadav, Dr. Zahid Parwez, and Dr. R. Sree Parimala [2] proposed in this paper that an Industry 5.0 is poised to advantage from a wide array of machineries and claims that have the potential to revolutionize production processes and enable the on-demand supply of customized products. This article focuses on exploring the diverse requests of Industry 5.0 across various sectors, including information and infrastructures, real estate, business services, business, finance/insurance, public administration, automotive, and healthcare. To gain insights into the subjects and investigation trends associated with Industry 5.0, text mining tools and procedures are employed to analyze and assess relevant information. Before delving into further research using text mining techniques such as keyword withdrawal and frequency analysis, the data undergoes meticulous cleaning and preprocessing. The subsequent sections of this article will delve into the technologies that form the foundation of Industry 5.0, facilitating its progressions and enabling its transformative potential.

In the study [3] by Jakub Pizoń, Michał Cioch, and Lukasz Kanski, the approach described in this article lays the groundwork for innovative solutions that combine organizational, management, and technical aspects in the context of utilizing collaborative robots (cobots). This approach presents an occasion to progress money-making implementations of modern cobotic system technology, aiming to reduce waste and align with lean organization principles. It also raises important questions regarding the implementation of cobotic systems within organizations, particularly for companies operating in specific industries and employing specific technologies and work organization methods. By addressing these questions, organizations can explore the potential benefits and challenges associated with adopting cobotic systems, paving the way for efficient and optimized operations.

In the study by Aslı Kılıç and Yavuz Selim Balcıoğlu [4], this article delves into the evolution from Industry 4.0 to Industry 5.0, shedding light on the transformative shift toward a more human-centered approach. The focus is on fostering collaboration, adaptability, and creative problem-solving. The piece further explores the application of Industry 5.0 knowledge across various sectors, such as smart factories, agile manufacturing, supply chain management, logistics optimization, quality control, predictive maintenance, workforce training, upskilling, customer appointment, and service revolution. Additionally, it addresses the opportunities and challenges arising from these technologies, highlighting the importance of collaboration among businesses, politicians, and enlightening institutions to tackle potential risks, ethical apprehensions, and skill gaps. The article advocates for cultivating a culture of innovation and continuous learning in order to connect the full probable of Industry 5.0.

The study [5] by Zhen Gao, Tom Wanyama and Ishwar Singh aims to revisit the significance of co-operative and intellectual robotic systems and their enabling machineries, which encompass ROS (Robot Operating System) and ROS2, integrated drive systems, mechanical sensors, horizontal incorporation of robotic networks, human–robot welcoming and natural interaction, and deep learning in robots. The research highlights the seamless incorporation of motion, computing, awareness, and reasoning intelligences in Robotics 4.0 to effectively address the diverse requirements of industries and society. The abstract provides a roadmap and presents case studies to showcase ongoing efforts in realizing the concept of Robotics 4.0.

The survey by Shah Zeb, Sunder Ali Khowaja, and Sunder Ali Khowaja [6] focuses on the junction of developing computational intelligence (CI) methods with softwarized next-generation wireless networks (NGWNs) to meet the demanding communiqué and calculation needs of Industry 5.0. The article reviews and analyzes current concepts and technologies such as CI tools and agendas, network-in-box architecture, open radio access networks, softwarized service architectures, and possible enabling amenities. These elements are essential for designing CI-NGWNs that align with the requirements of Industry 5.0. The study also discusses ongoing initiatives, demos, and frameworks associated with Industry 5.0 while offering insights into lessons learned, research challenges, and open questions that need to be lectured for the realization of Industry 5.0 in CI-NGWNs.

Carayannis and Morawska [7] proposed the concept of University 5.0 and Education 5.0, which emerges as a response to the continuing digital transformation and green changeovers, aiming to foster the social aspect of universities' assignments. This perspective encompasses various levels of analysis. At the micro level, it involves optimizing research and innovation processes within a single university. The meso level focuses on analyzing the innovation ecosystem surrounding the university, considering local characteristics to facilitate cooperative synergies. Lastly, the macro level involves examining external factors such as political, economic, demographic, sociocultural conditions, legal aspects, and technology to support innovation growth, aligning with the 5.0 version of a fully realized university (Carayannis *et al.*, 2022). This multi-level approach remains relevant in the transition from Industry 4.0 to Industry 5.0/Society 5.0. By starting with a clear apparition and assignment statement and translating them into approaches and operative plans, sustainability can be ensured for the entire innovation ecosystem while considering the ethical, legal, and social implications involved.

The paper by Bader Alojaiman [8] explores the possibilities, boundaries, and upcoming forecasts of Industry 5.0, which represents an example shift emphasizing collaboration between humans and machines rather than solely relying on technology. The objective is to examine possible employs of Industry 5.0 by reviewing existing research, identifying current progress and challenges, and developing a systematic analysis procedure. The discussion begins with an examination of the classifications of Industry 5.0 and the advanced technologies associated with this industrial revolution. Furthermore, the paper explores various application domains enabled by Industry 5.0, including healthcare, supply chain management, production growth, and cloud-based industrial production. Additionally, the research addresses challenges and issues related to the interaction between organizations, robotic systems, and individuals on production lines. By analyzing these factors, this paper contributes to a better understanding of the implications and complexities associated with Industry 5.0.

The research by Olesia Suntsova [9] employed various methodologies, including grouping, system analysis, historical methods to methodical research, synthesis and examination, and forecasting. The study aimed to redefine and structure the concept of the smart economy by incorporating new business technologies within the context of the pervasive digital transformation of Industry 4.0 and 5.0. Notably, novel organization approaches were developed to organize the smart reduced and facilitate the digital change of businesses within the frameworks of Industry 4.0 and 5.0. These approaches offer significant economic benefits by enabling economic agents to adapt to the challenges and opportunities presented by digital business conversion in the framework of Industry 4.0 and Industry 5.0.

In the survey by George and George [10], we examine the transformative potential of collaborative robots (cobots) by presenting case studies that showcase successful cobot implementations across diverse industries. These case studies serve to illustrate the significant impact of cobots in various applications. Moreover, we address the challenges and limitations associated with cobots, including technical constraints, ethical considerations, and legal and social implications. Additionally, we explore future research opportunities and potential applications of cobots in the manufacturing sector, highlighting the importance of continuous development and innovation to overcome current limitations and maximize their benefits. The survey concludes by emphasizing the pivotal role that cobots play in shaping the future of the manufacturing industry, with profound implications for productivity, workforce dynamics, and sustainability. By embracing and understanding the transformative power of cobots, manufacturers can leverage their potential to drive ongoing growth and enhance competitiveness in an increasingly interconnected and automated world.

7.3 Concepts of Intelligent Robots and Cobots

7.3.1 Definitions and Differentiation

Intelligent robots and cobots are two distinct but related concepts within the realm of robotics. While they both incorporate advanced technologies and automation, they differ in their functionality and the manner in which they interact with humans.

Intelligent Robots:
Intelligent robots Figure 7.4 refer to autonomous machines that possess advanced capabilities such as perception, decision-making, and problem-solving. [11] These robots are typically equipped with AI algorithms, machine learning models, and sensor systems that enable them to gather information from their environment, process it, and execute actions accordingly. Intelligent robots can operate independently and are often employed in manufacturing, logistics, healthcare, and other industries where precise and repetitive tasks need to be performed.

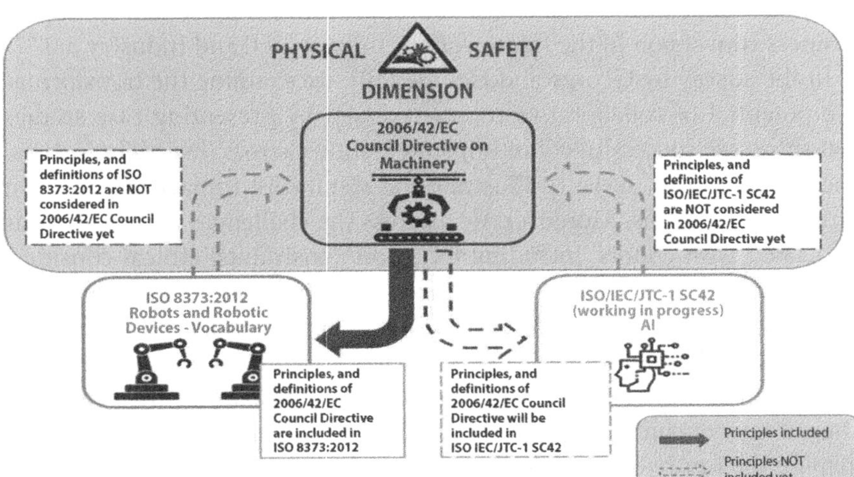

Figure 7.4 Frontiers | Redefining safety in light of human–robot interaction.

Cobots (Collaborative Robots):

Cobots, on the other hand, are designed specifically to collaborate and work alongside humans. Figure 7.5 is unlike traditional industrial robots, cobots are prepared with sensors and protection structures that allow them to operate in close proximity to creatures without posing a significant risk of injury. Cobots are programmed to perform tasks that complement human capabilities, such as assembly, material handling, or assistance in complex operations. They can be easily programmed or guided by human operators, enabling a flexible and interactive work environment.

7.3.2 Characteristics and Capabilities

Intelligent Robots:

- Autonomy: Intelligent robots have the ability to operate autonomously, making decisions and executing actions without constant human intervention.
- Perception and Sensing: They are equipped with sensors, cameras, and other sensing technologies to perceive and gather data from their environment.
- Decision-Making: Through the use of AI and machine learning algorithms, intelligent robots can analyze data and make informed decisions.
- Adaptability: These robots can learn and adapt to changing circumstances, allowing them to handle dynamic and unpredictable situations.
- Precision and Accuracy: Intelligent robots excel in performing tasks with high precision and accuracy, minimizing errors and enhancing efficiency.

Cobots:

- Safety: Cobots are designed with built-in safety features, such as force and collision sensors, to ensure safe interaction with human workers.
- Collaboration: They are meant to work collaboratively with humans, assisting them in tasks that require strength, precision, or repetitive actions.
- Flexibility: Cobots can be easily programmed or taught by human operators, allowing for quick task reconfiguration and adaptation to changing needs.

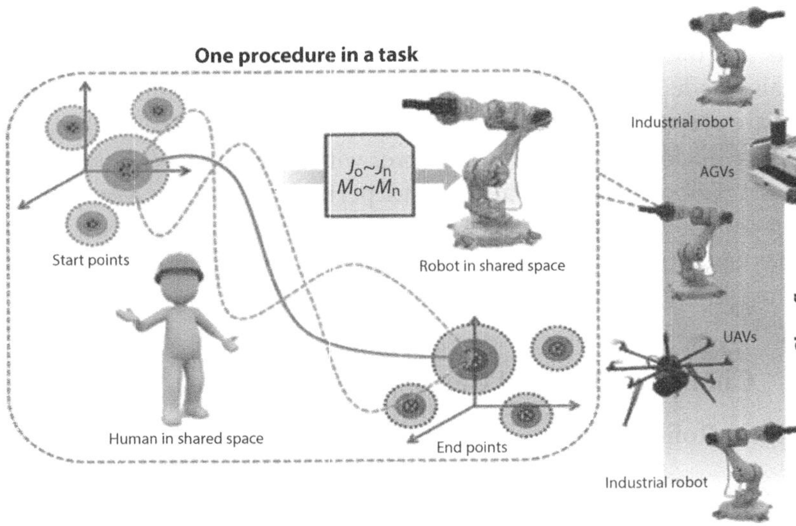

Figure 7.5 Robot learning toward smart robotic manufacturing.

- Ease of Use: User-friendly interfaces and intuitive controls make it convenient for non-experts to operate and interact with cobots.
- Human–Machine Interaction: Cobots are designed to have natural and intuitive communication with humans, facilitating seamless collaboration and coordination.

7.3.3 Human-Centric Design Principles

Human-centric design principles are crucial in the development and deployment of intelligent robots and cobots. [12] These principles ensure that the machines are designed to enhance and support human capabilities while considering factors such as safety, usability, and user experience. Some key principles include the following:

- Safety: Ensuring that the machines have robust safety features, are reliable, and adhere to safety standards to prevent harm to human workers.
- Ease of Use: Designing intuitive interfaces, controls, and programming methods that enable non-experts to interact with the machines easily.

- Ergonomics: Considering the physical and cognitive well-being of human workers by designing machines that align with human capabilities, minimizing physical strain and mental workload.
- Transparency: Promoting transparency in the machines' decision-making processes and behaviors, enabling human workers to understand and trust their actions.
- Collaboration: Enabling effective communication and coordination between humans and machines through natural language interfaces, visual cues, and feedback mechanisms.
- Ethical Considerations: Incorporating ethical guidelines and frameworks into the design and programming of the machines, addressing issues such as fairness, accountability, and privacy.

7.4 Benefits of Intelligent Robots and Cobots

7.4.1 Enhanced Productivity and Efficiency

One of the significant benefits of intelligent robots and cobots is their ability to enhance productivity and efficiency in various industries. These machines can perform tasks with high precision, speed, and accuracy, resulting in increased output and improved quality of products or services. By automating repetitive and physically demanding tasks, they free up human workers to focus on more complex and value-added activities. The integration of intelligent robots and cobots in manufacturing, logistics, and other sectors streamlines operations, reduces cycle times, and optimizes resource utilization, leading to overall productivity gains.

7.4.2 Improved Safety and Risk Mitigation

Intelligent robots and cobots contribute to improved workplace safety by handling hazardous or dangerous tasks that pose risks to human workers. They can operate in environments with high temperatures, toxic substances, or heavy machinery, minimizing the exposure of humans to potentially harmful conditions. Cobots, specifically designed for collaboration with humans, incorporate safety features such as force sensors,

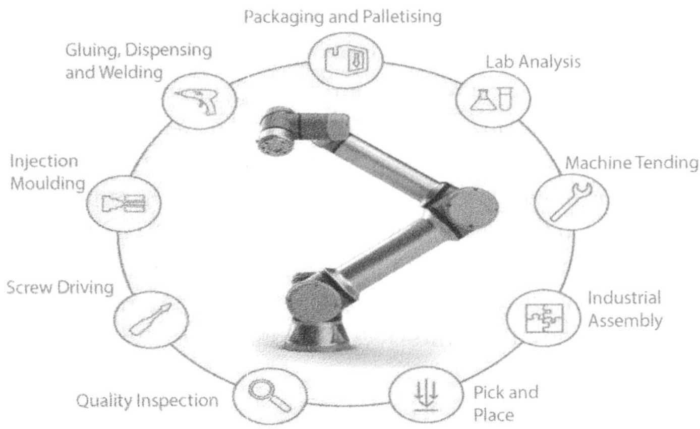

Figure 7.6 Benefits of collaborative robots.

collision detection, and speed limitation to ensure safe interaction. The above Figure 7.6 with their ability to accurately perform tasks without fatigue or distraction, robots can significantly reduce human-related errors and accidents, creating a safer work environment.

7.4.3 Workforce Augmentation and Skill Enhancement

Rather than replacing human workers, intelligent robots and cobots augment their capabilities and enable skill enhancement [13]. By collaborating with these machines, human workers can leverage their precision, speed, and efficiency to achieve higher levels of performance. The machines can handle repetitive or physically demanding tasks, freeing up humans to focus on more cognitive, creative, or supervisory roles. This collaboration enhances human skills such as problem-solving, critical thinking, and decision-making, leading to increased innovation and productivity. Intelligent robots and cobots also serve as valuable tools for skill development and training, providing opportunities for upskilling and reskilling the workforce.

7.4.4 Flexibility and Adaptability

Intelligent robots and cobots offer flexibility and adaptability to meet changing production needs and market demands. These machines can be easily reprogrammed or reconfigured to perform different tasks, allowing for quick setup and redeployment. In dynamic manufacturing environments, cobots excel at handling small batch production or customizations, where frequent task switching is required. Their versatility enables businesses to respond rapidly to market fluctuations, scale production up or down, and introduce new products or variants with minimal disruption. The ability to quickly adapt to changing requirements enhances operational agility and competitiveness.

Overall, the benefits of intelligent robots and cobots include enhanced productivity, improved safety, workforce augmentation, skill enhancement, and flexibility. By harnessing the strengths of both humans and machines, these technologies pave the way for increased efficiency, innovation, and a better working environment. However, it is essential to address potential challenges and ethical considerations to ensure a balanced and equitable integration of these technologies.

7.5 Application Areas

7.5.1 Manufacturing and Production

Automation has had a significant impact on manufacturing and production processes. Here are some key areas where automation is widely used.

Assembly Line Automation:
Assembly lines are often automated to improve efficiency, speed, and accuracy. Robots and automated systems can perform repetitive tasks such as picking and placing components, welding, painting, and quality control. Automation allows for increased production rates, reduced errors, and improved overall product quality [14].

Material Handling and Logistics:
Automation plays a crucial role in material handling and logistics within manufacturing facilities. Automated guided vehicles and robotic systems can efficiently move materials and components throughout the production floor, reducing manual labor, minimizing errors, and optimizing inventory management.

Quality Control and Inspection:
Automation systems, including machine vision and AI algorithms, are used for quality control and inspection processes. Automated systems can detect defects, measure dimensions, and ensure product consistency and compliance with quality standards. This helps improve product quality, reduce waste, and increase customer satisfaction.

Packaging and Labeling:
Automated packaging and labeling systems are commonly used in manufacturing. These systems can handle tasks such as filling, sealing, labeling, and palletizing products. Automation in packaging and labeling enhances efficiency, reduces errors, and enables customization and scalability.

Inventory Management:
Automation is utilized in inventory management to optimize stock levels, track inventory movement, and facilitate efficient supply chain operations. Automated systems can monitor inventory levels, generate purchase orders, and ensure timely replenishment. This helps reduce inventory costs, minimize stockouts, and improve overall supply chain efficiency.

Maintenance and Predictive Analytics:
Automation systems enable predictive maintenance by collecting and analyzing data from machines and equipment. By monitoring various parameters such as temperature, vibration, and performance indicators, automated systems can predict maintenance needs and schedule repairs proactively. This approach minimizes unplanned downtime, reduces maintenance costs, and extends the lifespan of machinery.

These are just a few examples Figure 7.7 of how automation is applied in the manufacturing and production sector. Automation technologies continue to evolve, enabling greater flexibility, adaptability, and integration across various processes, leading to improved productivity, cost-efficiency, and competitiveness in the manufacturing industry.

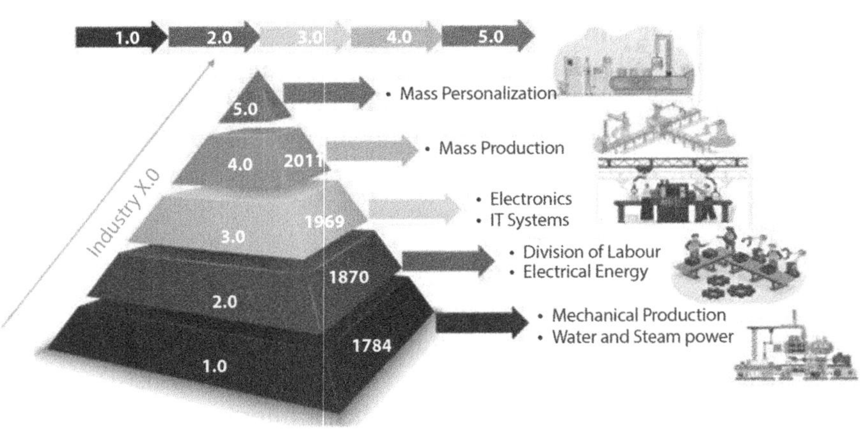

Figure 7.7 Enabling technologies and potential applications.

7.5.1.1 *Collaborative Assembly and Manufacturing Processes*

Intelligent robots and cobots play a significant role in manufacturing and production. [15] They can collaborate with human workers in assembly lines, assisting in tasks that require precision, dexterity, and repetitive actions. By working in tandem, these machines enhance the efficiency of production processes, reduce errors, and increase throughput.

7.5.1.2 *Quality Control and Inspection*

Intelligent robots equipped with advanced sensors and vision systems can perform quality control and inspection tasks in manufacturing. They can identify defects, measure dimensions, and conduct thorough inspections with high precision and speed. This improves the overall quality of products and reduces the likelihood of faulty items reaching the market.

7.5.2 Healthcare and Medical Assistance

7.5.2.1 *Surgical Assistance and Rehabilitation*

In healthcare, intelligent robots and cobots assist surgeons in performing complex procedures with precision and accuracy. They can be used in minimally invasive surgeries, providing surgeons with enhanced visualization, dexterity, and control. These machines also contribute to rehabilitation processes by assisting patients in regaining mobility, conducting repetitive exercises, and monitoring progress.

7.5.2.2 *Elderly and Patient Care*

Intelligent robots and cobots have applications in elderly and patient care. They can provide companionship, monitor vital signs, and assist with daily activities for the elderly or individuals with disabilities. These machines can contribute to the delivery of personalized and attentive care, improving the overall well-being of patients and reducing the burden on healthcare providers.

7.5.3 Logistics and Warehouse Automation

In logistics and warehouse operations, intelligent robots and cobots automate tasks such as picking, packing, sorting, and inventory management. They can navigate complex environments, identify objects, and optimize logistics operations, increasing efficiency and accuracy in supply chain management.

7.5.4 Agriculture and Farming

Intelligent robots and cobots find application in agriculture and farming, where they can perform tasks such as seeding, planting, crop monitoring, and harvesting. They can navigate fields, collect data on soil conditions and crop health, and assist in precision farming practices. These machines contribute to increased efficiency, reduced resource consumption, and improved yields in agricultural operations.

7.5.5 Construction and Infrastructure

Intelligent robots and cobots are utilized in construction and infrastructure projects. They can assist in tasks such as bricklaying, concrete pouring, welding, and heavy lifting, reducing manual labor and increasing construction efficiency. These machines enhance safety, precision, and speed in building processes, leading to improved project timelines and quality.

The application areas of intelligent robots and cobots extend beyond the mentioned industries. Their versatility and adaptability allow them to find use in various sectors, including education, retail, hospitality, and more. As technology advances and new possibilities emerge, the integration of humans and machines continues to shape and transform industries in the era of Industry 5.0.

7.6 Challenges and Considerations

7.6.1 Safety and Risk Management

One of the primary challenges in implementing new technologies and automation systems is ensuring safety and managing associated risks. With increased automation and the integration of AI and robotics, it is crucial to establish robust safety measures. This involves designing systems that can detect and mitigate potential hazards, implementing fail-safe mechanisms, and conducting thorough risk assessments. Safety protocols should be put in place to protect both human workers and the technology itself. Ongoing monitoring and evaluation are necessary to identify and address any emerging safety concerns.

7.6.2 Ethical Implications and Human–Machine Interaction

The rise of automation raises important ethical considerations. As machines take on more tasks previously performed by humans, it becomes necessary to address questions of responsibility, accountability, and decision-making. Ethical frameworks and guidelines should be developed to ensure that automated systems adhere to societal norms and values. Additionally, human–machine interaction should be carefully designed to strike a balance between efficiency and human involvement. Maintaining transparency, fairness, and accountability in the decision-making processes of AI systems is crucial.

7.6.3 Workforce Transition and Adaptation

The introduction of automation technologies can lead to workforce displacement and changes in job requirements. Some jobs may become obsolete, while new roles may emerge. It is essential to consider the impact on the workforce and support affected individuals in transitioning to new employment opportunities. Investing in reskilling and upskilling programs can help workers acquire the necessary skills to adapt to the changing job landscape. Figure 7.8 discussed and collaboration between industries, educational institutions, and governments is essential to facilitate a smooth transition and mitigate potential social and economic disruptions.

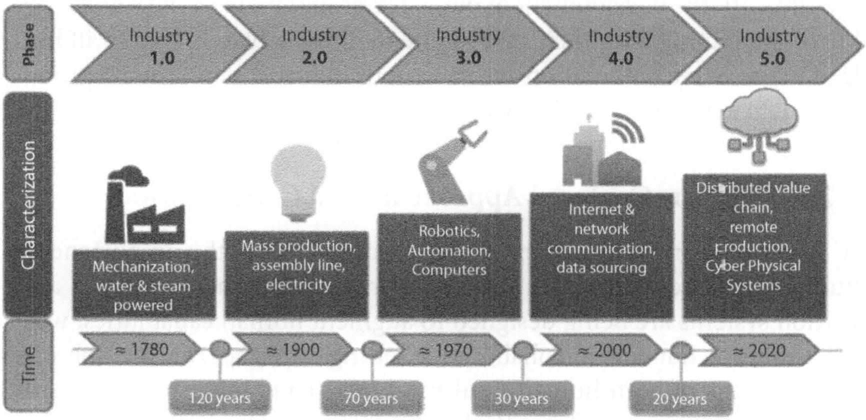

Figure 7.8 Industry 5.0 challenges and perspectives.

7.6.4 Legal and Regulatory Frameworks

The rapid advancement of automation technologies often outpaces the development of corresponding legal and regulatory frameworks. [16] It is crucial to establish comprehensive and up-to-date laws and regulations that govern the use of automation systems. This includes issues such as liability, data privacy, cybersecurity, and the protection of workers' rights. Policymakers and legal experts must work closely with industry stakeholders to develop flexible and adaptable frameworks that can keep up with the evolving technology landscape while ensuring public safety and ethical standards are met. International collaboration may also be necessary to address the global nature of automation technologies.

Addressing these challenges and considerations requires a multidisciplinary approach involving industry leaders, policymakers, ethicists, researchers, and the public. By proactively addressing these issues, we can maximize the benefits of automation while minimizing risks and ensuring a fair and sustainable future.

7.7 Future Prospects and Impacts

7.7.1 Advancements in Artificial Intelligence and Robotics

The future prospects of automation are closely tied to advancements in AI and robotics. AI algorithms are becoming more sophisticated, enabling machines to handle complex tasks, learn from data, and make decisions with minimal human intervention. Robotics technology is also advancing rapidly, with the development of more dexterous and versatile robots capable of interacting with their environment. These advancements will likely lead to increased automation across various industries, improving productivity, efficiency, and the ability to tackle challenging tasks.

7.7.2 Human-Centered Approaches and Collaboration

A key trend in automation is the recognition of the importance of human-centered approaches. Instead of replacing human workers, automation systems are being designed to augment human capabilities, working alongside humans in collaborative settings. This approach emphasizes the strengths of both humans and machines, combining human creativity, intuition, and empathy with the precision, speed, and data processing capabilities of automation technologies. Human–machine collaboration will likely lead to enhanced productivity, improved decision-making, and the ability to solve complex problems.

7.7.3 Socioeconomic Effects and Employment Landscape

The widespread adoption of automation technologies will undoubtedly have significant socioeconomic effects. While automation can lead to increased productivity and economic growth, it can also disrupt employment patterns. Some jobs may be replaced by automation, leading to job displacement and potential unemployment for certain segments of the workforce. [17] However, new jobs and industries may emerge as a result of automation, requiring different skill sets. The key lies in effective workforce transition strategies, including retraining programs, education initiatives, and job creation in emerging fields, to mitigate the negative impacts and ensure a smooth transition.

7.7.4 Potential Barriers to Adoption

Despite the potential benefits, there are several barriers that may hinder the widespread adoption of automation technologies. These include the high costs associated with implementing and maintaining automation systems, especially for small and medium-sized enterprises. Concerns about data privacy and security may also slow down adoption, as well as the need for regulatory clarity and standards. Figure 7.9 additionally, there may be resistance to change from workers and stakeholders who fear job

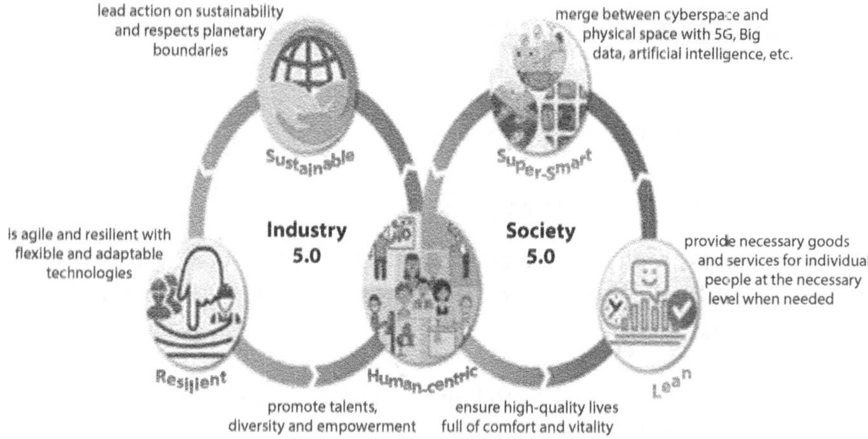

Figure 7.9 Industry 5.0 and Society 5.0.

Table 7.1 Modern society has reached its limits.

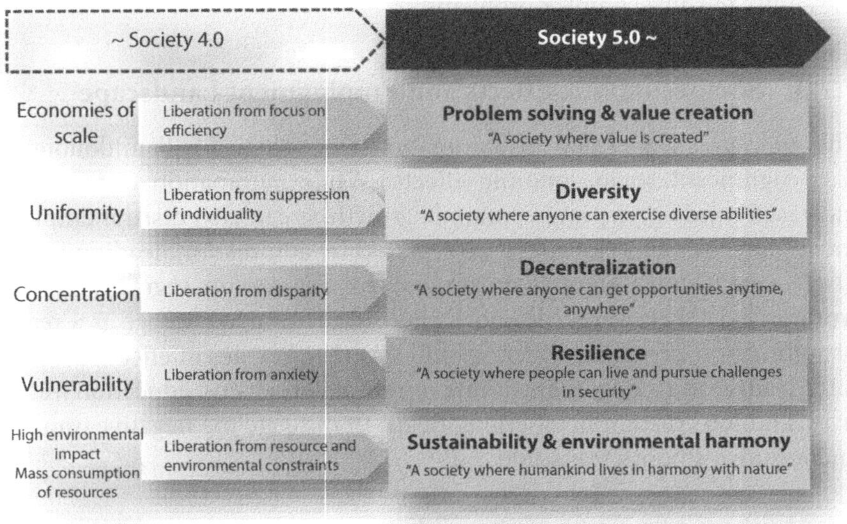

displacement or have concerns about the ethical implications of automation. Overcoming these barriers will require collaboration between industry, policymakers, and other stakeholders to address concerns, provide support, and create an enabling environment for automation adoption.

The future Table 7.1 mentioned impacts of automation will depend on how these prospects and challenges are addressed. [18] By focusing on advancements in AI and robotics, adopting human-centered approaches, managing the socioeconomic effects, and overcoming barriers to adoption, we can shape a future where automation enhances productivity, improves quality of life, and creates new opportunities for individuals and society as a whole.

7.8 Conclusion

7.8.1 Recapitulation of Key Points

Automation is rapidly transforming industries and societies, driven by advancements in AI and robotics. In this discussion [19], we explored several key points:

- Automation brings both benefits and challenges. It improves productivity, efficiency, and accuracy while also posing risks and ethical considerations.

- Safety and risk management are crucial to ensure the safe operation of automated systems and protect human workers.
- Ethical implications and human–machine interaction must be carefully addressed to ensure transparency, fairness, and accountability in decision-making processes.
- Workforce transition and adaptation programs are necessary to support individuals affected by automation, providing them with the skills needed for the changing job landscape.
- Legal and regulatory frameworks must keep pace with technological advancements to address issues such as liability, data privacy, and workers' rights.

7.8.2 Future Outlook and Industry 5.0 Transformation

Looking ahead, the future of automation holds immense potential. Advancements in AI and robotics will continue to drive innovation and increase the capabilities of automated systems. Collaboration between humans and machines, with a focus on human-centered approaches, will lead to enhanced productivity and problem-solving abilities.

The transformation toward Industry 5.0, which emphasizes the integration of automation and human skills, is on the horizon. This paradigm shift recognizes the importance of human creativity, intuition, and empathy alongside automation technologies. It aims to create a symbiotic relationship between humans and machines, where each complements the other's strengths [20].

While automation may lead to changes in the employment landscape, it also presents opportunities for new job creation and economic growth. By addressing the socioeconomic effects and investing in workforce transition programs, we can navigate this transition more smoothly and ensure a more inclusive and equitable future.

To fully realize the potential of automation, collaboration among industry leaders, policymakers, researchers, and the public is essential. By working together, we can shape the future of automation in a way that maximizes its benefits, mitigates risks, and creates a sustainable and prosperous society.

References

1. Emma-Ikata, D. and Doyle-Kent, M., Industry 5.0 Readiness – "Optimization of the Relationship between Humans and Robots in Manufacturing

Companies in Southeast of Ireland". *IFAC PapersOnLine*, 55-39, 419–424, 2022.
2. Yadav, R., Parwez, Z., Sree Parimala, R., Analysis and Prediction of Future Research Trends in the State of Industry 5.0. Published/ publié in. *Res Mil. (resmilitaris.net)*, 13, 3, 2331–2339, March Spring 2023.
3. Pizoń, J., Cioch, M., Kanski, L., Cobots Implementation in the Era of Industry 5.0 Using Modern Business and Management Solutions. *Adv. Sci. Technol. Res. J.*, 16, 6, 166–178, 2022, https://doi.org/10.12913/22998624/156222, ISSN 2299-8624, License CC-BY 4.0.
4. Kılıç, A. and Balcıoğlu, Y.S., EMBRACING THE FUTURE: A COMPREHENSIVE REVIEW OF INDUSTRY 5.0 RESEARCH TRENDS. *9th International Congress on Humanities and Social Sciences in a Changing World Proceedings Book.*
5. Gao, Z., Wanyama, T., Singh, I., From Industry 4.0 to Robotics 4.0 - A Conceptual Framework for Collaborative and Intelligent Robotic Systems. *13th International Conference Interdisciplinarity in Engineering (INTER-ENG 2019), Procedia Manuf.*, 46, 591–599, 2020.
6. Zeb, S., Khowaja, S.A., Khowaja, S.A., Towards Defining Industry 5.0 Vision with Intelligent and Softwarized Wireless Network Architectures and Services: A Survey. Special issue on "Edge Computing for the IoT", in: *Computers*, vol. 13, pp. 86--97, 2023, ISSN 2073-431X.
7. Carayannis, E.G. and Morawska, J., University and Education 5.0 for Emerging Trends, Policies and Practices in the Concept of Industry 5.0 and Society 5.0, ISBN 978-3-031-26231-9, https://doi.org/10.1007/978-3-031-26232-6, © The Editor(s) (if applicable) and The Author(s), under exclusive license to Springer Nature Switzerland AG, 2023.
8. Alojaiman, Technological Modernizations in the Industry 5.0 Era: A Descriptive Analysis and Future Research Directions, 16, 168–181, 2022, Springer © 2023 by the author. Licensee MDPI, Basel, Switzerland. This article is an open access article distributed under the terms andconditions of the Creative CommonsAttribution (CC BY) license (https://creativecommons.org/licenses/by/4.0/).
9. Suntsova, O., The definition of smart economy and digital transformation of business in the concepts Industry 4.0 and 5.0. *Technol. Audit Prod. Reserves*, 4, 66, 18–23, 2022, doi: http://doi.org/10.15587/2706-5448.2022.265105.
10. Shaji George, A. and Hovan George, A.S., The Cobot Chronicles: Evaluating the Emergence, Evolution, and Impact of Collaborative Robots in Next-Generation Manufacturing. *Partners Univers. Int. Res. J. (PUIRJ)*, 02, 02, April-June 2023, ISSN: 2583-5602 | www.puirj.com.
11. Universal Robots | the Benefits and Capabilities of Cobots, 6 Apr. 2020.
12. Shaji George, A. and Hovan George, A.S., Revolutionizing Manufacturing: Exploring the Promises and Challenges of Industry 5.0. *Partners Univers. Int. Innov. J. (PUIIJ)*, 01, 02, 22–38, 2023, https://doi.org/10.5281/zenodo.7852124.

13. C.E.T. and Martin, E., How Collaborative Robots Fit Into the Production Facilities of the Future, JAE Automation, 25 Feb. 2020, jaeautomation.com/how-collaborative-robots-fit-into-the-production-facilities-of-the-future.
14. Shaji George, A., Hovan George, A.S., Baskar, T., Gabrio Martin., A.S., Human Insight AI: An Innovative Technology Bridging The Gap Between Humans And Machines For a Safe, Sustainable Future, 2023.
15. Mady, T. and El-Sherbeeny, A.M., Collaborative Robots: A Review of the State-of-the-Art and Applications. *Int. J. Adv. Rob. Syst.*, 16, 6, 95–109, 2019.
16. Kock, S. *et al.*, The Potential of Collaborative Robots in the Agri-Food Industry. *Trends Food Sci. Technol.*, 94, 85–97, 2020.
17. Alosaimi, W., Ansari, M., Alharbi, A., Alyami, H., Ali, S., Agrawal, A., Khan, R.A., Toward a unified model approach for evaluating different electric vehicles. *Energies*, 14, 6120, 2021.
18. Alyami, H., Ansari, M., Alharbi, A., Alosaimi, W., Alshammari, M., Pandey, D., Agrawal, A., Kumar, R., Khan, R.A., Effectiveness Evaluation of Different IDSs Using Integrated Fuzzy MCDM Model. *Electronics*, 11, 859, 2022.
19. Carayannis, E.G., Draper, J., Bhaneja, B., Towards fusion energy in the Industry 5.0 and Society 5.0 context: Call for a global commission for urgent action on fusion energy. *J. Knowl. Econ.*, 12, 1891–1904, 2021.
20. Kasinathan, P., Pugazhendhi, R., Elavarasan, R., Ramachandaramurthy, V., Ramanathan, V., Subramanian, S., Kumar, S., Nandhagopal, K., Raghavan, R.R.V., Rangasamy, S. *et al.*, Realization of Sustainable Development Goals with Disruptive Technologies by Integrating Industry 5.0, Society 5.0, Smart Cities and Villages. *Sustainability*, 14, 15258, 2022.

8

Artificial Intelligence–Driven Cobots for Innovative Industry 5.0 Workforce

S. K. Manju Bargavi[1]*, K. R. Tejashvar[2] and K.R. Tejashwini[2]

[1]*School of Computer Science and IT, Jain Deemed-to-be University, Bengalore, India*

[2]*B. Tech Computer Science and Engineering, Christ (Deemed-to-be) University, Bengalore, India*

Abstract

"Cobots" are robots created expressly to work with humans. The goal of Industry 5.0 development is to improve computer science capabilities in order to foster a tighter bond between people and technology. Cobots are created to develop straightforward interactions with people. The objective of intelligent process automation is to assist cobots in establishing both production objectives and industrial plant safety measures. Intelligent cobots can then mimic these actions, assess where they can be of assistance, and support operators in their daily tasks. While both people and AI may play a variety of functions inside an organization, for collaboration to be successful, companies where both play the role of decision-making agents must place a strong emphasis on regulating the effects of technological engagement on human performance and vice versa. In order to help businesses and researchers alike determine the best approach for organizational decision-making in Industry 5.0, where human–machine collaboration is the main objective, many human–machine collaboration strategies that have been studied are also reviewed. The goal of Industry 5.0 is to change how people and technology interact through artificial intelligence.

***Keywords*:** Business, cobot, collaboration, human–machine, intelligent

**Corresponding author*: cloudbargavi@gmail.com

V. Ramasamy, S. Balamurugan and Sheng-Lung Peng (eds.) Intelligent Robots and Cobots: Industry 5.0 Applications, (159–176) © 2025 Scrivener Publishing LLC

8.1 Introduction

The introduction of new, innovative technologies such as Artificial Intelligence (AI) and collaborative robots (cobots) are revolutionizing the manufacturing process, allowing for faster, more accurate production of goods, and providing the opportunity to reduce human labor costs. AI-driven Cobots are being used in Industry 5.0 as cost-effective, efficient, and safe solutions to production problems. Cobots are designed to work alongside human workers, simplifying and streamlining repetitive tasks while allowing humans to focus on higher value-added activities. Cobots also enable workers to collaborate with machines in order to gain a deeper understanding of how to optimize process efficiency [2]. With AI-driven cobots, manufacturers can reduce costs and improve quality, as well as increase productivity and safety. Furthermore, AI-driven cobots can facilitate improved workplace communication, allowing for better collaboration between workers, managers, and machines.

The Web of Things is characterized and utilized by a notable specialist Kevin Ashton in the beginning of 2000. From Kevin Ashton's clarifications, the IoT is a framework/construction of material things to connection to the web through clever sensors. Ashton likewise considered the RFID innovation, which vigorously applied to transportation following administrations without any human intercessions. Presently, there are various definitions accessible based on their particular thought in reality situation. For example, from IEEE, "IoT is a system of associated gadgets through the web, for new machines and administrations empower the connection in the physical and virtual world as M2M communication" [3]. From Web Engineering Sheets' definition, "IoT is the systems administration of savvy objects, meaning numerous gadgets sensibly impart within the sight of web convention that not straightforwardly worked by people yet exist as parts in structures, vehicles, or the environment" [6].

By highlighting the crucial role of people and encouraging collaboration, Industry 5.0 will alter the labor market. "Collaborative robots" or "cobots" are a new breed of robots that allow humans and other robots to work together. These robots, or cobots, are made to operate alongside humans and, more significantly, to assist them in carrying out everyday duties in the real world. They are easy to use, and their main purpose is to physically aid operators by carrying out unpleasant and dangerous tasks [1]. Cobots should eliminate the worry of the production line being lost to automation, which has been a significant Industry 4.0 concern, and will provide the smart factory improved adaptability.

Robotics, ergonomics, and cognatics come together in the field of cobotics, a transdisciplinary technology. This new field of study focuses on understanding a cobot and human interacting.

As a result of ergonomic and cognitive engineering studies, several components of cobotics—such as the human representation of the robot, the ergonomics of the workstation, or operator acceptance—are not included in robotics.

At the end of the 1990s, a passive effort-assistance machine directly operated by an operator was referred to as a "cobot." The definition of the phrase has changed since then. It only became well-known in the 2010s. Modern robotics allows for the extension of human gestures or behaviors, setting it apart from basic robots by allowing for in-person, direct, or remote connection.

Researchers are putting a lot of effort into figuring out how to train algorithms using less data, and they have high hopes of succeeding [4]. Consequently, AI specialists advocate for a paradigm shift towards understanding the significance of "big data". The AI growth equation's inclination will be reversible. Big data will be replaced by little data as the cutting-edge AI engine of Industry 5.0. The five industrial revolutions are described in Figure 8.1.

This chapter aims to provide an innovative method for cobot training using tiny quantities of data. A framework based on common sense and quick multitasking methods is something that we highly value. The current state of cobotics research is therefore summarized in this article.

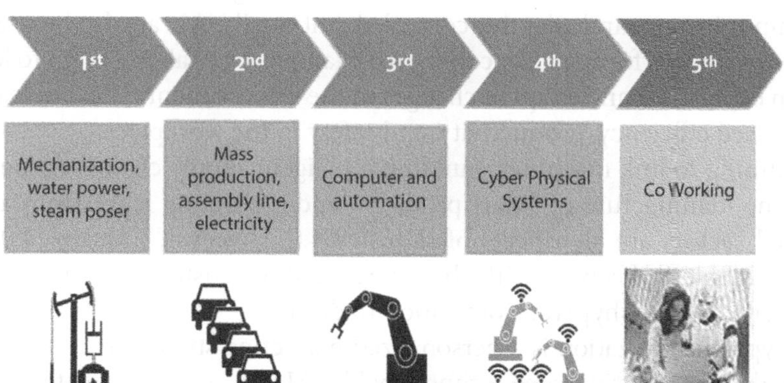

Figure 8.1 The five industrial revolutions.

8.2 Literature Review

A literature survey on AI-driven cobots for innovative Industry 5.0 workforce would likely cover several key areas, including the following:

The benefits and challenges of implementing cobots in Industry 5.0 environments include increased efficiency, flexibility, and safety, as well as concerns around cost, complexity, and safety risks.

Cobots, or collaborative robots, are designed to work alongside humans in Industry 5.0 environments. They offer several benefits such as increased efficiency, flexibility, and safety. For example, by taking on repetitive or dangerous tasks, cobots can free up human workers to focus on more complex or value-added activities. Additionally, cobots can work in close proximity to humans without the need for safety barriers, which can save space and improve overall safety.

However, there are also challenges to implementing cobots in Industry 5.0 environments. One major concern is cost, as cobots can be expensive to purchase and maintain. Additionally, implementing cobots may require significant changes to existing processes and infrastructure, which can be complex and time-consuming. Finally, there are safety risks to consider, as cobots can be dangerous if not properly programmed and maintained.

The use of AI in cobots includes the integration of machine learning, computer vision, and natural language processing to enable more advanced capabilities and improved performance.

Cobots are robots designed to work alongside humans in a shared workspace [5]. The integration of AI technologies such as machine learning, computer vision, and natural language processing can enable cobots to perform more advanced tasks, such as object recognition, natural language communication, and adaptive control. Additionally, AI can also be used to improve the performance of cobots, for example, by enabling them to learn from experience and adapt to changes in the environment. This can lead to increased efficiency, productivity, and safety in the workplace.

Aiming to link machines, building intelligent supply chains, encouraging the manufacture of smart products, and separating labor from automated sectors are all aspects of Industry 4.0. However, Industry 4.0 has been unable to keep up with the rising need for customization, whereas Industry 5.0 uses hypercustomization to do so.

Hypercustomization is a personalized marketing strategy that uses real-time data and cutting-edge technology like AI, ML, cognitive systems, and computer vision to offer each customer more specialized goods, services, and content [7]. Robotics and human intelligence combined allows for mass customization of items by producers. To accomplish this, various functional material variations are communicated with other staff members with the goal of customizing the product with various variations for consumer choice.

Industry 4.0 intended for massive production with minimal waste and maximum efficiency, but Industry 5.0 wants to achieve mass customization at the lowest possible cost with the highest level of precision. The industries can coordinate the manufacturing processes to incorporate client requests and market changes thanks to the interaction of humans, robots, and cognitive systems. Making the switch to an agile supply chain and production process is the first step in hyper personalization [8]. Moreover, the production team, consumer preferences, and human intervention are required. Also, the viability of hypercustomization is highly dependent on how economically viable the generated items are.

Case studies and examples of cobots are being used in various Industry 5.0 applications, such as manufacturing, logistics, and maintenance.

Cobots are being used in a variety of Industry 5.0 applications, including manufacturing, logistics, and maintenance.

In manufacturing, cobots are being used for tasks such as assembly, packaging, and material handling. For example, in the automotive industry, cobots are being used to assist with tasks such as welding and painting, whereas, in the electronics industry, they are being used for tasks such as soldering and circuit board assembly.

In logistics, cobots are being used for tasks such as order picking, packing, and palletizing. For example, in warehouse environments, CoBots are being used to assist with the pick and pack process, which can help to improve efficiency and accuracy.

In maintenance, CoBots are being used for tasks such as inspection, cleaning, and repair. For example, in the aerospace industry, CoBots are being used to assist with tasks such as aircraft engine maintenance and inspection, whereas, in the oil and gas industry, they are being used for tasks such as inspection of pipelines and drilling equipment.

One of the most popular cobot, universal robots, is widely used in various industry for their flexibility and ease of programming [11].

The role of cobots in the broader context of Industry 5.0 and the Fourth Industrial Revolution, including the potential impact on the workforce and the future of work.

8.3 Revolution of Industry 5.0

Technology and innovation frequently experience rapid growth, resulting in creative developments that seriously affect entire industries.

Businesses that seek to resist change or cannot keep up with it suffer the same fate of going out of business. Businesses that welcome disruption have a higher chance of riding the perceived chaotic wave and will be better positioned to take advantage of the state of the industry that will come next. There have been four significant changes in the industrial landscape so far; the most recent is known as Industry 4.0. Industry 5.0 is already starting to make noise, but, for some context, let us update you on where we are right now.

The Industrial Internet of Things (IIoT) and intelligent devices and systems have assumed a central role in Industry 4.0. The IIoT enables businesses to fully leverage the "smart factory" concept, where devices speak to one another and offer a new level of visibility and interactivity at every stage of the manufacturing process. This is in relation to the numerous devices involved in manufacturing and automation that have grown increasingly intertwined.

The insurgency of Industry 5.0 implies that people and machines are cooperating, working on the effectiveness of modern creation. Human laborers and widespread robots are supporting the efficiency of the assembling business [5]. Every one of the leader groups of the fabricate organization is expected to characterize the creation line, then, at that point, to follow the key presentation markers, and also to guarantee that the cycles are working easily. The future bearing of Industry 5.0 is the assembling of robots and modern robots. The progression of man-made consciousness and mental processing innovations is taking the assembling scene to a fast and expanding business productivity [15].

Aside from the benefits in the assembling industry, Industry 5.0 likewise helps in maintainability as it means to foster a practical framework that sudden spikes in demand for sustainable power. To embrace industry 5.0 for the organizations, the work force are required appropriate cooperation among the machines as well as administrators. Figure 8.2 shows that industrial evolution from 1.0 to 5.0. It is information in the fields like advanced mechanics as well as computerized reasoning [3, 14].

The job of the business association depends on settling on choices around the high level elements. Preparing representatives is expected with virtual schooling to diminish the expense for the organizations, as it is not expected creation to stop for giving preparation to the workers. It gives safe preparation that can keep the specialists from being uncovered to unnecessary issues during the instructional courses.

Correspondence and worker inspiration are helped by bringing about intuitive information conditions. The work positions are connected with

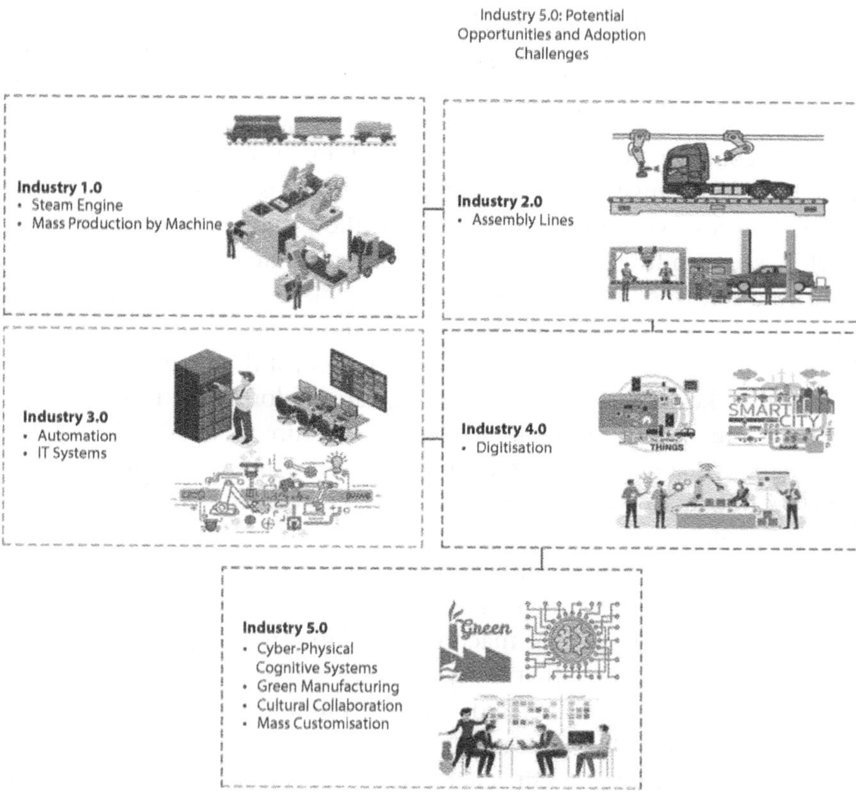

Figure 8.2 Industrial evolution from 1.0 to 5.0 [2].

correspondence with the advanced mechanics frameworks as well as fake knowledge.

The emergence of cobots is seen as a key part of Industry 5.0 and the Fourth Industrial Revolution. Cobots are designed to work alongside humans and can be used to automate processes, free up employees for higher-value roles, and reduce costs. The potential impact of cobots on the workforce is twofold. On one hand, cobots can help to alleviate the burden of mundane and repetitive tasks, allowing employees to focus on more creative and higher-value activities. This, in turn, can lead to enhanced job satisfaction, improved productivity, and better job security [10]. On the other hand, cobots can also lead to job displacement, as they are increasingly being used to replace human labor in certain tasks. The future of work is likely to be more automated, with cobots playing an increasingly important role. In the near future, cobots will become more intelligent and capable and will be used in a variety of industrial and commercial applications.

The challenge for employers will be to ensure that CoBots are used to augment, rather than replace, human labor. Ultimately, cobots will have a significant impact on the nature of work in the Fourth Industrial Revolution. While the potential benefits are considerable [13], employers will need to carefully consider how best to integrate cobots into their existing workforce in order to ensure that the human workforce is not displaced.

The business 5.0 transformation means the conjunction of people and machines, expanding the adequacy of modern result. The assembling area is turning out to be more useful thanks to human work and widespread mechanical technology. Every one of the assembling organization's chief groups should lay out the creation line and, then, at that point, stick to the KPIs and make sure that the methodology are all done effortlessly.

The production of industrial robots is the future focus of Industry 5.0. The manufacturing industry is advancing quickly, and company efficiency is rising, thanks to the development of AI and cognitive computing technologies [16].

Industry 5.0 is a new production paradigm that emphasizes interaction between people and machines. Industry 5.0 focuses on maximizing cooperation between more precise machinery and the creative potential of people. It creates procedures that reuse and recycle the resources in order to make production sustainable. Reduced environmental effects in the manufacturing sector are also necessary [17].

To maximize resource efficiency and waste, personalization must be increased using additive manufacturing. Industry 5.0 is transforming production processes all over the world by relieving human workers of monotonous jobs. Intelligent robots and systems have reached an unprecedented level of penetration into manufacturing supply chains. By storing design files in the cloud with strict access controls and utilizing the production resources across many locations, smart manufacturing enables designers to protect the design elements of manufactured goods.

8.4 Robotic Collaboration

Automation procedures are being greatly simplified by the use of robots, yet the powerful machines that make this advancement possible continue to pose serious risks to both their human teammates and them on the shop floor [12]. With a focus on collaborative robots, or cobots for short, the discipline of collaborative robotics is tackling this. Although the level of automation in use today gives factories incredible capabilities, it can only carry out instructions from its human operators.

8.4.1 Widespread Personalization

Mass personalization may seem contradictory, but Industry 5.0 ideas are bringing it into the real world. Human laborers will play a more specialized role in the production process as they work on the new functionality and process improvement that collaborative robotics will offer [11]. Humans will be able to quickly and accurately mass-produce individualized and customized products for customers alongside their intelligent machine coworkers.

8.4.2 Productivity and a Novel Human–Machine Connection

By now, it should be abundantly clear that Industry 5.0 is shifting back to a human focus from Industry 4.0, which had a heavily automated focus. The dangers posed by huge, powerful robots will decrease as more articulated and nimbler cobots are incorporated into the production process, and humans will have more control over the robots there.

Productivity rises as a result. It was determined that this was the case when Toronto's Paradigm Electronics added a polishing machine in addition to a human worker to their loudspeaker manufacturing process. Paradigm polished their speaker cabs to a mirror-like brilliance using the UR10 robotic arm from universal robots, which increased production by 50%. This not only frees up human employees from tedious and oftentimes dangerous chores, but it also saves the business money on the liability and medical costs related to such tasks.

Hyper-personalization is made possible by empowering human intellect with cognitive computing and intelligent automation. The use of technology like machine learning, robotic automation, and others is enabling staff to become more business-savvy and provide clients with high-quality services more quickly. The enterprise resource planning (ERP) system controls the supply chain for the business organization, including raw material delivery, transactions, and transportation. The technology to support the digital supply chain is being created and implemented by the next generation of supply chain solutions. It entails introducing personalization into the supply chain, raising customer happiness, and managing corporate productivity and profit margins. Based on the business' current information, risks associated with waste and the supply chain are being reduced.

8.4.3 New Employment

In the current world, there is a pervasive although largely hypothetical fear that robots will displace people from their employment. Thanks to collaborative robots, manufacturing floor workers can still accomplish their jobs but with less filthy, dangerous, and tedious tasks. Their already indispensable and difficult-to-replace skill sets gain value from learning how to operate, manage, and program a new fleet of robotic workers. Amazon has already made great strides toward educating its current employees to become the upcoming generation of robot experts.

Efficiency and profit maximization are the main drivers of business today, not resilience. Resilience will take the place of development, profit, and efficiency as the organizations' main priorities if it genuinely becomes one of the three pillars of Industry 5.0. However, the focus will be on developing "anti-fragile" organizations, which are able to foresee, respond to, and learn systematically from any crises in order to assure steady and long-lasting performance.

The Chief Robotics Officer post is the newest C-suite position to emerge, along with the growing robot workforce. The Chief Robotics Officer Research Scenario states that the primary focus of the position is on robotics and intelligent operational systems technology described in Figure 8.3. As manufacturing activities grow more entrenched with developing technology, businesses will witness new divisions appearing to correspond with their expanding mechanized workforces.

There is hardly any need to introduce the concept of sustainability given the widely held concerns about climate change. But fully integrating

Figure 8.3 Robotics by 2050 [encompass.com].

sustainability into a company's strategy involves doing considerably more than is being done right now. Companies are moving away from being a part of the problem and toward being a part of the solution in Industry 5.0.

8.5 Technological Issues with AI in the Cobot Age of Industry 5.0

Researchers today are pushing the envelope to develop more sophisticated or complex kinds of interaction by giving cobots' comprehension and anticipatory abilities with the help of AI.

Future cobots must be able to discern between intended and accidental gestures made during collaborative work, as well as identify human signals, actions, and intents. The ability of cobots to precisely record, process, and comprehend human desires is essential for the natural collaboration between humans and cobots.

The technological difficulties with AI, however, are in the interaction modalities like speech, gaze, or gesture planning.

8.5.1 Real-Time Applications

The application of AI-driven cobots for innovative industry 5.0 workforce can revolutionize the way industrial processes are carried out. By allowing AI-driven cobots to perform tasks that are too complex or too dangerous for human workers, they can increase efficiency and safety in industrial settings. AI-driven cobots can be deployed to automate repetitive tasks, manage inventory, and optimize production lines [9]. Additionally, AI-driven cobots can be used to collect data and provide predictive analytics to help managers make better decisions. This technology can also be used to monitor processes and alert operators of any potential issues, ensuring that safety protocols are being followed. AI-driven cobots can also be trained to interact with human workers, providing an extra layer of collaboration and communication. Finally, AI-driven cobots can enable virtual training and remote monitoring, allowing for more efficient management of industrial workforces.

Cobots, or collaborative robots, are a type of robot that is designed to work alongside human operators in manufacturing and other industrial settings. They are increasingly being equipped with AI capabilities to make them more flexible, efficient, and capable of handling a wider range of tasks. In Industry 5.0, Cobots with AI capabilities will be able to learn from human operators, adapt to changing conditions, and make decisions on their own, which will help to improve the overall productivity and efficiency of the workforce. Additionally, by working alongside human operators, cobots with AI can help to improve worker safety and reduce the risk of injury.

8.5.2 Current Trends

Current trends and future directions for the development and deployment of AI-driven CoBots in Industry 5.0 include advances in robotics and AI technology, as well as regulatory and ethical considerations.

AI-driven cobots are being deployed in many industrial settings, such as manufacturing, logistics, healthcare, and retail. Cobots are becoming increasingly intelligent, as AI technologies are quickly becoming more sophisticated.

Robotics and AI-driven cobots are becoming more cost-effective, allowing for greater adoption in industry.

Companies are increasingly investing in collaborative robots to increase efficiency and reduce costs. Figure 8.4 shows applications of Industry 5.0.

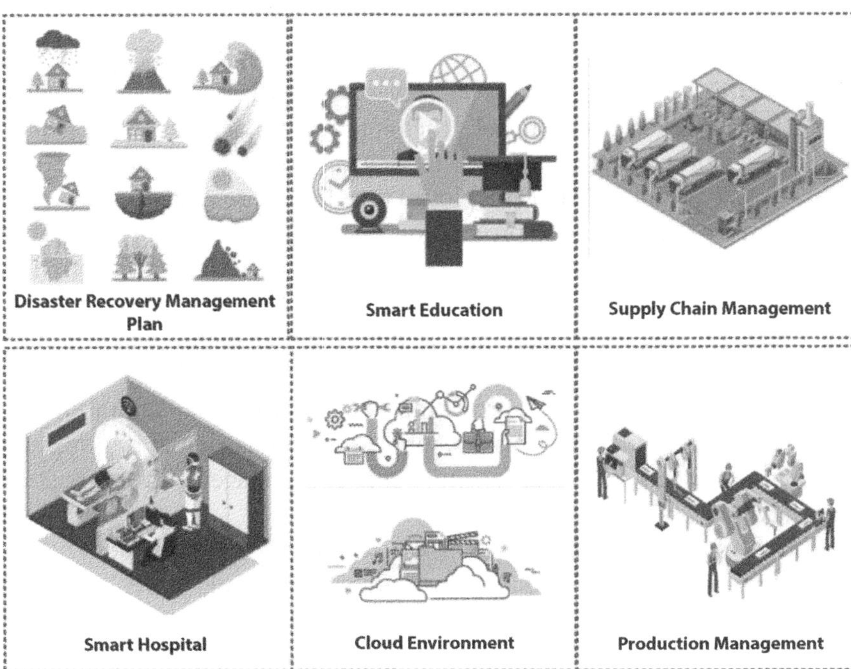

Figure 8.4 Applications of Industry 5.0.

Robots are necessary for the management of the supply chain in high production quantities of standardized operations, added this to each product, and it is a challenge where the robots need correct instruction. Redoing and customizing things do not be guaranteed to require the human touch. Nonetheless, it likewise ensures that the inventory network's start to finish activities run as expected, including picking the unrefined components in the wake of figuring out the necessities of every client's exceptional personalization and customization. Industry 5.0 aims to combine the human touch with automated and intelligent digital ecosystems. The use of human aspects in this process allows for the customization of end-user experiences and improved workflows.

Hyper-personalization is made possible by empowering human intellect with cognitive computing and intelligent automation. Technologies including robotic automation, machine learning, and others are assisting staff members in boosting firm productivity and providing high value to clients more quickly [12]. The ERP framework controls the store network for the business association, including exchanges, transportation, and the conveyance of natural substances. The innovation to help the advanced inventory network is being created and carried out by the up and coming age of store network arrangements. It entails improving customer happiness, adding customization to the supply chain, managing business efficiency, and managing market margins.

8.5.3 Future Directions

The possibility of autonomous drones has increased over the past 5 years from "maybe possible" to "definitely possible" to "inevitable" to "how did anyone ever think this was not inevitable?" to the present of "now commercially available." In contrast, Industry 5.0 presents us with the idea of self-driven autonomous drone system fed with a mapping application that makes them capable of taking the necessary response actions on their own in the given circumstances. Industry 4.0 introduced us to this drone concept, which has almost become a daily chore for different activities.

Let us analyze the hypothetical instance of a drone watching over a fence in this case of aerial operations. Instead of controlling the drone manually, a program that allows the drone to choose the timing and scope of the surveillance on its own can be fed. These drones have the ability to identify any questionable individuals, monitor their behavior, and send a message alert as a warning of trespassing. Here, the role of Industry 5.0 and its products is crucial in changing the traditional drone flight procedures and revolutionizing their functionality.

1. There will be a greater focus on developing AI-driven cobots that can interact with humans in a safe and efficient manner.
2. AI-driven cobots will become increasingly capable of performing complex tasks, such as decision-making, planning, and problem-solving.
3. Regulatory and ethical considerations will play an increasingly important role in the development and deployment of AI-driven cobots.
4. AI-driven cobots will become more widely adopted in industry, as they become more cost-effective and reliable.

Overall, while cobots offer many potential benefits, they also pose significant challenges that must be carefully considered before implementation.

8.5.3.1 Artificial Intelligence

AI is the simulation of human intelligence processes by machines, especially computer systems. AI technologies have been used in various industrial applications such as manufacturing, logistics, and healthcare. AI can automate tasks and optimize processes, leading to increased efficiency and productivity. AI-driven cobots are increasingly being used in Industry 5.0 to assist workers and make processes more efficient.

8.5.3.2 Cobots

Cobots are a type of robot designed to be used in collaboration with human workers. They are programmed to interact with humans in order to complete tasks that may be too complicated or dangerous for humans to do. Cobots can be used for a range of tasks, from simple assembly to more complex tasks such as welding. Cobots can be used for automation and to optimize processes, leading to increased efficiency and productivity.

8.5.3.3 Cobots in Industry 5.0

Industry 5.0, also known as the digital transformation era, is characterized by the use of advanced technologies such as AI, robotics, and automation. AI-driven cobots can be used to assist workers in a variety of tasks and to reduce errors, leading to improved efficiency and productivity. Cobots are particularly well suited for tasks that require precision such as welding, painting, and assembly. Cobots can also be used in medical applications to help with tasks such as surgery and diagnostics.

8.5.3.4 *Benefits of Using AI-Driven Cobots*

AI-driven cobots can provide a number of benefits to Industry 5.0. They can reduce errors, improve safety, and increase efficiency and productivity. AI-driven cobots can also help to reduce costs associated with labor and materials, leading to greater profitability. Additionally, cobots can be used to automate tasks that are too complex or dangerous for humans to do, thus freeing up human workers to focus on more creative tasks.

8.6 Conclusion

AI-driven cobots are increasingly being used in Industry 5.0 to assist workers and make processes more efficient. Cobots can reduce errors, improve safety, and increase efficiency and productivity. They can also be used to automate tasks that are too complex or dangerous for humans to do. AI-driven cobots have the potential to revolutionize the industry and create a more innovative and productive workforce.

The use of AI-driven cobots in industry 5.0 is a growing trend and one that is set to revolutionize the way businesses operate. Cobots are collaborative robots that are capable of working alongside humans, and they are becoming increasingly popular as a way to augment human workers and increase productivity. AI-driven cobots are specifically designed to collaborate with humans, and they can be programmed to take on a variety of tasks. This can include tasks that range from simple repetitive tasks to complex ones that require a greater degree of cognitive input. AI-driven cobots are also capable of learning from their environment and adjusting their behavior accordingly, which can help to increase efficiency and reduce errors. The advantages of using AI-driven cobots in industry 5.0 are numerous. For example, they can help to reduce the need for manual labor, which can help to reduce costs. Additionally, they can help to improve safety, as they can be programmed to take appropriate safety measures. Finally, they can help to improve overall productivity, as they are capable of taking on more complex tasks that may otherwise have been too difficult or time-consuming for humans to complete. In this way, they can help to optimize processes and make them more efficient.

References

1. Lee, S.K., Bae, M., Kim, H., Future of IoT Networks: A Survey. *Appl. Sci.*, 7, 1072, 2017.
2. Garcia, C.G., Nunez-Valdez, E.R., Garcia-Diaz, V., Cristina PelayoGBustelo, B., Lovelle, J.M.C., A Review of Artificial Intelligence in the Internet of Things. *Int. J. Interact. Multimedia Artif. Intell.*, 5, 4, 9–20, 2018.
3. Al-Fuqaha, A., Guizani, M., Mohammadi, M., Aledhari, M., Ayyash, M., Internet of Things: A Survey on Enabling Technologies, Protocols, and Applications. *IEEE Commun. Surv. Tutor.*, 17, 4, 2347–2376, 2015.
4. Lin, J., Yu, W., Zhang, N., Yang, X., Zhang, H., Zhao, W., A Survey on Internet of Things: Architecture, Enabling Technologies, Security and Privacy, and Applications. *IEEE Internet Things J.*, 4, 5, 1125–1142, 2017.
5. Samie, F., Bauer, L., Henkel, J., From Cloud Down to Things: An Overview of Machine Learning in Internet of Things. *IEEE Internet Things J.*, 6, 3, 4921–4934, June 2019.
6. Altexsoft, IoT Architecture: the Pathway from Physical Signals to Business Decisions, 2020, https://www.altexsoft.com/blog/iotarchitecture-layers-components.
7. Chander, G.B. and Kumaravelan, G., Introduction to Wireless Sensor Networks, in: *Soft Computing in Wireless Sensor Networks*, 2018.
8. Chander, B. and Kumaravelan, G., Internet of things: foundation, in: *Principles of Internet of Things (IoT) Ecosystem: Insight Paradigm*, pp. 3–33, Springer, Cham, 2020.
9. Gopalakrishnan, K., Security vulnerabilities and issues of traditional wireless sensor networks in IoT, in: *Principles of Internet of Things (IoT) Ecosystem: Insight Paradigm*, pp. 519–549, Springer, Cham, 2020.
10. Lin, Y.C., Development of Advanced Manufacturing Cloud of Things (AMCoT) - A Intelligence Manufacturing Platform. *IEEE Rob. Autom. Lett.*, 2, 1, 1809–1816, 2017.
11. Chen, C.C., A Novel Automated Construction Scheme for Efficiently Developing Cloud Manufacturing Services. *IEEE Rob. Autom. Lett.*, 3, 3, 1378–1385, 2018, DOI 10.1109/LRA.2018.2799420.
12. Gupta, H., Dastjerdi, A.V., Ghosh, S.K., Buyya, R., iFogSim: A toolkit for modeling and simulation of resource management techniques in the Internet of Things, Edge and Fog computing environments. *J. Softw.: Pract. Exper.*, 47, 9, 1275–1296, 2015.
13. Zhou, B. and Buyya, R., Augmentation techni9u es for mobile cloud computing: A taxonomy, survey, and future directions. *ACM Comput. Surv. (CSUR)*, 51, 1, 1–38, 2018.
14. Ranjan, R., Rana, O., Nepal, S., Yousif, M., James, P., Wen, Z., The next grand challenges: Integrating the Internet of Things and data science. *IEEE Cloud Comput.*, 5, 3, 12–26, 2018.

15. Gubbi, J., Buyya, R., Marusic, S., Palaniswami, M., Internet of Things (IoT): A vision, architectural elements, and future directions. *Future Gener. Comput. Syst.*, 29, 7, 1645–1660, 2013.
16. Chen, H., Simoska, O., Lim, K., Grattieri, M., Yuan, M., Dong, F., Lee, Y.S., Beaver, K., Weliwatte, S., Gafne, E.M., Minteer, S.D., Fundamentals, applications, and future directions of bioelectrocatalysis. *Chem. Rev.*, 120, 23, 12903–12993, 2020.
17. Deepa, N., Pham, Q.V., Nguyen, D.C., Bhattacharya, S., Prabadevi, B., Gadekallu, T.R., Pathirana, P.N., A survey on blockchain for big data: approaches, opportunities, and future directions. *Future Gener. Comput. Syst.*, 131, 2, 209–226, 2022.

9

Cobot Collaboration in the Healthcare Industry

V. Gopala Krishnan[1]*, R. Jayamurugan[2], P. Sakthivel[3] and M. Gowtham[4]

[1]Department of Physics, Dr. N. G. P. Arts and Science College, Coimbatore, Tamil Nadu, India
[2]Department of Physics, Sri Ramakrishna Mission Vidyalaya College of Arts and Science, Coimbatore, Tamil Nadu, India
[3]Department of Physics, Dr. N.G.P. Institute of Technology, Coimbatore, Tamil Nadu, India
[4]PG and Research and Department of Physics, Kongunadu Arts and Science College, Coimbatore, Tamil Nadu, India

Abstract
Universal robot is the prominent creator in the recent era. Collaborative robots register their place in the automated Industry 5.0 revolutions, which is now driving the health sector. However, in recent years, a vision-guided robotic device known as a "cobot" has been developed that may be used for a variety of precise tasks. The cobot's ease of programming, installation, and collaboration allows maximizing throughput and ensuring consistent quality. As a result, it can be readily redeployed for new product lines. Cobots can be used for hygienic and precise manipulation and assembly of medical equipment or implants, reducing the danger of human contamination, penetrating illnesses, and maintaining clean environments. Cobots can also free workers from material handling jobs in dusty, noisy, and high-vibration environments, allowing them to focus on higher-value tasks and quality inspections. Configuring robots for medical purposes is frequently estimated in hours, days, or even weeks with straightforward implementation. Lightweight cobots on mobile carts can be quickly transferred from process to process, and a single robot can frequently run much equipment. Similarly, three universal robots' robotic arms have been integrated into this technology so that engineers may quickly modify the program to unique patient needs and essential duties. Because of the ease of programming, installation, and collaborative

**Corresponding author*: vgowrigopal@gmail.com

V. Ramasamy, S. Balamurugan and Sheng-Lung Peng (eds.) Intelligent Robots and Cobots: Industry 5.0 Applications, (177–202) © 2025 Scrivener Publishing LLC

nature of industrial robotic arms, surgeons, nurses, operating room employees, and robots can work together to care for patients.

Keywords: Robots, medical industry, cobots, working progress, man-handling collaboration

9.1 Introduction

Computers, heavy machinery, and medical technologies are becoming increasingly intelligent as smart technology advances. New artificial intelligence (AI) algorithms are continually adapting to our needs and acting in our best interests. Industrial "cobots," or collaborative robots, are meant to cooperate astutely and carefully with people and are predicted to be crucial in Industry 4.0. Collaborative robots are those that can work with humans.

This partnership aims to foster the growth of human talent in a secure environment. In contrast, robotic solutions that lack human–robot collaboration often operate independently of humans and are frequently enclosed within cages. Moreover, they can be programmed to halt their operations when someone enters the area where the robot is working. This can result in unwanted operational or production delays, which can be mitigated by utilizing collaborative robots or cobots. By actively monitoring their surroundings and engaging with nearby humans, cobots can maintain performance and safety without any setbacks. The ongoing progress in various domains characterizes continuous innovation in the field of cobot industrial development. Enhanced homing capabilities, more precise movement patterns, faster response times, and the ability to mimic human behavior are all attributes driving the advancement of cobot technology [1–3].

> ***"Cooperative robots, or Cobots, eliminate some of the environmental and spatial hazards that traditional robots may pose, allowing them to work alongside humans. Their relationship with humans is more natural, they work in the same place, and Cobots can be readily reprogrammed."***

9.2 Cobots and Their Role

Cobots are the most modern robotic technology, and they have transformed the world of automation. A "cobot" is an abbreviation for a "collaborative robot." These robots are collaborative because they can work safely with people. It is outfitted with sensitive detectors which enable the robot to "hear." As a result, these robots are designed for continuous interaction with humans in a shared environment or near people. In 1996, Northwestern University professors J. Edward Colgate and Michael Peshkin created the Cobot for the first time [4, 5]. "Cobots" are large manufacturers who have benefited from automation for decades by deploying regular industrial robots, according to their 1997 US patent. However, unlike the low-volume, high-diversity manufacturing of small businesses, these enormous, expensive, and complicated robots were designed for high-volume, standardized manufacturing processes. Participatory robotics has developed into an adaptable, cost-effective, and simple-to-implement technology that helps organizations of all sizes to increase productivity, improve quality, and respond to changing consumer needs more quickly [6, 7]. Cobots are very helpful when a person is required to be present near the robots. Human–robot collaborations include instances, in which a human directs the robots, observes the operation, and, sometimes, teaches from the robots.

There is a clear risk when cobots interact with 5G networks. Quality-of-service requirements will be one of the most significant aspects of cobots, with greater demands for high communication, significant bandwidth, low latency, and quick ability to make decisions through efficient processing. Much research is being done to investigate how robots might mimic human motions and thoughts. On the contrary, AI systems have demonstrated to be able to outperforming human-discovered answers, and this in-game strategy and design optimization challenges are examples of this. It is fascinating to employ cobots to augment human labor, and human–robot collaboration is more efficient [8].

9.3 Impact of Cobot

When an invention becomes reliable, it gains popularity. When users lack trust in the gadget or application, the cobot demonstrates its worth. Moreover, the search space can be exceptionally vast as cobots make numerous real-time decisions. This demands the navigation of extensive knowledge graphs, which, in turn, requires substantial processing power, a reliable network, and specialized software designs. Recent technological breakthroughs, on the other hand, such as networked data, processing in parallel, computing at the edge, and dispersed intelligence (AI), empower cobots to make intelligent decisions, causing robust effective execution [9, 10].

9.4 The Challenges of Deploying Cobots at Scale

Because industries must build use cases for cobots, large installations of any novel technology are tough. Traditionally, all robotic machinery was installed and validated during the design phase. It was data-drivenly adapted throughout runtime at its best. In this situation, cobots give process developers the ability to teach cobots how to complete jobs as well as evolve throughout runtime, allowing for AI breakthroughs. Cobot technology includes hardware layout, actuators and sensors, effective data processing, video handling, planning, and several areas of AI scenarios, as well as technology that assure the system's safety, stability, and safety [11].

Collaborative robots, commonly known as cobots, are automated devices that have been employed across various industries for many years. They perform pre-programmed tasks to enhance human productivity and safety. The technology is known as a collaborative robot because it collaborates with humans and is equipped with force-feedback sensors to ensure that it does not harm anyone. Contrary to common belief, the primary objective of healthcare cobots is to serve as an extension of healthcare professionals (HCPs) and assist them in enhancing patient health outcomes [12]. Cobots are highly useful in removing monotonous or time-consuming duties from their shoulders of previously overburdened healthcare personnel. They have previously demonstrated to be useful in managerial duties, laboratory sample selection, surgical aid, and so on. Cobots are also in great need in tracking patients, electronic authorization, invoicing, claim processing, and other sectors.

Along with current industrial automation, terms like "Cobot" exist on the web more than ever before. Because the industrial revolution, robots have replaced human labor in critical and dangerous production jobs while operating under safety procedures. Cobots, or collaborative robots, operate with humans to eliminate some of the environmental and space risks that traditional robots might bring. Their connection with people in the same workstation is more natural, and cobots can be quickly reprogrammed [13, 14].

9.5 Cobot Background

Cobots were conceived in the 1990s, although they have only been commercially accessible for approximately a decade. General motors provided research funding between 1994 and 1995 to develop robots capable of collaborating with humans. The inception of cobots occurred in 1996, thanks to the efforts of J. Edward Colgate and Michael Peshkin at Northwestern University. General motors coined the term Intelligent Assist Device (IAD), although the word "Cobot" or "collaborative robot" was more widely used. In 2002, a draught of the IAD/cobot safety standards and procedures was produced, which was later upgraded in 2016.

9.6 Benefits of Cobots

Industry in the 21st century is intensely competitive, and lagging behind competitors can spell doom for a company. Industries can profit from collaborative robots in a variety of ways, including enhanced efficiency [15].

Here are five advantages that cobots provide straight out of the box.

- **Improved Robot–Human Interface and Performance:** In a typical assembly line configuration, human workers load the necessary components onto automated robots, and these robots then manage the remaining tasks. If human intervention is required, then it results in either a complete halt in the production process or a lengthy wait until the machines have completed their tasks. In a collaborative robot setup, people work alongside robots in the same workspace, increasing efficiency significantly. Multiple investigations show that cobots can reduce human rest time spent manufacturing by 85% [16].
- **Ideal for Small- and Medium-Scale Production:** Small- and medium-sized enterprises often resort to time-consuming methods in order to reduce expenses, whereas larger corporations, dealing with higher production volumes, often opt for robotics-oriented solutions. However, the landscape concerning cobots has evolved. Nowadays, most collaborative robots are user-friendly, offering a range of capabilities while being simple to configure and deploy. Cobots are commonly available for small to medium-sized organizations for duties like as cutting, forming, sanding, finishing, pulling, arranging, and packaging [17].
- **Higher Return on Investment (ROI):** After implementing cobots, industries profit from enhanced efficiency, adaptability, and manufacturing capacity. As a result, a higher ROI cannot be avoided. The cobots can be placed and moved as

needed, efficiently removing any ineffective activity during working hours. In general, this increases your ROI. Labor and maintenance expenses are reduced, leading to a higher profit margin [18].

- **A Secure Working Environment:** Because the cobots are developed with human participation and collaboration in mind, each unit is extensively tested to assure safety from the start. Cobots create a more secure place to work, especially in instances where human error might stymie advancement, such as healthcare scenarios, where the robots are able to acquire command and carry out their obligations without error [19].
- **Cobots Are the Future:** Increasing production automation is unavoidable. It is tough to forecast what human jobs will look like in a few decades from now when machines take control. Cobots, on the other hand, mix mechanical and human labor. By the year 2020, it is projected that cobot sales will reach $3.1 billion, and, with substantial investment in research and development, this figure is expected to steadily increase. Therefore, it is advisable for the industry to embrace cobots as early as possible.

In recent years, productivity in the manufacturing industry has been gradually growing. Cobots, or collaborative robots, are robots that work alongside humans in a shared office. As a result, in this section, we will discuss seven advantages of cobots and process automation.

9.6.1 Fast Installation

Typically, it takes a few days (or even weeks) to get a robot up and running. Cobot configuration, on the other hand, can be done in half an hour, and, with straightforward software, cobot can perform its first duty in an hour. As a result, a cobot can be utilized as a transitory or adaptable worker who can do a variety of tasks in several departments.

9.6.2 Quickly Programmed

Traditional robots take a great deal of knowledge and time to program tasks and may frequently only execute one. A cobot, unlike standard industrial robots, is simple to program. A modern operator can also manually position cobots in desirable locations and save them in specialized software.

9.6.3 Can be Used in Different Departments

Cobots are adaptable robots, and adding additional jobs to the robot is simple. Cobot may do the same thing every day for 40 h, but it could also do something different every day. Every day, for example, users can look for spots in the manufacturing process where the robot can be useful. This adaptability enables users to hire cobot as a temporary employee. For example, during peak production hours, hiring a robot for 2 weeks is more feasible than hiring an individual for 2 weeks.

9.6.4 More Consistent and Accurate Than Humans

Cobots, unlike humans, are extremely exact. These cobots can conduct activities with 0.1-mm precision. Furthermore, cobot will never depart from its actions and will always carry out duties with the same level of power. For example, while tightening screws, the tension and pressure will always stay the same, ensuring quality consistency.

9.6.5 Positive Effect on Employees

A cobot is intended to collaborate with humans rather than to replace them. Cobots, also known as human-centered robots, may assist people more easily and refine the task to be done. Employees can focus on other duties by having the robot execute dirty, risky, dull, monotonous, or repetitive work. Consider quality checks and maintenance, but product assembly might also be done in partnership with the cobot. For example, the cobot might be assembling a product so that the employee can finish it one more step in the process.

9.6.6 Employees Become Happier in Their Work

Employees will like their jobs because they are supported in their efforts. Boring and boring tasks will be delegated to robots so that employees can do their duties in a more "human" manner. The task requires greater innovation, solution-oriented thinking, and initiative. In this way, robots contribute to a desirable work environment. Not to mention the prevention of RSI (robotic system integrators), this occurs when trading the same trade over an extended period of time. In the long run, this means fewer absenteeism, greater employee engagement, job enrichment, a high level of job satisfaction, and a contribution to the personal development of employees.

9.6.7 Increased Productivity and Process Optimization

The primary goal of robotization is, of course, to optimize a process. In other words, attain the highest possible goal at the lowest possible expense. Errors are reduced, staff have more time for other things, productivity rises, and the end result is a better operating result. Setting up and outfitting a cobot clearly needs an investment, but it saves money and also converts into profit in the long run. Figure 9.1 depicts the use of cobots and their importance in various tasks.

9.7 The Need for Cobot Regulation Frameworks

The Three Laws of Robotics, introduced by the renowned science fiction author Isaac Asimov in 1942, served as the foundational principles for robotics for many years. However, significant advancements have occurred since then. Robots are now ubiquitous, coming in a wide range of sizes and designs. As robots continue to gain popularity and intelligence, there is a need for contemporary and relevant regulations to govern and oversee their behavior. Cobots have the potential to revolutionize the future of

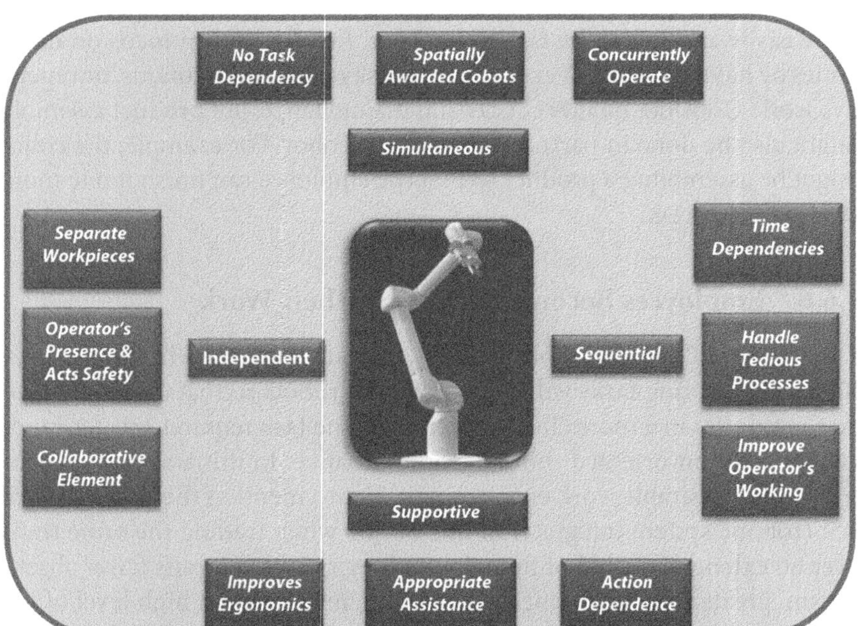

Figure 9.1 Relationships and dependencies during cobot utilities [20].

healthcare delivery. To achieve this, users must first meticulously develop a widely accepted regulatory framework for cobots that ensures the utmost care and safety of individuals when interacting with them. Only then can we usher in a new era of cobots that elevate medical and healthcare capabilities to unprecedented heights.

Brooklyn, 9 April 2021: In a press release, Global Market Estimates' latest industry research reveals that the collaborative healthcare robots market is expected to experience rapid growth between 2021 and 2026. Collaborative healthcare robots, which are automated devices, perform various tasks within the medical sector. They are involved in office work, laboratory tasks, patient care, and even assist in surgical procedures. These cobots have the potential to address the current strain and workforce shortages in the medical field effectively. The demand for healthcare automation has been on the rise, driven by the need to reduce infection exposure risks for frontline healthcare workers and ongoing technological advancements in hospital care services. Market penetration will be driven by rising demand for services such as patient monitoring, medicine dispensing, patient record keeping, electronic prescribing, billing, insurance claims, and patient care. Furthermore, cobot adoption will encourage precision in work, reduced stress for nurses, and increased production [21].

9.8 The Hardware Sector will be Led by Robotic Arms and Sensors

The upcoming period is expected to rely heavily on the use of robotic arms and sensors. The increasing demand for repairing factory computer systems through automated processes will facilitate sensor and robotic arm integration. Demand will be driven by the rising demand for surgical applications and the goal of gaining more precision throughout the procedure. The use of cobots will also eliminate the need for a big number of employees, making the operation more profitable [22–24].

9.9 Testing in Laboratories and Care for Patients Have a Lot of Possibilities

Cobots provide for exceptionally efficient laboratory test jobs. High accuracy, fast response times, and less reliance on manual operations will benefit laboratory testing applications. Another promising major application is patient care, encompassing tasks such as medication administration, swab collection, monitoring of blood pressure, temperature, and conducting various spot tests. These activities have proven beneficial in easing the burden on patients [21].

9.10 APAC Significant Gains

In the future years, the Asia Pacific region [Asia Pacific Accreditation Cooperation (APAC)] will have the highest CAGR (compound annual growth rate). The growth of the regional industry will be driven by the expanding medical sector and the wide range of applications to meet the increasing automation needs in healthcare. Japan, China, South Korea, and Singapore are expected to be the primary contributors to the region's growth, thanks to their significant investments in the medical equipment industry. The main strategies will be technological improvement and expanded reach. COVID-19, the global epidemic, caught everyone off guard, and the healthcare sector is now facing one of the most difficult difficulties on a global scale. Healthcare personnel labor around the clock, and the high strain on the front lines has led to tiredness in many, virus infection in others, and the tragic death of some in the line of duty [25].

The global healthcare system is collapsing under the weight of this pandemic. Nonetheless, a recent analysis found that, even in the absence of a worldwide pandemic, the world would need 6 million more nurses to meet global health standards. This is a concerning figure, and it is unlikely that the demand can be supplied in a short period of time.

9.11 Human Factors and Errors

Sleep deprivation, shift work, repetitive jobs, physically demanding duties, and, of course, stress are all elements that lead to human errors within the boundaries of human factors, according to the research of human factors and exhaustion. Aside from a shortage of nurses, the medical sector faces other challenges such as a high number of human errors due to a pressurized and non-automated environment, and work shifts with a significant amount of time dedicated to administrative tasks, which diverts valuable time away from patient care. All of these things contribute to nurse weariness and stress. However, technological progress has a silver lining. Collaborative robots (cobots), the agile and smart cousins of industrial robots, are gaining traction in the healthcare industry [26–29].

9.12 Robots Lending a Hand During the COVID-19 Outbreak

During this difficult time of the COVID-19 outbreak, the healthcare industry is looking to robotics for assistance. Robots lessen the risk to HCPs by reducing face-to-face interaction with isolated patients. Researchers at Tsinghua University, one of China's leading universities, created a cobot to assist health personnel in the perilous duty of virus testing. The growth of the regional industry will be driven by the expanding medical sector and the wide range of applications to meet the increasing automation needs in healthcare. Japan, China, South Korea, and Singapore are expected to be the primary contributors to the region's growth, thanks to their significant investments in the medical equipment industry. Cobots are also assisting in the fight against the spread of COVID-19 throughout the universe. A Texas-based firm, DetectaChem, is developing a low-cost COVID-19 test kit that uses a smartphone to analyze data in 10 to 30 min. DetectaChem will be able to grow production to 50,000 kits per day after the kit has been approved by the Food and Drug administration [30, 31].

9.13 Universal Robots Cobots in Healthcare

Prior to the COVID-19 outbreak, the healthcare industry was already using cobots to assist with numerous duties. Simple duties like drawing blood, monitoring patients' vital signs and conditions, and taking care of patient hygiene can be performed by the robots. The robots can also manufacture and dispense prescriptions in drug labs, as well as assist paraplegics in moving about and administering physical therapy, decreasing therapists' burden and physical strain. Gentofte University Hospital in Copenhagen, Denmark, has enlisted the assistance of two Universal Robot UR5 collaborative robots (cobots) to aid in the processing and organization of blood samples for testing. Despite a 20% increase in samples arriving for examination, the cobots allow the lab to retain its objective of delivering more than 90% of results within 1 h. In Denmark, the Odense University Hospital Neurorehabilitation Centre in Ringe collaborated with Universal Robots to establish the Patient@Home project, which aided people in the rehabilitation of blood clots and stroke-related ailments [32].

Rainer is a service cobot that is designed to assist patients with repetitive functional actions as part of their rehabilitation process. This technology can assist therapists by allowing them to create customized training programs that patients can complete safely using the cobot. In Singapore, NovaHealth's Traditional Chinese Medicine (TCM) facility features a physio-massage robot that rubs patients' backs and knees. The robot masseuse, named Emma (Expert Manipulative Massage Automation), was developed by AiTreat, a startup company backed by Nanyang Technological University. Emma has a robotic arm that extends from a device and uses soft tips for massage, this has been made from silicone to generate heat between 38°C and 40°C to stimulate human touch. After consultation, the TCM doctor prescribes the style of massage, and the robot is calibrated with a camera to accurately target the acupuncture site. During the massage, the patient can adjust the pressure using his or her palm. The robot assesses muscular stiffness before and after the massage using built-in sensors, allowing the doctor to precisely diagnose the patient's condition. Cobot, a vision-guided robotic device, is being developed in Spain [33]. Various sorts of physiotherapy can be applied automatically. While a skilled physician continues to diagnose the patient's illness, the robotic system takes over the duty of the physiotherapist, relieving overburdened clinics of their labor. Three universal robots' robotic arms were put into this piece of equipment so that developers could quickly adjust the program to the patient's individual demands and the duties necessary. Because of the cobot arms' ease of programming, setup, and collaborative nature, surgeons, nurses, operating room employees, and robots can work together to care for patients.

Furthermore, robotic physician assistants can monitor patients' vital data and warn nurses when human intervention is required, allowing a single nurse to watch numerous patients at the same time. Furthermore, these assistants automatically enter the data into the patient's electronic medical record. Everything from industries to hospitals for decades, automation has been an essential component of the global industrial industry, resulting in tremendous increases in industry efficiency and production. Robotic technologies have evolved through time into cobots, a more user-friendly version of the massive, heavy industrial robots of yesteryear, and they are now ready to help the healthcare system, providing health professionals more time to do what is most important: care for our loved ones [21].

The global market for collaborative healthcare robots is highly competitive, primarily due to the presence of significant industry players. These include ABB Ltd., Yaskawa Electric Corporation, KUKA AG, Kawasaki Heavy Industries Ltd., Bosch Rexroth AG, Universal Robots, Festo, Techman Robot Inc., Intuitive Surgical, Paro Robots, Accuray Incorporated, MOV. AI, Teradyne Inc., OMRON Corporation, AUBO Robotics Inc., Fetch

Robotics Inc., and FANUC Corporation Inc. An exceptional illustration of how cobots have seamlessly integrated into the modern healthcare system was witnessed during the COVID-19 pandemic. In 2020, numerous hospitals opted for cobots to disinfect beds and rooms instead of assigning this task to their staff. Cobots were effectively entrusted with potentially hazardous medical procedures, thus safeguarding healthcare personnel from unnecessary exposure to the deadly infection. The increasing interest in telemedicine during the epidemic is likely to fuel the emergence of cobots as well. Visualize, let us call him Boymax, Baymax's cousin. Boymax's humanoid design simplifies interactions with people, including children. With its voice interface, there is no need for manual commands. Its visual camera and sensors can detect changes in body temperature and blood oxygen levels, as well as other indicators of a patient's need for medical attention. Boymax can also serve as a telemedicine platform, enabling users to remotely consult with doctors. Doctors do not need to be concerned about patients misinterpreting prescriptions, as they are directly communicated to the Cobot. Boymax can receive orders, dispense medications, assist patients in taking the correct dosage at the prescribed times, provide adequate nutrition, relay real-time updates to the doctor, and alert them in case of emergencies. In urgent situations, they can even perform emergency procedures. In certain critical scenarios, having Boymax could be the distinction between survival, severe injury, or even fatality.

9.14 What Role Can Cobots Play in Enhancing Healthcare Brand Experiences?

Cobots can play a role in enhancing the brand experiences of hospitals and pharmaceutical companies for stakeholders such as patients, HCPs, and payers. Collaborative surgical robots have been assisting doctors for a considerable time. There are different types of medical devices: proactive ones with pre-programmed components that can operate independently, semi-active systems that allow surgeon-guided components alongside pre-programmed ones, and master–slave systems entirely controlled by the surgeon. The da Vinci Surgical System, as a master–slave cobot, conducts surgeries [12]. It employs robotic arms to translate the surgeon's hand movements into precise, smaller movements for minimally invasive surgical procedures [17]. By improving the patient and caregiver experience during surgical interventions, cobots can contribute to higher success rates and shorter hospital stays. Newer advancements, such as the tactile glove, can offer surgeons with nearly the same experience as they are accustomed to while using their hands [25]. This has the ability to make operations even more exact, which, in turn, can assist boost brand value and employee productivity, as shown in Figure 9.2.

Hospital supervision cobots have been used to monitor hospital beds in Japan. "Nurse" cobots accompany doctors on their rounds and present patient histories, vital signs, and so on, saving the doctor time that would otherwise be spent examining charts. These cobots can also serve to ease acute nursing staff shortages by relieving the load on present people, resulting in improved productivity and more time for individual interactions with patients and professional advancement. Accompaniment for the elderly is critical in aiding the elderly. Chronic loneliness has been proven in research to have an effect on older people's recall, physical wellness, psychological well-being, and life expectancy. The growing older population and the global rise of the nuclear family, together with rising life expectancy and the burden of chronic disease, make elderly companionship a key role that cobots could serve in the future. In the hospital setting, cobots can fulfill two functions in senior care: first and foremost, make life easier for the elderly; second, and most importantly, provide bedside companionship by speaking with patients in the absence of nurses or doctors. After release, the cobots may offer medical services and monitoring of patients in the patient's residence, reducing the need for hospital beds and equipment.

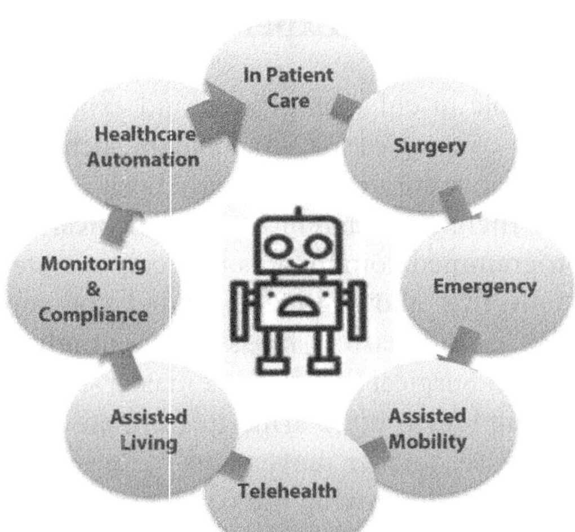

Figure 9.2 Categorized work in cobot.

Recovery from illness and assistance pet therapy have been investigated for hospitalized patients with mental illness and for older individuals. The mere existence of a being (even if it is not human) aids in a person's psychological healing or improvement. Voice assistants such as Alexa already exist in current circumstances. Cobots will soon provide medical care as well as personalized talks for patients, boosting greater mental wellness throughout illness. Across the globe, there is a shortage of doctors, nurses, and other skilled HCPs. This makes it difficult to afford assisted living, which may be required for some medical problems. While cobot ownership remains prohibitively expensive, mass manufacture and wider acceptance can help shrink the gap. Another area where cobots can be incredibly useful is in assisted mobility. Patients with medical illnesses such as paraplegia, paralysis, or congenital abnormalities are confined to their beds or wheelchairs for an extended period of time, or perhaps for life. While wheelchairs are very beneficial and reasonably inexpensive, cobots can provide these individuals with independence of movement. The development of wearable cobotic exoskeletons is aimed at enabling patients of all ages not only to regain the ability to walk but also to run and dance. As technology advances, cobotic exoskeletons have the potential to significantly enhance the quality of life for individuals with mobility challenges [34–36].

9.15 The Rise of the Health Companion

According to a study conducted by the India Brand Equity Foundation, the healthcare industry in India is experiencing significant growth and is projected to reach $372 billion by 2022. Despite the abundance of physicians, surgeons, and specialists in the country, there remains a substantial shortage. In a nation with an estimated population of 1,350,000,000 people (as per a 2019 United Nations report), a single doctor may be responsible for the medical care of 1,457 individuals. This, coupled with increased investment in technical advancements within previously non-automated sectors, has led to the introduction of robotic assistants, or cobots, in the healthcare industry. Cobots, equipped with advanced robotic arms, can effectively and safely operate alongside humans in the same environment. Traditional robots are larger, bulkier, and even dangerous to humans. Enhanced security features and simple installation of cobots enable human–robot collaboration, prompting medical industry deans to gradually explore with using cooperative robotics technologies to apply limited automation to medical challenges. Because of their sophisticated safety features, cobots are the ideal answer for securely assisting doctors and HCPs in operating rooms and other medical applications, subject to risk evaluation. Furthermore, all of this is accomplished while maintaining exactness and truth.

While operation is the most contentious application of robotics in the medical field, it is not the only one that is controversial. Aurolab, a division of Arvind Eye Care Systems located in Madurai, has been at the forefront of cobot technology adoption. They have integrated eight cobots from the Danish pioneer, universal robots, to advance their mission of eradicating blindness [37]. Previously, the NGO was importing intraocular lens implants (IOLs), a critical component in vision restoration, at exorbitant prices, which led to soaring costs for their cataract kits. Aurolab, with a workforce primarily composed of high school-educated women and a commitment to women's empowerment, sought a solution that would not only reduce expenses and boost production capacity but would also be safe and user-friendly, even for individuals with no prior experience in robot programming. As a result, the focus turned to cobots, which now oversee the lens manufacturing process. Thanks to the assistance of cobots, Aurolab has achieved a remarkable feat—tripling IOL production from 150 to 10,000 units per day. The improved quality has enabled the NGO to export their lenses to over 130 countries. Furthermore, this substantial increase in output has led to a workforce expansion from 15 to 700 employees, providing a remarkable success story for India's healthcare sector.

Another important application of cobots in the healthcare business is laboratory automation. For example, the University Hospital in Copenhagen, Denmark, had a quandary when the amount of blood samples coming at their laboratory increased by 20%, and they wanted to retain their target of testing at least 90% of them in an hour. In surgical settings, the presence of cobots is becoming increasingly common. While cobots do not perform surgeries themselves, they do play a crucial role in assisting surgeons and enhancing the precision of procedures. An early adopter of collaborative robots in the operating room was the Arora Neuroscience Innovation Institute (ANII) in the United States. They utilized an optical device designed for neurosurgeons to precisely identify the location of brain tumors, facilitating their removal by the surgeon. Cobots enable surgeons to locate tumors, even those situated deep within the body, which was previously a challenging task. Synaptive Medical, based in Canada, has performed marvels with the aid of cobots in neurological surgery, most notably in doing minimally invasive microsurgery with unparalleled clarity. These operations, according to Synaptive, frequently take less time than normal surgeries and require people to remain in a medical facility for a shorter period to recover.

The battle against the highly contagious COVID-19 virus, a global pandemic that is currently impacting the entire world, serves as a more recent and widely recognized example. As can be seen, numerous individuals in the healthcare profession are seeking assistance from robots.

In order to protect nurses and doctors worldwide from contracting the coronavirus, which is a major concern for HCPs, researchers at Tsinghua University have proposed the use of cobot arms for digitally screening and treating coronavirus patients. While the technology is still in the research phase, cobots are expected to provide a cost-effective and high-tech solution to this global problem. Essentially, cobots enable safe and precise collaboration between humans and machines, allowing them to leverage their respective strengths. It is evident that the adoption of collaborative robots in the medical field has gained global recognition and acceptance. However, despite the fact that cooperative robots in India have proved their usefulness in other sectors, India continues to be in its early phases of development when it involves cobots in the field of healthcare. The utilization of this specialized technology in the healthcare sector offers substantial advantages, including increased productivity rates, enhanced efficiency within the medical industry, support for HCPs to perform at their best, and a reduced risk of infectious disease transmission, among others. Cobots are poised to assist healthcare and pharmaceutical companies in enhancing their performance on the production floor, ultimately contributing to the "Make in India" initiative by delivering high-quality products. India's healthcare industry is undergoing a technical transformation (Version 4.0) to enable global competitiveness. Collaborative robots, or cobots, are gradually finding their way into this traditionally non-automated sector, with applications ranging from pharmaceutical and medical device manufacturing to hospital environments. Their versatility and user-friendliness are making cobots an increasingly valuable addition to the entire healthcare industry.

Cobots are sophisticated arms made of robots that function in the same natural setting as humans. Traditional robots are heavier, bulkier, and possibly dangerous to humans. With a risk assessment, cobots are the right solution for securely aiding healthcare providers and physicians in surgical centers and other healthcare environments while improving precision and accuracy. In industrialized countries such as Canada and the United States, for example, cobots are employed in robotic-assisted surgery and are equipped with cameras to assist doctors in identifying malignancies for removal. Although surgery is one of the most obvious applications for robotics, the adaptability of cobots makes them appropriate for numerous applications in the healthcare sector. Similarly, packaging in the pharmaceutical industry is another field where cobots are being examined, as they allow providers to achieve high criteria while maintaining hygiene standards. While the science is still being assessed, cobots are expected to provide an inexpensive, highly sophisticated way to addressing this worldwide issue.

Cobots enhance human–robot interdependence by allowing each to perform what it knows how to do, but in a safe and precise environment. The utilization of this one-of-a-kind technology is gaining traction every day and is already gaining traction internationally. However, India appears to be in its infancy when it comes to examining the possibilities of cobots in healthcare. In difficult times like these, robotic automation will be a blessing to many industries, including healthcare, by increasing efficiency and scalability. Cobots are obviously critical in combating a pandemic of this magnitude by promoting human–robot collaboration, a key notion of 4.0 Industry, but they will also be critical in the future as the economy changes toward deploying cobots to take over. Humans are accustomed to jobs that involve mental effort and make human creativity tedious, monotonous, or risky. The potential of cobot technology will assist the country in revolutionizing the sector, projecting it into the future, ensuring the present, and protecting people's future [38–41].

9.16 Cobots are Revolutionizing the Field of Medicine

Many in the medical device and healthcare industries have hastened to embrace the future of collaborative robotics, citing the clear promise for faster operations, repeated safety measures, and a more efficient workforce. Transformation of cobots is most visible in the following areas.

9.16.1 Lab Testing

The University Hospital of Copenhagen employs cobots to collect, separate, and load blood samples onto an input system for analysis. These cobots utilize visual programming, gathering samples, examining them with a barcode scanner, and subsequently sorting them based on cap color. Another robot places samples with matching colors into a tray and inserts it into an analyzer. Thanks to the robots processing approximately seven to eight blood tubes per minute, equating to around 3,000 tests daily, 90% of test results are typically available within an hour. This rapid turnaround has greatly benefited patients. Instead of returning to the hospital, outpatients can now get findings on their initial appointment. Furthermore, on shifts in the morning, hospitalized patients receive same-day findings, allowing clinicians to evaluate current and accurate information.

9.16.2 Patient Care

Researchers at Japan's Toyohashi University of Technology have created Terapio, a cobot designed to monitor patients at their bedside and continuously update electronic medical records, thereby saving valuable time for busy nurses. By tracking medical personnel as they move around each patient's room and automatically updating vital signs in patient files, Terapio relieves nurses of administrative duties, allowing them to focus on patient care.

Another companion robot, known as RIBA (Robot for Interactive Body Assist), features a bear-like appearance and is designed to visit patients at their bedsides. RIBA was developed by the RIKEN-SRK Collaboration Centre for Human-Interactive Robot Studies and Sumitomo Riko Company. Its purpose is to assist Japan's growing elderly population by lifting patients from beds and mobility aids, thereby reducing the risk of caregivers sustaining injuries during the frequent process of transferring patients, which can occur up to 40 times per day.

9.16.3 Surgery

The da Vinci system, equipped with state-of-the-art robotic arms and cameras to assist surgeons during procedures, stands as one of the most renowned cobots in the realm of surgery. The precision of the da Vinci's robotic arms, which translate the surgeon's hand movements into smaller and more accurate actions, enables less invasive treatment options. In 2018, nearly 100,000 patients underwent procedures using the da Vinci system, highlighting the early adoption of cobots in the medical field, encompassing a wide range of surgical procedures. Various types of surgical robots are currently in operation. The CyberKnife device, developed by Accuray, employs a pre-programmed treatment plan to deliver radiation therapy to cancerous tumors and possesses the capability to make real-time corrections for patient movements. CyberKnife therapy has demonstrated superior performance compared to traditional radiation therapy in two multicenter studies involving prostate cancer patients, achieving success rates of 97%–100% for low-risk patients and 88%–97% for intermediate-risk patients, respectively.

In the demanding field of brain and spinal surgery, robotic applications have recently made significant strides. One such advancement is the Modus V device from Synaptive Medical Inc., which is currently assisting Dr. Sebastian Koga, the Physician Director of the Ochsner Health System Neurological Institute. This cobot operates by providing a high-resolution image of the surgical area, detecting the surgical tools in use, and automatically adjusting its powerful magnification to match the surgeon's position. Additionally, Stryker/Mako has begun implementing cobot technology in total knee arthroplasty procedures to enhance axis alignment correction. The technique delivers final element positions that are 3.1 times more accurate overall and five times more precise in bone cuts. Furthermore, the Mako Total Knee robotic arm-assisted surgery enhances the surgeon's visibility during cutting, addressing concerns such as ligament rupture. This approach reduces the necessity for tibial subluxation or patella eversion, leading to improved and quicker functionality while ensuring tissue safety.

9.17 Manufacturing of Medical Devices

Aurolab, an Indian manufacturer of cataract surgical kits and intraocular lenses, has partnered with Universal Robots in order to deploy collaborative robots to streamline its production procedures. Material handling as well as the precise selection and delivery of elements are among the operations done by the cobots, which are critical to the sensitive lens manufacturing process.

9.18 Applications Outside of Healthcare

Even small- and medium-sized enterprises are quickly discovering methods to automate operations with the help of cobots, thanks to lower costs and less setup expertise required. Here are some of the fascinating uses outside of healthcare that are gaining traction.

9.18.1 Automotive

Automakers are early users of industrial robotics as an industry with numerous potentially harmful activities. They were also among the first to incorporate cobots into their operations. Cobots, for example, are doing visual checks for flaws in machine parts using better vision technology, proving to be more accurate than the human eye.

9.18.2 Fulfillment Centers

Cobots are being used to fill orders at Amazon fulfilment centers and other online grocery retailers, eliminating the need for people to move from shelf to shelf in factories. In a similar vein, the Chinese e-commerce giant "jd. com" has recently constructed a cobot-driven warehouse capable of handling 200,000 orders per day with only four employees on site.

9.18.3 Food Service

Medium-sized food company Axium Food Inc. has adopted cobots to perform repetitive tasks that previously led to health issues like carpal tunnel syndrome among human workers. In response to the increasing demand in the food industry, companies are now incorporating cobots into their cooking lines. One of the most renowned cobots, Flippy, designed for flipping hamburgers, made its debut at Caliburger in March.

9.18.4 Industry Predictions

As cobots continue to satisfy the demands of the pharmaceutical and health industries, providing high rates of success and excellent patient outcomes, adoption rates are expected to grow. Product owners may expect machine learning and AI to play a larger role as a result of creative industry alliances like Google's life sciences division Verily and Johnson & Johnson's surgical division Ethicon.

9.19 Next Steps of Cobot

9.19.1 Service Providers

Service suppliers must be ready to deliver a consistent and distinct level of service throughout their networks. Because application privacy and predictability needs (in this case, cobots) propagate over network infrastructure, suppliers of services will be required to deliver a specific level of service in critical security scenarios.

9.19.2 Technology Vendors

System reliability must be ensured by technology providers in terms of certainty, safety, and defense. It will also have to ensure legislative and standard conformity.

9.19.3 Enterprises

Businesses will need to develop applications that take benefit from human–robot collaboration as this technology matures. Human–robot cooperation is on the horizon in the near future, and everyone from producers to service providers will be able to participate. The precision, greater efficiency, diverse operations, and lowered risk factors related to human mistake indicate that this technology will be used extensively in the not too distant future.

9.20 Conclusion

Robots that collaborate will play an important part in the upcoming medical and technological developments. Cobots are extremely useful when humans require close proximity with their robots. When a human directs, oversees, or even teaches from robots, this is referred to as a human–robot partnership. Service quality requirements, which evolve with time and adapt to changing environments, will be among the most vital aspects of cobots. Unlike "conventional" robots, cobots exhibit "human-like" characteristics such as adaptability, flexibility, and awareness of their surroundings. Cobots are able to be aware of the environment and work safely alongside people thanks to security sensors and scanners. Utilizing cobots offers numerous benefits to businesses, including increased productivity, improved site efficiency, and enhanced operator well-being, all contributing to higher-value manufacturing. This term is commonly employed by those who heavily rely on the Internet of Things (IIoT).

Cobotics and Industry 4.0 as a whole are having a significant global impact. Many people believe that the combination of cobots and the Industrial Web of Things (IIoT) will be one of our clock's most disruptive forces. Programming and integrating are simplified and can be done manually. This means that an operator can demonstrate the appropriate positions/paths to the robot, and the robot will swiftly reproduce them. Overall, the team-based robotics approach has a chance to improve application usability by decreasing the amount of space required for a robotic unit. Businesses, manufacturing facilities, and employees all profit from this collaborative technology. These are frequently used in production, warehouses, medical care, and other areas to perform tasks that would otherwise be performed by humans. Robots used in factories, on the other together, frequently need separation from true human touch for them to function properly without endangering humans. This includes working in completely vacant areas or within cages to avoid unintentional human encounters.

Cobots, also known as collaborative robots, are designed to perform tasks in close proximity to and in collaboration with humans. They represent a technological counterpart to AI, as they do not replace humans with autonomous substitutes but instead enhance human capabilities with their superior strength, precision, and data processing capacities. This enables employees to achieve more and provide greater value to their organizations. Cobots are trained and guided by human operators who control their movements, in contrast to being programmed for a predefined set of actions. These robots are specifically engineered for interaction with people in work settings and can be operated by individuals without extensive expertise. Furthermore, cobots are generally lighter than their counterparts available in the market. When combined with mobile robots, they can work in dynamic conditions. Cobots have gained in popularity as user-friendly robotics technology has improved. These are meant to be used in conjunction with a range of top modules, such as cobot arms, moving parts, lift tables, racks, and other one-of-a-kind solutions that improve human–robot collaboration. To avoid contact damage, cobots possess rounded corners and gentler, flexible surfaces than industrial robots. These help to improve safety at work in an array of methods. Smarter cobots can work safely with people, minimizing the risk of significant damage. In the future, this equipment will be utilized to do any difficult task, resulting in less lost time. Cobots are a critical driver of creativity and development in the robotics industry, as well as an essential part of Industry 4.0 advancement. Cobotics is the next natural step in the evolution of industrial technology that will combine specialized jobs. Also incorporated and integrated into the workplace will be reconfigurable science and technology, multifunction, printing with 3D printers, virtual realities, and self-manufacturing.

References

1. Simões, A.C., Soares, A.L., Barros, A.C., Factors influencing the intention of managers to adopt collaborative robots (cobots) in manufacturing organisations. *J. Eng. Tech. Manage.*, 57, 101574, 2020.
2. Galin, R., Meshcheryakov, R., Kamesheva, S., Samoshina, A., Cobots and the benefits of their implementation in intelligent manufacturing. *IOP Conf. Ser. Mater. Sci. Eng.*, 862, 3, 032075, 2020.

3. Frijns, H.A. and Schmidbauer, C., Design Guidelines for Collaborative Industrial Robot User Interfaces, in: *IFIP Conference on Human-Computer Interaction*, Springer, Cham, pp. 407–427, 2021.
4. Colgate, J.E., Edward, J., Peshkin, M.A., Wannasuphoprasit, W., Cobots: Robots for collaboration with human operators, 1996.
5. Colgate, J.E. and Peshkin, M.A., U.S. Patent No. 5,952,796. U.S. Patent and Trademark Office, Washington, DC, 1999.
6. Fast-Berglund, A., Palmkvist, F., Nyqvist, P., Ekered, S., Åkerman, M., Evaluating cobots for final assembly. *Procedia CIRP*, 44, 175–180, 2016.
7. Gil-Vilda, F., Sune, A., Yagüe-Fabra, J.A., Crespo, C., Serrano, H., Integration of a collaborative robot in a U-shaped production line: a real case study. *Procedia Manuf.*, 13, 109–115, 2017.
8. Xu, Y., Liu, X., Cao, X., Huang, C., Liu, E., Qian, S., Liu, X., Wu, Y., Dong, F., Wei Qiu, C., Qiu, J., Hua, K., Su, W., Wu, J., Xu, H., Han, Y., Fu, C., Yin, Z., Liu, M., Roepman, R., Zhang, J., Artificial intelligence: A powerful paradigm for scientific research. The Innovation, 2, 4, 100179, 2021.
9. Yuvethieka Sri, G.V., Balancing Assembly Line Using Collaborative Robots in Modern Manufacturing Industry under Improvements of Efficiency and Ergonomics Study. *Turk. J. Comput. Math. Educ. (TURCOMAT)*, 12, 12, 538–546, 2021.
10. Cohen, Y., Shoval, S., Faccio, M., Minto, R., Deploying cobots in collaborative systems: major considerations and productivity analysis. *Int. J. Prod. Res.*, 60, 6, 1815–1831, 2022.
11. Ji, S., Pan, S., Cambria, E., Marttinen, P., Yu, P. S., A Survey on Knowledge Graphs: Representation, Acquisition, and Applications. *IEEE Trans. Neural Netw. Learn. Syst.*, 33, 2, 494–514, Feb. 2022, doi: 10.1109/TNNLS.2021.3070843.
12. Kriakous, S.A., Elliott, K.A., Lamers, C., Owen, R., The Effectiveness of Mindfulness-Based Stress Reduction on the Psychological Functioning of Healthcare Professionals: a Systematic Review. *Mindfulness*, 12, 1–28, 2021.
13. Simões, A.C., Lucas Soares, A., Barros, A.C., Drivers impacting cobots adoption in manufacturing context: A qualitative study, in: *International Scientific-Technical Conference Manufacturing*, pp. 203–212, Springer, Cham, 2019.
14. Matheson, E., Minto, R., Zampieri, E.G., Faccio, M., Rosati, G., Human-robot collaboration in manufacturing applications: a review. *Robotics*, 8, 4, 100, 2019.
15. Villani, V., Pini, F., Leali, F., Secchi, C., Survey on human–robot collaboration in industrial settings: Safety, intuitive interfaces and applications. *Mechatronics*, 55, 248–266, 2018.
16. Galin, R.R. and Meshcheryakov, R.V., Human-Robot Interaction Efficiency and Human-Robot Collaboration, in: *Robotics: Industry 4.0 Issues & New Intelligent Control Paradigms*, vol. 272, pp. 55–63, 2020.

17. Faccio, M., Granata, I., Menini, A. *et al.*, Human factors in cobot era: a review of modern production systems features. *J. Intell. Manuf.*, 34, 85–106, 2023.
18. Lambrechts, W., Klaver, J.S. *et al.*, Human Factors Influencing the Implementation of Cobots in High Volume Distribution Centres. *Logistics*, 5, 2, 32, 2021.
19. Palleschi, A., Hamad, M., Abdolshah, S., Garabini, M., Haddadin, S., Pallottino, L., Fast and Safe Trajectory Planning: Solving the Cobot Performance/Safety Trade-Off in Human-Robot Shared Environments. *IEEE Robot Autom. Lett.*, 6, 3, 5445–5452, 2021, https://doi.org/10.1109/LRA.2021.3076968.
20. Javaid, M., Haleem, A., Singh, R.P., Rab, S., Suman, R., Significant applications of Cobots in the field of manufacturing. *Cognit. Rob.*, 2, 222–233, 2022.
21. Mourtzis, D., Angelopoul, J., Panopoulos, N., A Literature Review of the Challenges and Opportunities of the Transition from Industry 4.0 to Society 5.0. *Energies*, 15, 17, 6276, 2022.
22. Zaatari, S.E., Wang, Y., Hu, Y. *et al.*, An improved approach of task-parameterized learning from demonstrations for cobots in dynamic manufacturing. *J. Intell. Manuf.*, 33, 1503–1519, 2022, https://doi.org/10.1007/s10845-021-01743-w.
23. Hashemi-Petroodi, S.E., Thevenin, S., Kovalev, S., Dolgui, A., Operations management issues in design and control of hybrid human-robot collaborative manufacturing systems: a survey. *Annu. Rev. Control*, 49, 264–276, 2020.
24. Aydin, Y., Sirintuna, D., Basdogan, C., Towards collaborative drilling with a cobot using admittance controller. *Trans. Inst. Meas. Control*, 43, 8, 1760–1773, 2021.
25. Heymann, D.L., Prentice, T., Reinders, L.T., *The world health report 2007: a safer future: global public health security in the 21st century*, World Health Organization, Switzerland, 2007.
26. Djuric, A.M., Urbanic, R.J., Rickli, J.L., A framework for collaborative robot (CoBot) integration in advanced manufacturing systems. *SAE Int. J. Mater. Manuf.*, 9, 2, 457–464, 2016.
28. Salimbeni, S. and Mamani, D., Framework for Cobots´ incorporation in manufacturing lines. *Podium*, 38, 159–180, 2020.
29. Ma, R., Chen, J., Oyekan, J., A review of manufacturing systems for introducing collaborative robots, in: *3rd UK-RAS Conference for PhD Students & Early Career Researchers*, pp. 71–73, 2020.
30. Djuric, A., Rickli, J., Sefcovic, J., Hutchison, D., Goldin, M.M., Integrating collaborative robots in engineering and engineering technology programs. *ASME International Mechanical Engineering Congress and Exposition*, vol. 52064, American Society of Mechanical Engineers, p. V005T07A013, 2018.
31. Khan, Z.H., Siddique, A., Lee, C.W., Robotics utilization for healthcare digitization in global COVID-19 management. *Int. J. Environ. Res. Public Health*, 17, 11, 3819, 2020.
32. Knudsen, M. and Kaivo-Oja, J., Collaborative robots: Frontiers of current literature. *J. Intell. Systems: Theory Appl.*, 3, 2, 13–20, 2020.

33. Antuña, V. and Olmo, Transferring human skills to a Baxter robot. Master's thesis, Escola Tècnica Superior d'Enginyeria Industrial de Barcelona, Universitat Politècnica de Catalunya, Spain, 2019.
34. Ronzoni, M., Accorsi, R., Botti, L., Manzini, R., A support-design framework for Cooperative Robots systems in labor-intensive manufacturing processes. *J. Manuf. Syst.*, 61, 646–657, 2021.
35. El Makrini, I., Elprama, S.A., Van den Bergh, J., Vanderborght, B., Knevels, A.J., Jewell, C.I., Jacobs, A., Working with Wait: How a cobot was developed and inserted on an auto assembly line. *IEEE Rob. Autom. Mag.*, 25, 2, 51–58, 2018.
36. Kopp, T., Baumgartner, M., Kinkel, S., Success factors for introducing industrial human-robot interaction in practice: an empirically driven framework. *Int. J. Adv. Manuf. Technol.*, 112, 3, 685–704, 2021.
37. Mishra, D.K. and Paul, D., Industry 5.0: Human touch and the future, in: *2022 OPJU International Technology Conference on Emerging Technologies for Sustainable Development (OTCON)*, vol. 8, pp. 1–6, IEEE, 2023.
38. Fager, P., Sgarbossa, F., Calzavara, M., Cost modelling of onboard cobot supported item sorting in a picking system. *Int. J. Prod. Res.*, 59, 11, 3269–3284, 2021.
39. Galin, R.R., Mamchenko, M.V., Romanov, N.A., Business Processes Automation in a Production Facility Through the Use of Collaborative Robots, in: *2020 International Multi-Conference on Industrial Engineering and Modern Technologies (Far East Con)*, pp. 1–5, IEEE, 2020.
40. Schmitt, J., Hillenbrand, A., Kranz, P., Kaupp, T., Assisted Human-Robot Interaction for Industrial Assembly: Application of Spatial Augmented Reality (SAR) for Collaborative Assembly Tasks, in: *Companion of the 2021 ACM/IEEE International Conference on Human-Robot Interaction*, pp. 52–56, 2021.
41. De Simone, V., Di Pasquale, V., Giubileo, V., Miranda, S., Human-Robot Collaboration: an analysis of worker's performance. *Procedia Comput. Sci.*, 200, 1540–1549, 2022.

10

Robotic Arm for Industry Automation

Sanjeev Vyaas S.*, Shamsudeen Babu G., Raghulnath S. S., Manikandan P. and Joel E.

Department of Mechanical Engineering, KPR Institute of Engineering and Technology, Coimbatore, India

Abstract
Robotic arm plays a vital role in Fourth Industrial Automation. It has various applications like welding and assembly of various components, and it enhances automation in manufacturing. It has high productivity and reduces the manufacturing time and the cost. It also benefits for producing/manufacturing components in small industries by taking advantage of robotic arms. Optimization of the design through analysis is very essential. This chapter comprises the design, construction, application, and their role in industry. Response surface method is used to optimize the process variables. The computer-aided design model of a robotic arm is developed using CREO/Solidworks and is analyzed using ANSYS. It is pictured that the material properties play a major role in different loading conditions. This chapter's motive is to produce a robotic arm with high accuracy and precision, less in mass, and easily transportable.

Keywords: Robotic arm, design, mechanism, working, control, industrial automation

10.1 Introduction

Due to expeditious demand for the robotic arm, the end number of research has been carried out. In manufacturing industry, precision and repeatability have been demanded by several production tasks have been taken control by robot. In the 1950's modern industrial robot was introduced, since then, there has been a significant innovations and technological

**Corresponding author*: sanjeevvyaas2003@gmail.com

V. Ramasamy, S. Balamurugan and Sheng-Lung Peng (eds.) Intelligent Robots and Cobots: Industry 5.0 Applications, (203–220) © 2025 Scrivener Publishing LLC

development in robotic technology. The robotic arm can be utilized 24 h with its various special features and makes it better than a human arm. This robotic arm helps to increase productivity within a short span of time this robotic arm increases productivity and manufacturing flexibility. To top it all, the quality of the product and working environment is highly enhanced, and, in addition, it minimizes the floor space usage and focuses on the safety of the workers. Robots are considered as key element in automation; thus, they are widely used in different industrial applications: medical, transport, underwater, and entertainment [1]. A robotic arm is a mechanically controlled device with multiple usage with ample degree of freedom in either stationary or mobile state. Yusoff *et al.* have designed a mobile robotic arm, which can be controlled wirelessly and will perform functions like pick or place, which is being controlled by ps2 controller. The prospective mobile robotic arm can move forward and backward and can turn right and left [2]. It is at ease to do functions like grabbing and dropping harmful objects in a quick manner [3]. Kadir *et al.* have developed a robotic arm controlled by the IoT (Internet of Things), which minimizes the day-to-day work of the humans at home [4]. Gawli *et al.* have designed a robotic arm controlled with wire and IoT using android smart phone remotely [5]. This chapter comprises the following:

- This chapter comprises role of the robotic arm in industries, how they are automated, and their future scope.
- This chapter also comprises the design of robotic arm with a flowchart, construction of robotic arm, mechanism of a robotic arm briefing about the kinematics, actuators, sensors, and the controls and also comprises the working of the robotic arm with an flowchart.

10.2 Robotic Arm and Their Role in Industry

Robotic arms are mechanical devices that are designed to perform repetitive or complex tasks, typically in industrial settings. They consist of multiple segments connected by joints, which can be programmed to move in a specific way to accomplish a task. Robotic arms have various applications, comprising manufacturing, assembly, material handling, and welding.

- One example of a robotic arm in industry is the FANUC M-2000iA/2300 robotic arm, which is used in automotive manufacturing. This robot has a payload capacity of up to 2.3 tons and can perform heavy-duty tasks such as lifting and moving large automotive components. It is also equipped with advanced sensors and controls, which allow it to operate safely and accurately in a manufacturing environment [6].

- Another example is the KUKA KR QUANTEC series of robotic arms, which are used in a variety of industrial applications, including welding, painting, and material handling. These robots are designed to be highly flexible and can be programmed to perform a wide range of tasks. They are also equipped with advanced safety features, such as collision detection and avoidance, to ensure that they operate safely in an industrial environment [7].

Overall, robotic arms play a crucial role in industry by increasing productivity, improving efficiency, and reducing costs. They can perform tasks that are dangerous, repetitive, or require a high degree of precision, freeing up human workers to focus on more complex and creative tasks. With the continued development of robotics technology, it is likely that robotic arms will play an even larger role in industry in the years to come. Robotic arms are commonly used in industrial applications, such as manufacturing, assembly, and material handling. They offer many benefits, including increased productivity, accuracy, and safety [8]. This chapter provides a through explanation of the use of robotic arms in various industrial applications, including automotive manufacturing, food processing, and electronics assembly. The authors also discuss the advantages and disadvantages of using robotic arms in these applications and provide recommendations for future research. It highlights the key role robotic arms play in improving industrial productivity, efficiency, and quality control. For example, in the automotive industry, robotic arms are commonly used for welding, painting, and assembly tasks. Robotic arms also play a critical role in the food processing industry, where they can handle delicate or hazardous materials and improve food safety by reducing the risk of contamination. In conclusion, the use of robotic arms in industry has become increasingly widespread and has revolutionized many industrial processes. The above-mentioned robotic arms are few examples of robotic arm for automotive industry and can be used in variety of industrial application. Figure 10.1 shows a graph about the number of robotic arms used in various industries from 2011 to 2021.

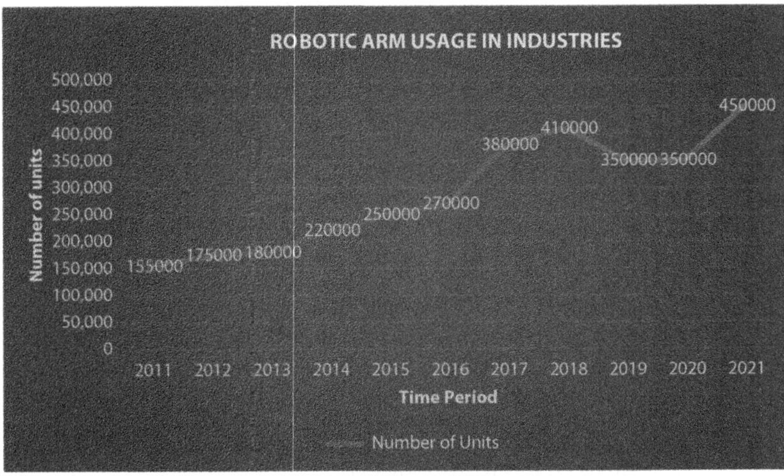

Figure 10.1 Statistics of robotic arm.

10.3 Desof Robotic Arm

Designing and constructing a robotic arm involves several key components, including mechanical structure, actuators, sensors, and control systems. Some important keywords related to this process include kinematics, dynamics, trajectory planning, and feedback control. One reference that covers the development and assembly of a robotic arm is the book "Introduction to Robotics: Mechanics and Control" by John J. Craig. This book provides a comprehensive introduction to robotics, covering topics such as kinematics, dynamics, and control systems. It also includes a detailed example of the design and construction of a six-degree-of-freedom robotic arm [9]. The robot arm has four degrees of freedom and can-do simple duties such as light material handling. It will be integrated onto a transportable platform that will function as an assistance for industrial labor. The robot arm is outfitted with numerous servo motors that connect the arms and conduct arm motions. Because the servo motors have encoders, no controller was required. To operate the robot, we utilized LabVIEW, which conducts inverse kinematic calculations and serially sends the correct angles to a microcontroller, which drives the servo motors and may change position, speed, and acceleration. The robot arm was tested and validated, and the findings demonstrate that it works perfectly [10]. The design and construction of a robotic arm can vary depending on the specific application and requirements. However, some common components of a robotic arm include joints, actuators, sensors, and a control system.

A programmable microcontroller–based brick capable of operating three stepper motors designed to form an anthropomorphic structure is the key component of the robotic arm. The original design was for testing on a human-sized industrial robot arm dubbed PUMA 560, which was utilized to investigate challenges in diverse item handling and compliance control in grab motions (Bejczy and Jau, 1986) [11]. In the Figure 10.2 the flowchart outlines the process for designing a robotic arm. The design and construction of the arm involve the following steps:

1. Design the mechanical structure of the arm, including the joints and linkages.
2. Select the appropriate actuators, such as servo motors or stepper motors, to control the movement of the arm.
3. Install sensors, such as potentiometers or encoders, to provide feedback on the position and orientation of the arm.
4. Develop a control system, which includes a microcontroller and software, to coordinate the movement of the arm based on user input and sensor feedback.

10.4 Construction of Robotic Arm

The construction of robotic arm focuses on the design, development, and operation of a low-cost, adaptable robotic arm that offers the benefits of flexibility and customization, allowing it to perform a range of jobs in various applications such as education, medical care, and daily living assistance. A budget-friendly prototype was created, and this document outlines its design and construction. This robotic arm is intended for use in educational and research settings, such as for teaching robotics concepts, conducting experiments in various fields, and developing new applications. The arm is designed to be simple and easy to build, with a focus on using readily available off-the-shelf components. This provides detailed information on the materials and components used, as well as the assembly and programming process [12]. The robotic arm is composed of three servo motors, a microcontroller, and an aluminium frame. The authors also provide a step-by-step guide to programming the arm using the Arduino integrated development environment. The arm is capable of basic movements, such as gripping and lifting objects, and can be controlled using a remote or a computer. It highlights the importance of developing low-cost robotic arms for educational and research purposes, as they can provide a valuable learning tool for students and researchers. The authors

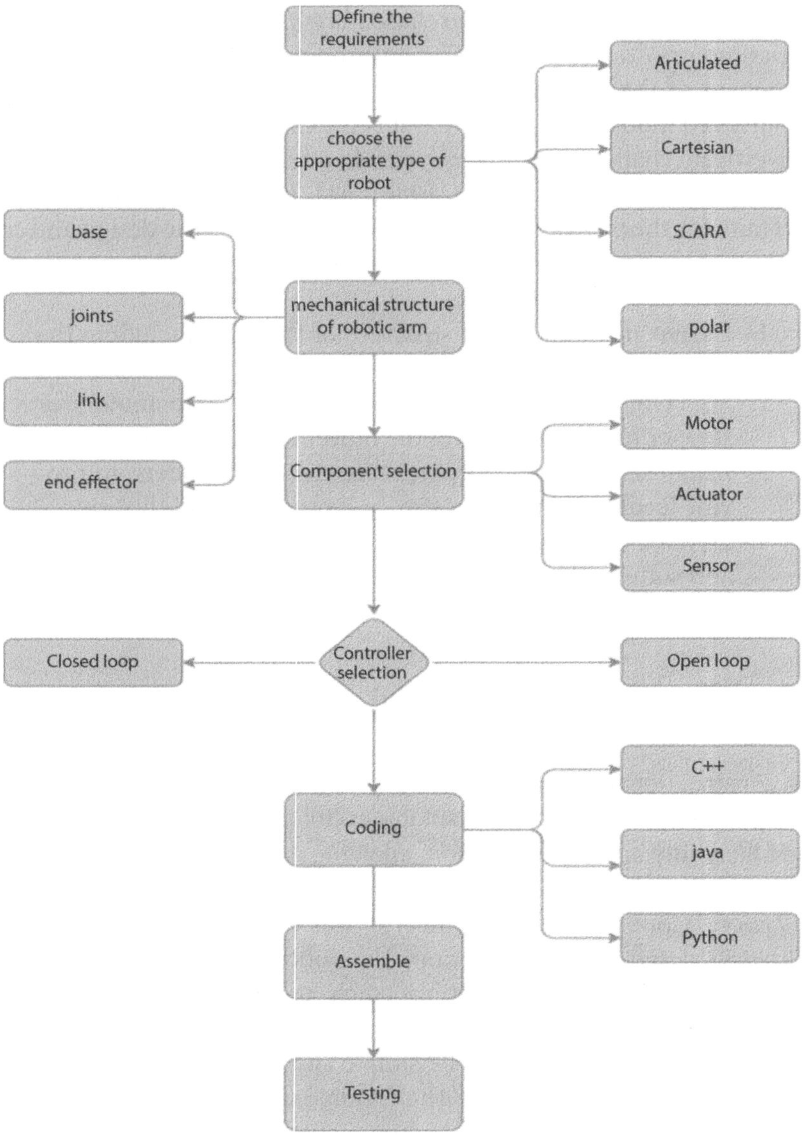

Figure 10.2 Specification of a robotic arm.

also discuss the potential for further development and improvement of the arm, such as the addition of more degrees of freedom or the incorporation of sensors for more advanced control. In conclusion, this chapter provides a valuable guide regarding the development and assembly of an economical robotic arm, which can be used for educational and research purposes.

The authors provide detailed information on the materials, components, and programming process and highlight the potential for further development and improvement of the arm [13].

10.5 Mechanism of a Robotic Arm

The typical structure of a robotic arm is composed of multiple connected links or segments that are joined together by joints, which are capable of moving in various directions and performing a range of tasks. The mechanism and working of a robotic arm involve several key components, including,

1. Joints: The joints of a robotic arm can be of different types such as revolute, prismatic, spherical, etc., which allows the arm to rotate, move up and down, and move forward and backward.
2. Actuators: These are the motors that drive the motion of the arm. Electric motors, hydraulic actuators, or pneumatic actuators can be used as actuators in a robotic arm.
3. Sensors: Sensors such as encoders, potentiometers, or accelerometers furnish information regarding the placement and alignment of the robotic arm.
4. Control system: A control system consists of hardware and software that coordinates the movement of the robotic arm based on input from the user or from sensors.
5. End effector: It is the tool or mechanism that is attached to the end of the robotic arm to perform a specific task.

10.5.1 Robotic Arm Kinematics

Robotic arm kinematics refers to the study of the motion of a robotic arm without considering the forces or torques that cause the motion. Kinematics is important for designing, controlling, and programming robotic arms for various applications.

The kinematics of a robotic arm can be divided into two categories: forward kinematics and inverse kinematics.

1. Forward kinematics: This involves calculating the position and orientation of the end effector of the robotic arm based on the joint angles or joint displacements. The

forward kinematics can be solved analytically or numerically, depending on the complexity of the arm's structure.

2. Inverse kinematics: This involves determining the joint angles or joint displacements required to achieve a desired position and orientation of the end effector. Inverse kinematics can be solved using numerical optimization techniques, geometric methods, or algebraic methods.

There are various methods and algorithms for solving the forward and inverse kinematics of a robotic arm, including the Denavit-Hartenberg method [14], the Jacobian method [15], and the iterative numerical methods [16]. Understanding the kinematics of a robotic arm is essential for tasks such as motion planning, trajectory generation, and obstacle avoidance. It also plays a critical role in applications such as robotic surgery, manufacturing, and space exploration.

10.5.2 Robotic Arm Control

Robotic arm control refers to the process of controlling the motion of a robotic arm to achieve a specific task or behavior. The control of a robotic arm can be achieved through various methods, including open-loop control, closed-loop control, and intelligent control [17].

1. Open-loop control: In open-loop control, the motion of the robotic arm is pre-programmed and executed without feedback from sensors. Open-loop control is simple and fast but is not suitable for tasks that require accuracy or adaptability.
2. Closed-loop control: In closed-loop control, feedback from sensors is used to adjust the motion of the robotic arm in real-time. This allows for greater accuracy and adaptability but can be more complex and slower than open-loop control.
3. Intelligent control: Intelligent control involves the use of artificial intelligence techniques such as fuzzy logic, neural networks, and genetic algorithms to optimize the control of the robotic arm. This can lead to more efficient and adaptive control but requires more computational resources.

The control of a robotic arm can also involve various levels of control, including joint-level control, task-level control, and motion planning [18].

- Joint-level control involves controlling the individual joints of the robotic arm to achieve a desired motion.
- Task-level control involves specifying a desired task or behavior, and the robotic arm will autonomously plan and execute the necessary motion to achieve the task.
- Motion planning involves generating a path or trajectory for the robotic arm to follow, considering constraints such as collision avoidance and joint limits.

Overall, the control of a robotic arm is an essential aspect of robotic arm design and is critical for various applications such as manufacturing, automation, and space exploration.

10.5.3 Robotic Arm Actuators

Actuators are a critical component of robotic arm systems, as they are responsible for providing the force or torque necessary to move the robotic arm. The choice of actuators for a robotic arm depends on various factors, such as the size and weight of the arm, the desired speed and precision of motion, and the power requirements of the application.

There are several types of actuators commonly used in robotic arm systems, including the following:

1. Electric motors: Electric motors are commonly used as actuators in robotic arms due to their high torque output and precise speed control. Common types of electric motors used in robotic arm systems include DC motors, stepper motors, and servo motors.
2. Pneumatic actuators: Pneumatic actuators use compressed air or gas to generate linear or rotary motion. They are commonly used in robotic arms for applications that require high speed and low precision.
3. Hydraulic actuators: Hydraulic actuators use pressurized fluid to generate force or torque. They are commonly used in applications that require high force output, such as heavy lifting or pressing operations.
4. Shape memory alloys (SMAs): SMAs are a type of actuator that can change shape when heated or cooled. They are often used in applications that require precise motion control and small form factors.

5. Piezoelectric actuators: Piezoelectric actuators use a piezoelectric material to generate motion when subjected to an electric field. They are commonly used in applications that require precise positioning and small form factors.

The choice of actuator for a robotic arm system depends on various factors, such as the size and weight of the arm, the desired speed and precision of motion, and the power requirements of the application. It is essential to consider the trade-offs between performance, size, cost, and reliability when selecting an actuator for a robotic arm [19].

10.5.4 Robotic Arm Sensor

Sensors are an essential component of robotic arms, as they provide feedback to the control system about the position, orientation, and motion of the arm. This feedback is critical for ensuring accurate and precise movement of the robotic arm. There are several types of sensors commonly used in robotic arm systems, including the following:

1. Encoders: Encoders are sensors that measure the position and speed of the motor shaft. They are commonly used in robotic arms to provide feedback about the position and motion of the arm [20].
2. Inertial measurement units (IMUs): IMUs are sensors that measure the orientation and motion of the robotic arm by combining data from accelerometers, gyroscopes, and magnetometers. They are commonly used in applications that require accurate tracking of the arm's orientation and motion [21].
3. Force/torque sensors: Force/torque sensors are used to measure the forces and torques applied to the robotic arm. They are commonly used in applications that require precise force control, such as assembly or machining operations [22].
4. Vision sensors: Vision sensors, such as cameras and 3D scanners, are used to provide visual feedback to the control system about the position and orientation of the arm and the objects it interacts with. They are commonly used in applications that require object recognition or tracking [23].
5. Proximity sensors: Proximity sensors, such as ultrasonic or infrared sensors, are used to detect the presence and proximity of objects in the arm's workspace. They are commonly

used in applications that require obstacle detection or collision avoidance [24].

The choice of sensors for a robotic arm system depends on various factors, such as the arm's size and weight, the desired accuracy and precision, and the nature of the application. It is essential to consider the trade-offs between performance, size, cost, and reliability when selecting sensors for a robotic arm.

10.6 Working of Robotic Arm

The working of a robotic arm involves a variety of complex mechanical and control systems, and there is a significant amount of research that has been conducted on this topic. In this answer, we will provide a brief overview of the working of a robotic arm and then highlight some journal references that provide more in-depth information on this topic.

A robotic arm typically consists of a series of joints and links that can move in multiple directions, allowing it to perform a wide range of tasks. The arm is controlled by a computer or electronic controller, which sends commands to the motors and actuators that move the arm. One of the key challenges in designing a robotic arm is ensuring that it can move with the necessary precision and accuracy. This requires careful design of the joints, links, and actuators, as well as the control system that manages the movement of the arm. There are many different types of robotic arms, each with their own unique design and set of capabilities. Some are designed for use in manufacturing or assembly lines, whereas others are used in medical applications, such as surgical procedures [25]. Here are a few journal references that provide more in-depth information on the working of robotic arms:

- "Design and control of robotic manipulators: A review" by S. Shah and D. Schmiedeler (Journal of Dynamic Systems, Measurement, and Control, 2013).
- "Modelling and control of robotic manipulators: A survey" by M. W. Spong, S. Hutchinson, and M. Vidyasagar (Proceedings of the IEEE, 1989).
- "Mechanics of robotic manipulators" by M. W. Spong and M. Vidyasagar (MIT Press, 1989).

These references provide detailed information on the mechanics, design, and control of robotic arms and can be useful for researchers and engineers working in this field [26]. Figure 10.3 as illustrated in the flowchart, designing a robotic arm involves defining its purpose.

10.7 How Robotic Arm are Automated

Robotic arms are typically automated using a combination of hardware and software components. The hardware components include the robotic arm itself, as well as sensors, actuators, and other devices used to control and monitor the arm. The software components include the control system that governs the arm's motion and the algorithms used to plan and execute its movements [27].

1. Design the robotic arm: The arm is designed to meet the specific requirements of the application, including the workspace, payload capacity, and range of motion.
2. Install the sensors and actuators: Sensors are installed on the arm to provide feedback about its position, orientation, and motion. Actuators, such as motors or hydraulic cylinders, are used to move the arm.
3. Develop the control system: The control system is designed to translate high-level commands into low-level motor commands. The control system may include a combination of hardware and software components, such as microcontrollers, motor controllers, and motion planning algorithms.
4. Calibrate the arm: The arm is calibrated to ensure that it moves accurately and precisely within its workspace. This may involve adjusting the sensor readings or tuning the control system.
5. Test and refine the system: The system is tested to ensure that it meets the requirements of the application. This may involve refining the control algorithms or adjusting the sensor placement to improve accuracy.

Once the robotic arm is fully automated, it can be integrated into a larger system, such as a manufacturing line or warehouse automation system. Overall, the automation of robotic arms involves the careful design and integration of hardware and software components to ensure that the arm can move accurately, precisely, and safely.

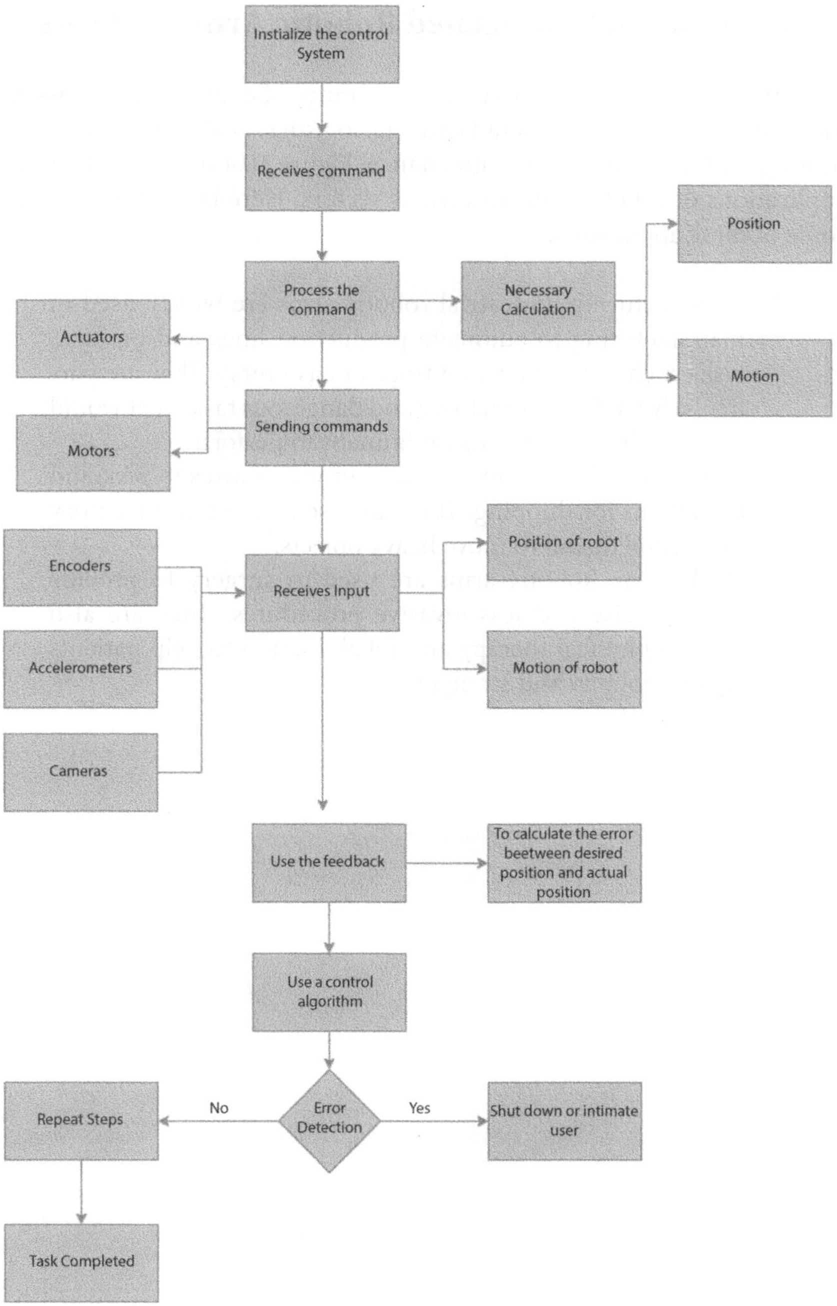

Figure 10.3 Procedure of robotic arm working.

10.8 Industrial Automated Robotic Arm Application

Industrial robotic arms are used in a wide range of applications across various industries, including manufacturing, logistics, healthcare, agriculture, mining, and inspection and maintenance. Figure 10.4 the chart showcases applications of robotic arms in various sectors. Here is a detailed view of some of these applications:

1. Manufacturing: Industrial robotic arms are widely used in manufacturing to automate production lines and perform tasks such as welding, painting, and assembly. They are particularly useful for repetitive and dangerous tasks that would be difficult or dangerous for humans to perform.
2. Logistics: Robotic arms are used in warehouses to pick and pack items for shipping. They can also be used for inventory management and to move heavy objects.
3. Healthcare: Robotic arms are used in surgery to provide more precise and less invasive procedures. They are also used in physical therapy and rehabilitation to help patients regain mobility and strength.

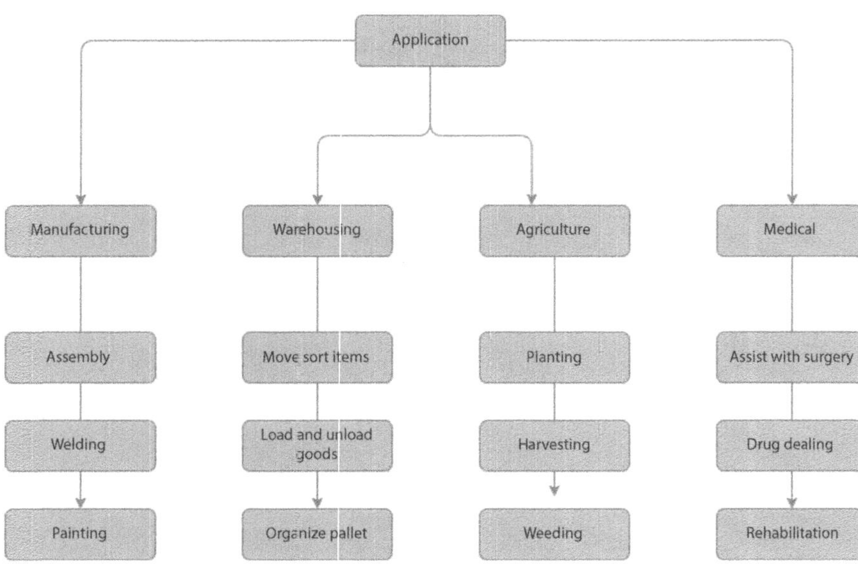

Figure 10.4 Robotic arm application.

4. Agriculture: Robotic arms are used in agriculture to perform tasks such as planting, pruning, and harvesting. They can help increase efficiency and reduce labor costs.
5. Mining: Robotic arms are used in mining operations to extract minerals and other resources. They can be used in hazardous environments where it is unsafe for humans to work.
6. Inspection and maintenance: Robotic arms are used for inspecting and maintaining equipment and infrastructure, such as pipelines and bridges. They can help reduce the need for human workers in dangerous or hard-to-reach locations.

Overall, industrial robotic arms offer a wide range of benefits, including increased efficiency, precision, and safety in a variety of applications across multiple industries [28].

10.9 Future Scope

There are numerous research and development projects being undertaken in the field of robotic arm technology, indicating the vast future scope of this technology. Here are some references that highlight potential areas of future development:

- "The Future of Robotics in Manufacturing" by Steve Lovass-Nagy and Susan Zuray, published in McKinsey & Company, discusses the potential for modular and adaptable robotic systems to help manufacturers be more flexible and responsive to changing market demands.
- "The Future of Robotics and Automation in Construction" by Hugh Mallett and Antony Wood, published in Construction Manager, discusses how the construction industry is beginning to adopt robotics and automation and the potential for these technologies to transform the industry.
- "Robotic Arms in Agriculture: Advancements and Applications" by Matias Garcia, published in Precision Ag, discusses the potential for robotic arms to revolutionize the agriculture industry and increase productivity while reducing labor costs and environmental impact.

- "The Future of Robotics in Healthcare" by Christophe Duguet, published in World Economic Forum, discusses how robotic arm technology is being used in healthcare and how it could transform the industry by improving patient outcomes and reducing costs.
- "The Future of Robotics in Logistics" by Steve Banker, published in Forbes, discusses the potential for robotic arm technology to revolutionize the logistics industry and enable faster, more efficient delivery of goods.

10.10 Conclusion

Robots have been one of the most talked about subjects during the talks of upcoming industrial revolution. Its importance has skyrocketed ever since the pandemic era has struck, due to its advantages where it does not require a human beings touch. Robots delivered food and manufactured components and various other works were still done even though humans were withheld indoors. Hence, it has become an irreversible force, which is going to decide the economic development of the world. In this book chapter, we have concluded the design, construction, working, and applications of a robotic arm. Industrial revolution 5.0 depends on artificial intelligence, but AI needs a physical form to perform a task, so it is crucial to have an efficient robot to implement the AI.

References

1. Wallach, W. and Allen, C., *Moral Machines Teaching Robots Right from Wrong*, Oxford University Press, Inc., New York. 2009, ISBN 978-0-19-537404-9.
2. Wang, W., Song, G., Nonami, K., Hirata, M., Miyazawa, O., Autonomous Control for Micro-Flying Robot and Small Wireless Helicopter X.R.B. *Proceedings of the IEEE/RSJ International Conference on Intelligent Robots and Systems*, Beijing, China, 2006.
3. Gawli, K., Karande, P., Belose, P., Bhadirke, T., Bhargava, A., Internet of things (IoT) based robotic arm. *Int. Res. J. Eng. Technol.*, 4, 03, 3–3.
4. Kamaril, Y.M.A., Ezuan, S.R., Kader, I.B.S., Wireless mobile robotic arm. *Procedia Eng.*, 41, 1072–8, 2012.
5. Agrawal, N.K., Singh, V.K., Parmar, V.S., Sharma, V.K., Singh, D., Agrawal, M., Design and Development of IoT based Robotic Arm by using Arduino, in: *2020 Fourth International Conference on Computing Methodologies*

and Communication (ICCMC), pp. 776–80, 2020, https://doi.org/10.1109/ICCMC48092.2020.ICCMC-000144.

6. Dryba, S., Vincon, A., Kloetzer, C., Extra-Large-Scale Robotics: Applications and Advances. *ISR Europe 2022; 54th International Symposium on Robotics*, VDE, 2022.
7. Buchstab, A., KUKA: Innovations in Industrial Robotics. *Rob| Arch 2012: Robotic Fabrication in Architecture, Art, and Design*, Springer Vienna, 2013.
8. Raza, K., Khan, T.A., Abbas, N., Kinematic analysis and geometrical improvement of an industrial robotic arm. *J. King Saud Univ. Eng. Sci.*, 30, 3, 218–223, 2018.
9. Merat, F., Introduction to robotics: Mechanics and control. *IEEE J. Robot. Autom.*, 3, 2, 166–166, 1987.
10. Elfasakhany, A. *et al.*, Design and development of a competitive low-cost robot arm with four degrees of freedom. *Mod. Mech. Eng.*, 1, 02, 47, 2011.
11. Olawale, J. *et al.*, Development of a microcontroller based robotic arm. *Proceedings of the 2007 Computer Science and IT Education Conference*, 2007.
12. Evliyaoğlu, Oltun, K., Elitaş, M., Design and development of a self-adaptive, reconfigurable and low-cost robotic arm, in: *Mechatronics and Robotics Engineering for Advanced and Intelligent Manufacturing*, Springer International Publishing, Berlin, 2017.
13. Sharma, A. and Noel, M.M., Design of a low-cost five-finger anthropomorphic robotic arm with nine degrees of freedom. *Rob. Comput. Integr. Manuf.*, 28, 4, 551–558, 2012.
14. Becerra, Y. *et al.*, A comparison among the Denavit-Hartenberg, the screw theory, and the iterative methods to solve inverse kinematics for assistant robot arm. *Advances in Swarm Intelligence: 10th International Conference, ICSI 2019, Chiang Mai, Thailand, July 26–30, 2019, Proceedings, Part I 10*, Springer International Publishing, 2019.
15. Hock, O. and Sedo, J., Inverse kinematics using transposition method for robotic arm. *2018 ELEKTRO*, IEEE, 2018.
16. Toquica, A. *et al.*, Kinematic modelling of a robotic arm manipulator using Matlab. *J. Eng. Appl. Sci.*, 12, 7, 1–3, 2017.
17. Megalingam, R.K., Boddupalli, S., Apuroop, K.G.S., Robotic arm control through mimicking of miniature robotic arm. *2017 4th International Conference on Advanced Computing and Communication Systems (ICACCS)*, IEEE, 2017.
18. Sun, L. and Fang, L., An approximation method for stiffness calculation of robotic arms with hybrid open-and closed-loop kinematic chains. *Adv. Mech. Eng.*, 10, 2, 1687814018761297, 2018.
19. de Jesús Rubio, J. *et al.*, Dynamic model with sensor and actuator for an articulated robotic arm. *Neural Comput. Appl.*, 24, 573–581, 2014.
20. Azizkhani, M. *et al.*, Dynamic control of soft robotic arm: An experimental study. *IEEE Rob. Autom. Lett.*, 8, 4, 1897–1904, 2023.

21. Schabron, B. *et al.*, Integration of forearm sEMG signals with IMU sensors for trajectory planning and control of assistive robotic arm. *2019 41st Annual International Conference of the IEEE Engineering in Medicine and Biology Society (EMBC)*, IEEE, 2019.
22. Li, Y.F. and Chen, X.B., On the dynamic behavior of a force/torque sensor for robots. *IEEE Trans. Instrum. Meas.*, 47, 1, 304–308, 1998.
23. Hofer, M., Sferrazza, C., D'Andrea, R., A vision-based sensing approach for a spherical soft robotic arm. *Front. Rob. AI*, 8, 630935, 2021.
24. Shimonomura, K., Nakashima, H., Nozu, K., Robotic grasp control with high-resolution combined tactile and proximity sensing. *2016 IEEE International Conference on Robotics and automation (ICRA)*, IEEE, 2016.
25. Tarn, T., Bejczy, A., Yun, X., Coordinated control of two robot arms. *Proceedings. 1986 IEEE International Conference on Robotics and Automation*, vol. 3, IEEE, 1986.
26. Spong, M.W., Hutchinson, S., Vidyasagar, M., *Robot modeling and control*, John Wiley & Sons, UK, 2005.
27. Banerjee, D., Yu, K., Aggarwal, G., Robotic arm-based 3D reconstruction test automation. *IEEE Access*, 6, 7206–7213, 2018.
28. Almurib, H.A.F., Al-Qrimli, H.F., Kumar, N., A review of application industrial robotic design. *2011 Ninth International Conference on ICT and Knowledge Engineering*, IEEE, 2012.

11

Cloud-Based Cobots for Industry 5.0: A Human-Centric Solution

Arvind Kumar[1]* and Richa Vijay[2]

[1]*Manav Rachna International Institute of Research and Studies, Faridabad, India*
[2]*Amity University, Noida, UP, India*

Abstract

Industry 5.0 means to deliver producing arrangements that are asset-proficient and client-liked by utilizing the imagination of human experts in blend with successful, smart, and exact machines. Industry 5.0 is expected to profit from different promising advances and applications empowering expanded creation and unconstrained conveyance of modified items. The fifth modern transformation will start when its three critical parts—clever gadgets, smart frameworks, and savvy computerization—completely join with the actual climate as a team with human insight. Independent robots are clever specialists that team up with individuals all the while in a similar work area, which is what the expression "robotization" alludes to. This article proposes the possibility of Industry 5.0, which expects to conquer these challenges by having robots team up with people as opposed to rival them. Chapter starts by presenting various new terms and phrasing connected with Industry 5.0 from the perspectives of different specialists and professionals in the area. The potential utilizations of Industry 5.0, including canny medical services, cloud fabricating, inventory network the board, and assembling creation, are then completely talked about. This chapter will discuss a few empowering innovations for Industry 5.0 and cobots, the Web of Everything, blockchain, and 6G and past organizations. To acknowledge Industry 5.0, we feature various exploration obstacles and annoying issues.

***Keywords*:** Cobots, intelligent robots, Industry 5.0, human interaction, cloud data

**Corresponding author*: akdangi@gmail.com

V. Ramasamy, S. Balamurugan and Sheng-Lung Peng (eds.) Intelligent Robots and Cobots: Industry 5.0 Applications, (221–246) © 2025 Scrivener Publishing LLC

11.1 Introduction

Industry 5.0, known as the "savvy processing plant," is the subsequent stage in the advancement of assembling [17]. It is described by the joining of trend-setting innovations like the Web of Things, man-made brainpower, and distributed computing. These advancements consider more prominent mechanization, information examination, and network in the assembling system. Industry growth with web versions depicted by Figure 11.1 illustrates the technological advancements.

Important Features of Various Web Versions

- The initial iteration of the worldwide digital communications network is known as Web 1.0. The Internet made up of static web pages that only allow for passive involvement is frequently referred to as the "read-only" Internet.
- Web 2.0, or the "read and write" Internet, was the next development in the web's history. The social web was developed as a result of users' ability to communicate with servers and other users. This is the internet as it exists now.
- The term "Web 3.0" is in use to describe the next iteration of the Internet, which is a "read-write-execute" web with decentralization at its core.

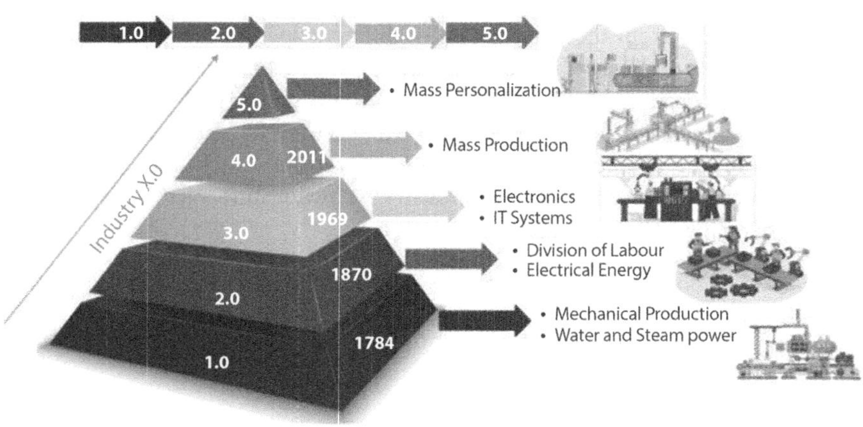

Figure 11.1 Industry growth with web versions (https://doi.org/10.1016/j.jii.2021.100257).

1. It refers to a digital environment created by utilizing blockchain technology, where individuals can communicate with one another directly.
2. Artificial intelligence (AI) and machine learning, where computers can understand data like people, will power it.

Table 11.1 here tabulates the comparative analysis of Web 1.0-3.0 thereby highlighting the peculiar details of these versions.

Table 11.1 Comparative study between Web 1.0-3.0.

WEB 1.0	WEB 2.0	WEB 3.0
1996–2004	2004–2016	2016+
The Hypertext Web	The Social Web	The Semantic Web
Tim Berners Lee	Tim O'Reilly, Dale Dougherty	Tim Berners Lee
Read Only	Read and Write Web	Executable Web
Millions of User	Billions of User	Trillions+ of Users
Echo System	Participation and Interaction	Understanding self
One Directional	Bi-Directional	Multi-user Virtual environment
Companies Publish Content	People Publish Content	People build application though which people interact and publish content.
Static content.	Dynamic content.	Web 3.0 is curiously undefined. AI and 3D, The web learning
Personal Websites	Blog and Social Profile	SemiBlog Haystack.
Message Board	Community portals	Semantic Forums
Buddy List, Address Book	Online Social networks.	Semantic Social Information

11.2 Web 5.0

As being made by Dorsey's Bitcoin specialty unit, The Nitwit, it is in its beginning phases. Building a more decentralized network that places clients in charge of their information and characters is the objective of Web 5.0. Users will actually want to "own their personality on the Web" and "deal with their information" because of Web 5.0, which consolidates Web 2.0 and Web 3.0. Web 3.0 and Web 5.0 both expect a future wherein there could be, at this point, not any gamble of huge blackouts or restrictions from enormous tech or governments. The applications and benefits of Industry 5.0 is depicted in Figure 11.2. The idea of Industry 5.0 was first presented in 2011 at the Hannover Messe modern fair in Germany and has since acquired boundless acknowledgment as a system for the fate of assembling. It expands on the past Business 4.0 stage, which zeroed in on the utilization of advanced advances, for example, robotization and information trade in assembling.

Cloud-based cobots, or cloud-connected robots, play a key role in Industry 5.0 by providing manufacturers with the ability to remotely monitor, control, and update their robots. This allows for greater flexibility and scalability in the manufacturing process. Additionally, cloud-based cobots can access data and resources stored in the cloud, such as machine learning models, which can be used to improve their performance and capabilities.

Overall, cloud-based cobots are helping to drive the Industry 5.0 revolution by enabling more efficient, flexible, and autonomous manufacturing

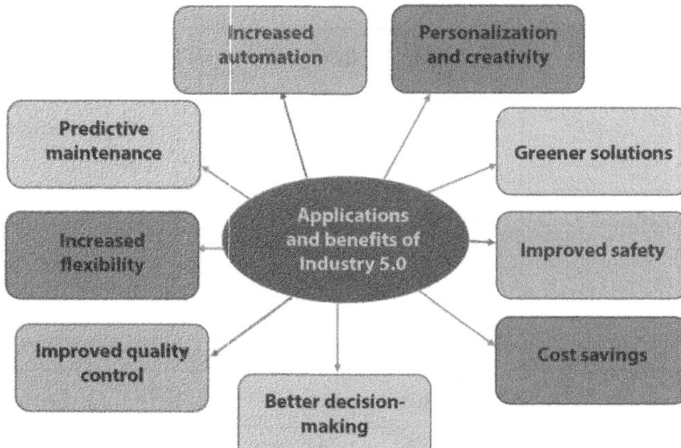

Figure 11.2 Applications and benefits of Industry 5.0.

operations, allowing for a more connected and intelligent manufacturing system. The integration of cloud-based cobots in Industry 5.0 allows manufacturers to improve efficiency, reduce costs, and increase competitiveness in the global market.

11.3 Applications and Benefits of Industry 5.0

Industry 5.0 [11] has an extensive variety of applications and benefits as depicted by Figure 11.3 in the manufacturing industry, some of which include the following.

Increased automation: Industry 5.0 technologies like AI, cloud computing, and IoT enable greater automation in the manufacturing process. This can lead to improved efficiency, reduced human error, and increased productivity [11].

Predictive maintenance: Industry 5.0 technologies allow for live monitoring of equipment and machinery, which can be used to predict when maintenance is needed and schedule it accordingly. This can reduce downtime and prolong the life of equipment.

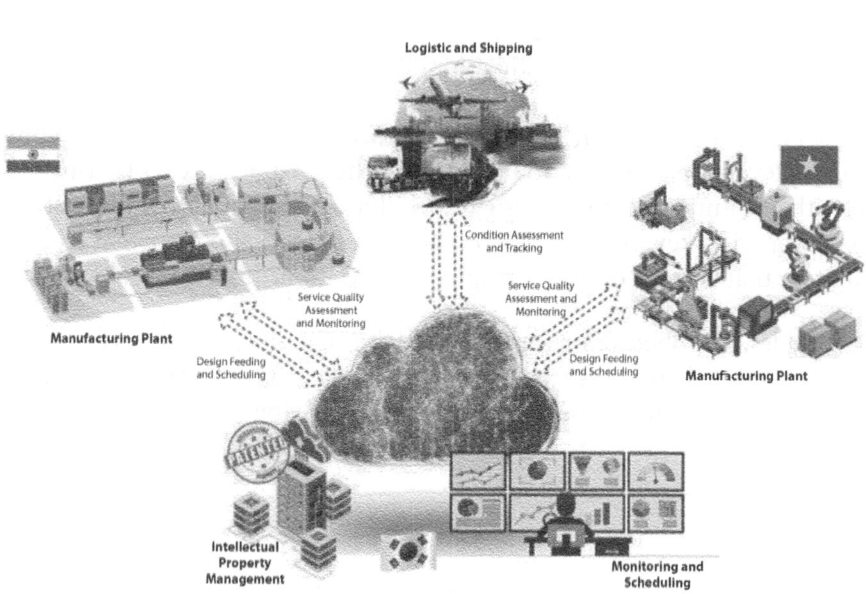

Figure 11.3 Applications of Industry 5.0 (https://doi.org/10.1016/j.jii.2021.100257).

Increased flexibility: Industry 5.0 technologies allow for greater flexibility in the manufacturing process. For example, cloud-based cobots can be controlled and updated remotely, allowing manufacturers to quickly adapt to changes in demand or product specifications.
Improved quality control: Industry 5.0 technologies such as IoT and AI can be used to monitor production in real-time and quickly identify and correct any issues that arise. This can lead to improved product quality and customer satisfaction.
Better decision-making: These data can be used to make better decisions about production and supply chain management.
Cost savings: By increasing automation, reducing downtime, and improving efficiency, Industry 5.0 can lead to significant cost savings for manufacturers.
Improved safety: The integration of robots in the manufacturing process can reduce the need for human labor in dangerous or hazardous environments, thus improving safety for workers.
Greener arrangements: Given that there are various decisions alluded to. Present-day changes have zeroed in on the security of the climate as from an overall perspective critical. With Industry 5.0, new corporate types of progress and responsive attributes are changing this model. This has prompted the improvement of practical procedures where, for instance, a unimportant time of waste and its association become focal, cross-cutting cycles, likewise making the connection all the more remarkable. This shift is according to what overall affiliations, casual guidelines, and clients have consistently referenced.
Personalization and imagination: Mechanical development needs to consider a level of personalization that fulfills the need of the clients. The faculty that is important for Industry 5.0 will use the capability of innovation however will likewise track down space to add their thoughts that lead to an item that is created considering personalization.

Besides, the robotization accomplished during Industry 4.0 permits laborers to liberate themselves from specific redundant undertakings, zeroing in on making all the more special procedures or applying their imagination.

11.4 Industry 5.0 Poses the Following Opportunities and Challenges

Industry 5.0 [11] is good to go to show up when Industry 4.0 is right now acquiring notoriety yet to get made. Industry 5.0 is viewed as a fifth current upheaval in which clients can fulfill their single necessities according to their tendencies and doubts. Be that as it may, the awful undertakings are finished by robots in Industry 4.0, which is at the mass customization level; Industry 5.0 desires to perform mass personalization with the assistance of Man-made thinking. Industry 5.0 should change the creation cycle with higher freedom to agreeable robots as depicted by Figure 11.4. Industry 5.0 is the state-of-the-art present-day agitation, which the majority would view as ordinary to get more noteworthy creative mind and improvement in the things by allowing robots to perform grim endeavors. It is typical to utilize the innovative insightful limit of individuals preferably. Moving from enormous scope assembling to custom collecting at present, the necessity for the current region is speedy headway in gathering methodologies and creation system digitization and intelligentization (Vaidya *et al.*, 2018).

Starting from the essential present-day change, moderate types of progress have driven in gathering, from water and steam-driven gadgets to electrical

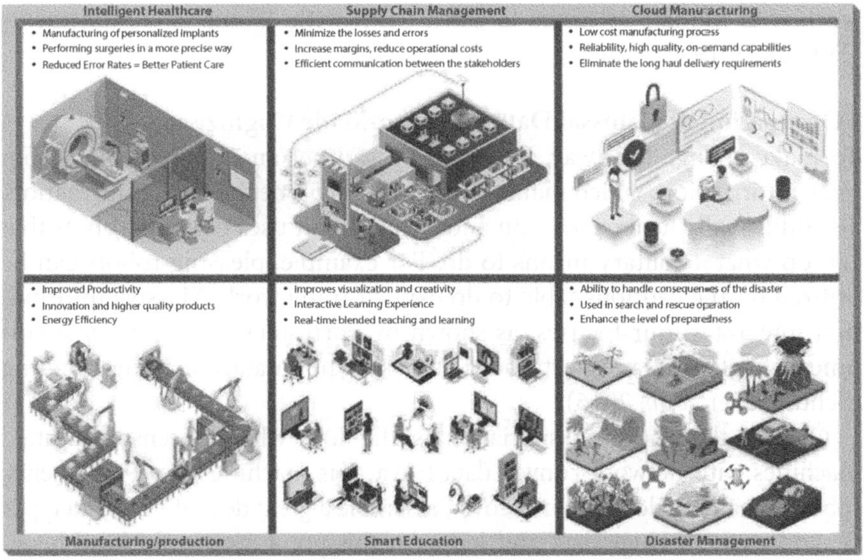

Figure 11.4 Different dimensions of Industry 5.0 in the surrounding of human (DOI: 10.1016/j.jii.2021.100257).

and computerized electronic creation, making fabricating processes seriously shocking, modified, and achievable so that machines can be worked easily, capacity, and consistency (Qin *et al.*, 2016). Individuals these days need the human bit of mass personalization, so Industry 5.0 assists them with changing from mass get-togethers to mass personalization (Haleem and Javaid, 2019). Industry 5.0 is making mass customization a reality, and the fast advances in collecting strategies, creation structure digitization, and information are the necessities of the current region (Vaidya *et al.*, 2018). Industry 4.0 as of late ordered mass customization, which was not sufficient. For instance, type 1 diabetes is trying to regulate considering the way that people have obvious levels of absorption and specific perspectives with specific skin thicknesses, approaches to acting, lifestyles, etc. Shift to Industry 5.0 grants giving people an application that follows their penchants and timetable, making a diabetes control creation technique and, in the end, a lower, more careful, and strong device modified to the person. The capacity to think up an Industry 5.0 procedure would as needs be totally historic for diabetes casualties.

- **Occupation of Robots**

Robots sincerely commit to works like stacking, unloading, painting, and welding. An autonomous robot is used to lead a more exact free strategy of gathering and to work where human delegates are confined to work. A critical example is a useful robot, or close called "cobot" expected to work with human delegates to assist them with countless commitments (Fukuda, 2019).

- **Occupation of Colossal Data and Man-Made Cognizance**

To make change headway, Industry 5.0 can get an understanding from huge information, which makes a relationship of mechanized information. To additionally foster precision and execution, it uses obliging robots that can do what a solitary means to do. For example, pleasing robots can be utilized on the cautious table to do explicit framework. Massive information integrates four focuses, as shown by Forrester's suspicion: data volume, data plan, data respect, speed of making new data, and understanding (Schumacher *et al.*, 2016).

One of the engaging specialists is the IoT, wherein sensor-outfitted machines with network convey data to various machines and PC systems, robotize many cycles, and, moreover, assemble a great deal of new data types (Fukuda, 2019). Huge data and IoT is at this point a one-of-a-kind benefit in Industry 4.0, and these substances make more conspicuous impact in Industry 5.0. Gigantic proportion of data is made, which is poor down using huge data assessment methodology and man-went with thinking estimations to seek after smart decisions. In like manner, more sensors and clever contraptions are used in Industry 5.0, which, for IoT, ensures top type, exactness, and productivity with higher customization. Industry 5.0 produces one more kind of open positions like creative and imaginative thinking, advancement interface, managing cobots, making counterfeit estimations, and some more as depicted by Figure 11.5.

- **Opportunities of Industry 5.0**

  - Cutting edge headway: Expanded robotization will impact work sincerely in different locales through the sending of best in class improvement.
  - Greator opportunity: Especially robotized fabricating structures give more critical doorway to customization to clients.
  - Progress of human productivity: Industry 5.0 gives more fundamental entrances to creative individuals to come and work which empowers the streamlining of human suitability.
  - Mindfully sponsorship to client: Machines are versatile as per specialist necessities to seek after an obvious level decision, and Industry 5.0 will help the client carefully in overseeing emphasized follow-up tasks (Haleem and Javaid, 2019).

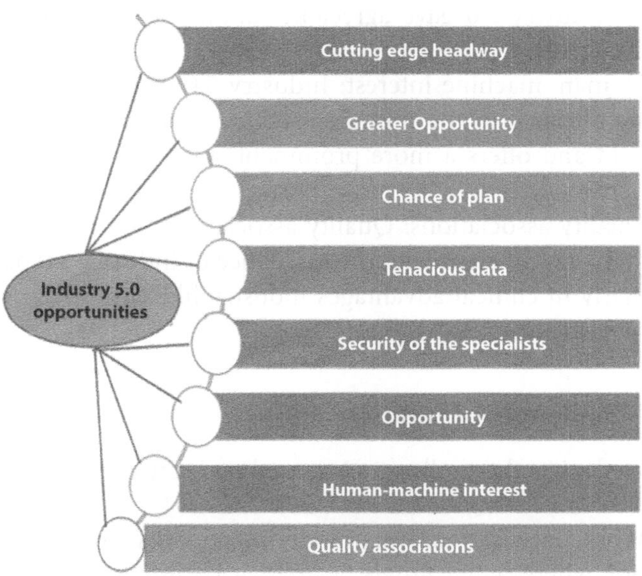

Figure 11.5 Opportunities of Industry 5.0.

- Higher-respect business: Industry 5.0 makes higher-respect work than prior to considering the way that this recommendations back the chance to individuals to be obligated for development (Rossi, 2018).
- Coordinating technique: In Industry 5.0, the director inside the creation cell gets more participated in the getting sorted out framework than in the basically modernized conveying system (Rossi, 2018).
- Chance of plan: It awards chance of plan to work and permits more changed and individual things.
- Tenacious data: It draws in the computerization of social occasion techniques better with Industry 5.0 by managing the consistent data from the area.
- Security of the specialists: Expanded success of the workers at the work floor considering the way that cobots can take up perilous and dangerous works.
- Increment purchaser commitment: More re-tried things and associations increment buyer faithfulness, devotion, and draws in new clients, which accomplishes broadened benefit and cut of the pie for the affiliations.
- Opportunity: It permits surprising open doors to new associations and business visionaries in imaginative and creative circles to devise new things and associations united to Industry 5.0 gave agreeable funding and foundation is accessible.
- Human–machine interest: Industry 5.0 gives stretched out significance to the human–machine joint exertion subject field and offers a more prominent stage for creative work here.
- Quality associations: Quality associations can be equipped at the far off areas with the assistance of Industry 5.0 particularly in clinical advantages industry like clinical activities in rustic districts by robots.

11.4.1 Challenges of Industry 5.0

Figure 11.6 depicts the challenges imposed on Industry 5.0.

- Work polarization: This model increments work polarization where center fitness business is lessening and the labor force is isolated into two associations: incredibly prepared

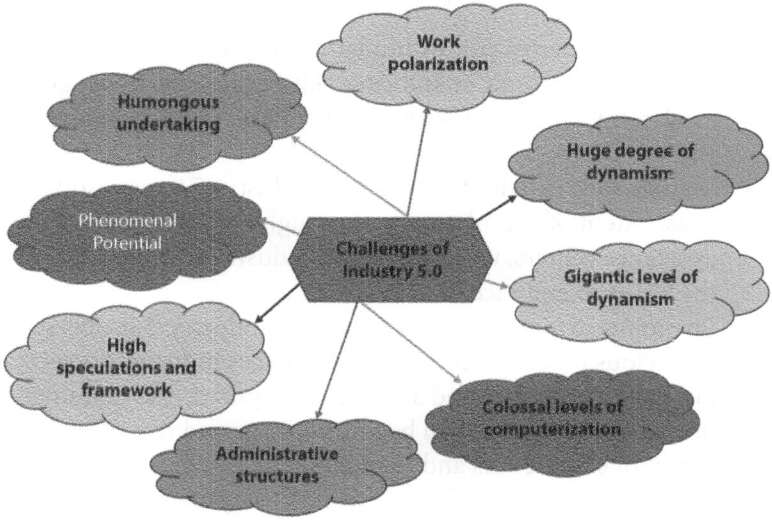

Figure 11.6 Challenges of Industry 5.0.

and qualified; and low-paid and lacking agents. This could work with the gifted and untalented opening in the general populace [11].

- Humongous undertaking: Due to essentially robotized conveying structures, expertise improvement is a humongous undertaking like arranging labor force for get-together of extreme front line and top-tier improvements, and provoking social change for talking with.
- Supportive significant level mechanics: Accommodating mechanical development is the system for robotization, which close by human accomplices similarly remains a tremendous bet on the shop floor (Rossi, 2018).
- Higher opportunity and sociality limits: Sharp assembling frameworks request higher independence and sociality limits as key parts of self-worked with structures. The shift from present setting to Industry 5.0 is irritating considering nonappearance of independence in the continuous designs, for example, created bearing.
- Testing to get great and constancy: From gathering structures, it is challenging to get shocking and uprightness of the information and it is attempting to oblige various information reports (Thoben, 2017).

- Expanded electronic security risk: There is a drawn out association prosperity risk in fundamental present-day designs and gathering lines at Industry 5.0 because of its lengthy availability and utilization of standard correspondences shows.
- Tremendous extent of experience: Industry 5.0 requesting a beast extent of speculation to thoroughly do the entire of its assistance centers, which is hard to industry and particularly the SMEs to embrace.
- Phenomenal potential: For example, Industry 5.0 offers a marvelous potential in clinical advantages industry yet serious level of accuracy and accuracy is required. The examination in this front is still in beginning stage and requests high extent of experiences and foundation.
- High speculations and framework: It is pursuing for new associations and monetary experts since Industry 5.0 requests high undertakings and foundation with top-tier progression necessities.
- Administrative structures: Testing to attract administrative instruments Industry 5.0 because of the high extent of computerization presence. For instance, whom to be seen as careful if there should arise an occurrence of bafflements and how much.
- Colossal levels of computerization: Considering extra basic levels of robotization in the undertakings, the continuous business method and procedures ought to be changed and patched up to meet the fundamentals of Industry 5.0. Because of mass personalization, business structure will zero in extra on client driven works out. Client subjectivity changes after some time and it is attempting to change the business frameworks and blueprints regularly.
- Gigantic level of dynamism: Business structures in Industry 5.0 requests more basic level of dynamism to help challenge because of differential client propensities.

11.5 What Does Industry 5.0 Mean for your Strategy?

The possibility of Industry 5.0 is not restricted to "industry." It applies to each area, and each association one can imagine. This implies that its relevance is fundamentally more extensive than Industry 4.0. In this way,

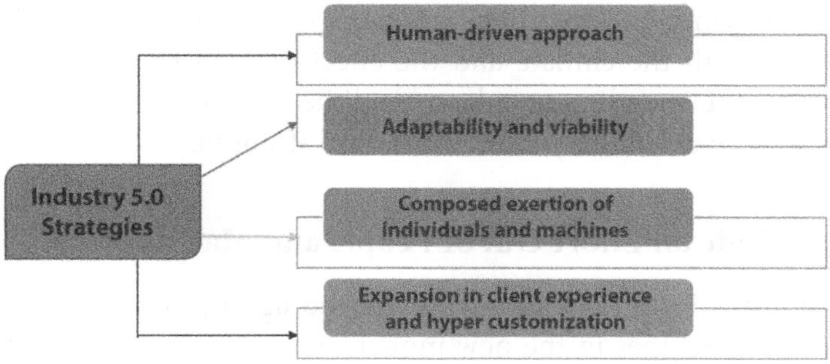

Figure 11.7 Industry 5.0 strategies.

while examining the ramifications of Industry 5.0 for a system, we want to take an expansive and general point of view applying to all enterprises. As the European Commission illuminates in this infographic, Industry 5.0 has three key points of support: human-driven, versatile, and manageable as depicted by Figure 11.7. Each of the three has critical ramifications for business technique.

11.5.1 Human-Driven Approach

If Industry 4.0 revolved around using mechanization frameworks, machines, and robots, interconnected with one another for most extreme execution improvement and productivity, then Industry 5.0 generally places that with regard to a more prominent human-driven approach. It raises the job of human ability and its significance to, by and large, outcome of tasks. With regard to executing robots and different machines, Industry 5.0 underscores the cooperation and associations between people and machines.

This approach is additionally, in some cases, called "society-driven" as it puts center human requirements and interests at the center of the creation cycle and advances adjusting modern mechanization advances to the necessities of industry laborers. Industry 5.0 perceives the significance of accomplishing high efficiency as well as establishing protected and comprehensive workplaces where human well-being and prosperity are fundamentally important.

11.5.2 Flexibility and Reasonability

As an expansion to engaging people, Industry 5.0 commits a ton of regard for different parts of versatility and manageability. The idea of Industry 5.0 gives a system that ought to adjust the seriousness and business productivity of the underlying vision with an emphasis on supportability and careful way to deal with the climate and the effect of modern computerization on our planet. Depending on adaptable, lithe, and versatile innovations is likewise a piece of Industry 5.0 drive toward strength.

11.5.3 Made an Effort Out of People and Machines

While Industry 5.0 is not making light of the significant job of robots and mechanized machines in the new modern unrest, the center is moved to cooperative collaborations among people and machines. As well as perceiving notable qualities and benefits of mechanical robotization, for example, being more exact, steady, and useful in their work contrasted with human representatives, Industry 5.0 additionally recognizes every one of the shortcomings of unnecessary mechanization, like low adaptability of exceptionally computerized arrangements and unfortunate flexibility to changing necessities and prerequisites.

11.5.4 Extension in Client Experience and Hyper-Customization

With regard to the client-confronting side of modern robotization—items and administrations produced by the Business 5.0 arrangements, the new idea is centered around conveying more noteworthy client encounters rather than simply accomplishing superior execution by interconnecting machines and programming. For buyers, this brings hyper-customization and hyper-personalization, which is the capacity of organizations to furnish clients with stunningly better decision and item customization choices while as yet lessening creation costs thanks to mechanical technology, computerization, and other inventive advancements.

11.6 Understanding the Architecture and Key-Components of a Cloud-Based Cobot System

A cloud-based cobot, or cloud-connected robot, typically has the following architecture:

- Hardware: The physical robot, which may be a collaborative robot (cobot) or a traditional industrial robot, is equipped with sensors and actuators for performing tasks.

- Connectivity: The robot is connected to remote communication and control.
- Cloud platform: The robot communicates with a cloud-based platform that manages data storage, computation, and communication between the robot and other devices.
- Application programming interface (API): The cloud platform provides an API that allows developers to create and run applications that interact with the robot, such as programming new tasks or monitoring the robot's status.
- User interface: A web-based or mobile interface is provided for users to interact with the robot and monitor its status.
- Machine learning: The cloud platform may also include machine learning algorithms to improve the robot's performance over time and make it more autonomous.
- Security measures: The system should be secure to protect against unauthorized access and data breaches.

11.7 A Cloud-Based Cobot System Typically Consists of Several Key Components

- Cobot hardware: This includes the physical robot itself, as well as any sensors or other devices that are required for the robot to function.
- Cloud infrastructure: The system runs on a cloud-based platform, such as Google Cloud Platform, Microsoft Azure, or Amazon Web Services. This provides the necessary computing power and storage for the system.
- Software: The system relies on a variety of software, including the robot operating system, which controls the robot's movements and communication with other devices, as well as software for tasks such as machine learning, computer vision, and data storage and management.
- Communication: The system uses various forms of communication, such as Wi-Fi, Ethernet, and cellular, to connect the cobot to the cloud infrastructure and other devices.
- User interface: A web-based interface allows users to interact with and control the cobot remotely. This may include

a dashboard for monitoring the robot's status and performance, as well as tools for programming and configuring the robot.
- Security: Security features such as encryption, authentication, and access control are implemented to protect the system from unauthorized access and ensure the integrity of data.

11.8 Robot Collaboration Rather Than Competition

Robot collaboration, also known as "cobotics," refers to the concept of robots working together with humans to achieve a common goal. This can take many forms, including the following:

- Human–robot collaboration: This refers to situations where a human and a robot are working together on a task, with the robot performing tasks that are either too dangerous or too tedious for the human.
- Multi-robot collaboration: This refers to situations where multiple robots are working together to accomplish a task. For example, a group of robots might be used to assemble a car or a group of drones might be used to survey a large area.
- Networked robots: This refers to robots that are connected to one another and able to share information and resources.
- Hybrid automation: This refers to systems where robots and humans work together in a shared workspace, with the robot performing certain tasks while the human performs others. This allows for a more efficient use of resources and can help increase productivity.
- Task allocation: Robots and humans can be used together more effectively by allocating tasks based on the strengths of each. For example, robots are well suited for performing repetitive tasks with high precision, whereas humans are better at tasks that require flexibility and creativity.
- Adaptability: Collaborative robots can adapt to the humans they are working with, adjusting their speed and force based on the humans' movements and abilities. This allows for a more natural and efficient working relationship.

Collaborative robots are designed to be safe to work alongside humans and can be programmed to perform a wide range of tasks, from simple repetitive tasks to more complex operations. By working together, robots and humans can achieve more than they would be able to alone, and the use of collaborative robots is expected to increase in a wide various industries like manufacturing, agriculture, logistics, and healthcare. Overall, the use of collaborative robots is expected to increase in a wide range of industries, as they offer many benefits over traditional robots, such as increased efficiency, productivity, and safety. Collaborative robots are also more adaptable and can be used in a wider range of applications. This is a new field of robotics, and research is ongoing to improve the technology and find new ways of collaboration.

11.9 Web 5.0–Enabled Technologies and Cloud-Connected Robots

Web 5.0 is the next age of the Internet, described by an emphasis on artificial consciousness, semantic comprehension, and the reconciliation of the physical and computerized universes. A portion of the innovations as depicted by Figure 11.8 that might be related to Web 5.0 include the following:

- Computerized reasoning (computer-based intelligence) and AI (ML): These advancements will assume a focal part in Web 5.0, empowering the improvement of additional clever and independent frameworks.
- Semantic web: The semantic web is a dream for the fate of the web where data is all the more effortlessly perceived and shared by machines. This will be empowered by innovations, for example, Natural Language Processing and ontologies.
- Web of Things: The reconciliation of the actual world with the computerized world will be a vital part of Web 5.0, and the IoT will assume a focal part in this. The IoT or Web of Things will interface with many gadgets and sensors, considering the consistent trade of information and the improvement of new applications and administrations.
- Edge figuring: Edge registering is a disseminated figuring model in which information is handled at the organization's edge, near the information's wellspring. This will be significant for handling a lot of information created by IoT gadgets.

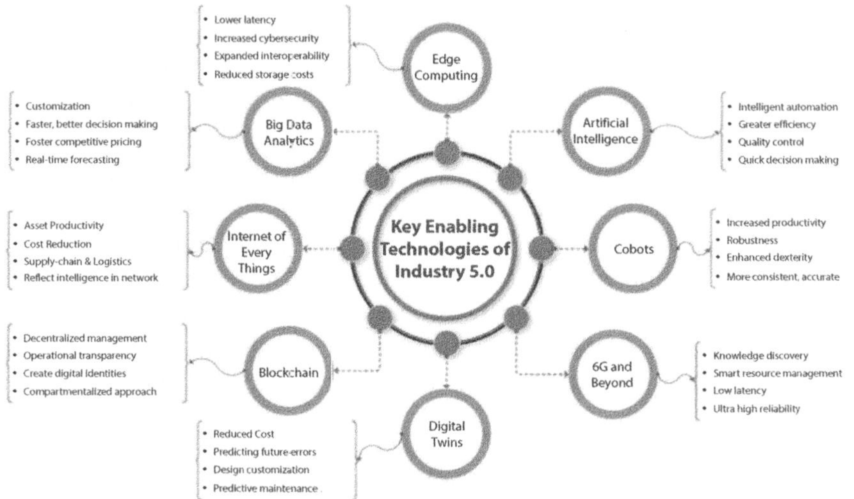

Figure 11.8 Web 5.0–enabled technologies (https://doi.org/10.1016/j.jii.2021.100257).

Cloud-connected robots are robots that are connected to the cloud [11], allowing them to access powerful computing resources and data storage, as well as to communicate with other devices and systems. With the help of Web 5.0 technologies, cloud-connected robots will be able to make more informed decisions, learn and adapt, and interact with the environment in a more natural and intuitive way. Additionally, the cloud infrastructure will provide a secure and scalable environment for the deployment and management of these robots.

11.10 Security and Data Privacy Concerns in Cloud-Based Cobot Systems

Security and information protection are the main pressing issues in cloud-based cobot frameworks as they include touchy data like the assembling process, creation information, and individual data of workers. Coming up next are a portion of the security and information protection worries that should be tended to in cloud-based cobot frameworks:

- Information encryption: Delicate information communicated and put away in the cloud ought to be scrambled to forestall unapproved access.

- Validation and access: Serious areas of strength for control and access control measures ought to be set up to guarantee that main approved clients approach the cobot framework and the information it creates.
- Network security: The cobot framework ought to be safeguarded from network-based assaults, like refusal of administration (DoS) assaults, by utilizing firewalls, interruption discovery, counteraction frameworks, and other safety efforts.
- Actual security: The cobot equipment and any information stockpiling gadgets ought to be genuinely able to forestall unapproved access and alter.
- Consistence: Cloud-based cobot frameworks might have to consent to different guidelines, like the Overall Information Assurance Guideline and the Medical Coverage Convenience and Responsibility Act on medical care information, which has severe prerequisites for information protection and security.
- Cloud supplier security: The security of the cloud framework ought to be consistently evaluated to guarantee that it meets the security necessities of the cobot framework.
- Episode reaction: An arrangement should be set up to conveniently and compellingly answer security occurrences, including information breaks and framework disappointments.

Security and data privacy are critical considerations in cloud-based cobot systems, and organizations need to take a comprehensive approach to address these concerns, including implementing strong security measures, monitoring for threats, and regularly assessing the security of the system.

In summary, cloud-based cobot systems offer many benefits such as increased efficiency and scalability, but they also raise important security and data privacy concerns. To ensure the security and data privacy of a cloud-based cobot system, organizations should take a comprehensive approach that includes implementing strong security measures, such as data encryption, authentication and access control, network security, and physical security; ensuring compliance with relevant regulations; regularly assessing the security of the cloud infrastructure; and having a plan in place to respond to security incidents. Organizations should also be aware of the potential risks and have a clear understanding of their responsibilities for protecting sensitive data. In addition, monitoring for potential threats and regularly testing the system for vulnerabilities should be a part of the security strategy.

11.11 6G Mobile Telecommunications

6G is the next generation of mobile telecommunications technology, which is expected to build upon the capabilities of 5G networks and provide even faster, more reliable, and more secure connectivity. Some of the technologies that may be associated with 6G include the following:

- Terahertz (THz) frequencies: 6G associations should use the THz range, which offers out and out higher information transmission than the frequencies used by current, flexible associations.
- Quantum exchanges: 6G associations could use quantum-based progresses, for instance, quantum key distribution, to give overhauled security to correspondences.
- Electronic thinking (PC-based insight) and AI (ML): 6G associations ought to have the choice to manage a ton of data continuously and use man-made consciousness and ML developments to foster organization execution and security further.
- Spatial figuring: 6G associations could have the choice to help spatial enrolling, which thinks about the compromise of virtual and augmented reality experiences with this current reality.
- Autonomous associations: 6G associations could have the choice to assist free associations that with canning self-direction and self-recover, chipping away at their adaptability and steadfast quality.

In terms of cobots, 6G networks could provide high-speed, low-latency connectivity to the cloud, enabling more advanced applications of cobots such as real-time monitoring and control, remote assistance, and real-time data processing. Moreover, it could enable new applications such as remote control of cobots by humans from a distance, providing more flexibility and new ways of collaboration. 6G networks will also facilitate the integration of cobots with other technologies such as IoT and 5G edge computing, to create a more connected and intelligent manufacturing environment.

11.12 Case Studies of Successful Implementation of Cloud-Based Cobots in Industry 5.0

There are several examples of successful implementation of cloud-based cobots in Industry 5.0 across different industries. Some examples include the following:

- Manufacturing: One example is a company that has implemented a cloud-based cobot system to automate the assembly of small electronic devices. The cobot is connected to the cloud and uses machine learning algorithms to optimize its performance and reduce errors.
- Healthcare: Another example is a hospital that has implemented a cloud-based cobot system to assist with patient care. The cobot is used to transport medications and medical supplies and is also equipped with cameras and sensors to monitor patients' vital signs and alert staff to any changes.
- Agriculture: A company has developed a cloud-based cobot system for precision agriculture. The cobot is equipped with sensors and cameras, which helps in collection of data on growth of the soil and crop. These data then analyzed with tools over the cloud to optimize crop yields.
- Retail: A large retail company has implemented a cloud-based cobot system to automate their warehouses. The cobots work alongside human employees to pick and pack items for shipment, increasing efficiency and reducing the number of errors.
- Logistics: A logistics company has implemented a cloud-based cobot system for loading and unloading cargo from ships. The cobots are connected to the cloud, and use computer vision and machine learning algorithms to identify and sort different types of cargo. Certainly, here are some additional examples of successful implementation of cloud-based cobots in Industry 5.0.
- Automotive Industry: A major car manufacturer has implemented a cloud-based cobot system to improve the efficiency of its assembly line. The cobots are connected to the cloud and use machine learning algorithms to optimize their performance and adapt to changing production requirements.

- Food and Beverage Industry: A food and beverage company has implemented a cloud-based cobot system for packaging and palletizing. The cobots are connected to the cloud and use machine learning algorithms to optimize the packaging process and reduce the number of errors.
- Pharmaceutical Industry: A pharmaceutical company has implemented a cloud-based cobot system for the handling and packaging of drugs. The cobots are connected to the cloud and use computer vision and machine learning algorithms to sort and package drugs accurately, ensuring patient safety.
- Construction Industry: A construction company has implemented a cloud-based cobot system for bricklaying. The cobots are connected to the cloud and use machine learning algorithms to optimize the bricklaying process and reduce the number of errors.
- Energy Industry: An energy company has implemented a cloud-based cobot system for the inspection and maintenance of wind turbines. The cobots are connected to the cloud and use machine learning algorithms to optimize their performance and reduce the number of errors.
- These examples show that cloud-based cobots can be successfully implemented in a variety of industries, and they can bring significant benefits such as increasing efficiency, improving quality, and reducing costs. The use of cloud-based cobots also allows companies to have real-time monitoring, analysis, and optimization of the production process, making data-driven decisions and improving the overall performance.

11.13 Challenges and Future Considerations for the Integration of Cloud-Based Cobots in Industry 5.0

While cloud-based cobots have the potential to bring many benefits to Industry 5.0, there are also challenges and considerations that need to be addressed for their successful integration:

- Technical challenges: One of the main challenges is the integration of different technologies such as cloud computing, robotics, and AI. This requires a high level of expertise in various fields and the ability to integrate different systems seamlessly.
- Security and data privacy: As with any cloud-based system, security and data privacy are major concerns, particularly when it comes to sensitive information such as production data and personal information of employees. Strong security measures and compliance with relevant regulations need to be in place to protect sensitive data.
- Scalability: As the number of connected devices and the amount of data generated increases, the ability to scale the system to handle these data becomes more important.
- Interoperability: Cloud-based cobots need to be able to communicate and interact with other systems and devices, both within the organization and outside of it, such as other robots, devices, and systems.
- Cost: The cost of implementing a cloud-based cobot system can be high, and it may require a significant investment in both hardware and software.
- Human factor: The integration of cobots in the production process may imply changes in the work environment and the roles of human workers, which may require training and education.
- Ethics and governance: The use of cobots raises ethical and governance questions, such as the impact on jobs and the responsibility of the companies for the actions of the cobots.

11.14 Conclusion

This chapter describes that the role of the human touch in Industry 5.0 cannot be overstated. It is a vital component of the future of manufacturing. Integrating the human touch into manufacturing will increase efficiency, productivity, and customer satisfaction. Industry 5.0 is about advanced technology and the human touch that makes it all possible. In the extended version of this chapter, authors can discuss the all dimensions of Industry 5.0 in more details, implementation challenges, and security issues of Industry 5.0.

References

1. Fukuda, K., Science, Technology and Innovation Ecosystem Transformation Toward Society 5.0. *Int. J. Prod. Econ.*, 2019.
2. Haleem, A., and Javaid, M., Industry 5.0 and its expected applications in medical field. *Curr. Med. Res.*, 9, 4, 167–169, 2019.
3. Qin, J., Liu, Y., Grosvenor, R., A categorical framework of manufacturing for industry 4.0 and beyond. *Procedia CIRP*, 52, 173–178, 2016.
4. Rossi, B., Manufacturing Gets Personal in Industry 5.0, Reconteur, 2018, https://www.raconteur.net/technology/manufacturing-gets-personal-industry-5-0.
5. Schumacher, A., Erol, S., Sihn, W., A maturity model for assessing Industry 4.0 readiness and maturity of manufacturing enterprises. *Procedia CIRP*, 52, 161–166, 2016.
6. Thoben, K.D., Wiesner, S., Wuest, T., Industrie 4.0" and smart manufacturing-a review of research issues and application examples. *Int. J. Autom. Technol.*, 11, 1, 4–16, 2017.
7. Vaidya, S., Ambad, P., Bhosle, S., Industry 4.0–a glimpse. *Procedia Manuf.*, 20, 233–238, 2018.
8. Carayannis, E.G., Draper, J., Bhaneja, B., Towards fusion energy in the industry 5.0 and society 5.0 context: Call for a global commission for urgent action on fusion energy. *J. Knowl. Econ.*, 12, 1–14, 2020.
9. King, G.S., Rameshwar, J.R., Syan, C.S., Industry 4.0 in a small commodity-based economy: A vehicle for stimulating innovation. *J. Ind. Integr. Manag.*, 5, 03, 365–391, 2020.
10. Choudhary, V. and Mishra, A., Analyzing the critical success enablers of industry 4.0 using hybrid fuzzy ahp–cocoso method. *J. Ind. Integr. Manag.*, 7, 2150018, 2021.
11. Maddikunta, P.K., Pham, Q.-V., B, P., Deepa, N., Dev, K., Gadekallu, T.R., Ruby, R., Liyanage, M., Industry 5.0: A survey on enabling technologies and potential applications. *J. Ind. Inf. Integr.*, 26, 100257, 2022, https://doi.org/10.1016/j.jii.2021.100257.
12. Weiss, Buchner, R., Tscheligi, M., Fischer, H., Exploring human-robot cooperation possibilities for semiconductor manufacturing. *2011 International Conference on Collaboration Technologies and Systems (CTS)*, Philadelphia, PA, USA, pp. 173–177, 2011, doi:10.1109/CTS.2011.5928683URL: http://ieeexplore.ieee.org/stamp/stamp.jsp?tp=&arnumber=5928683&isnumber=5928646.
13. Sadrfaridpour, B. and Wang, Y., Collaborative Assembly in Hybrid Manufacturing Cells: An Integrated Framework for Human–Robot Interaction. *IEEE Trans. Autom. Sci. Eng.*, 15, 3, 1178–1192, July 2018. doi:10.1109/TASE.2017.2748386,URL:http://ieeexplore.ieee.org/stamp/stamp.jsp?tp=&arnumber=8049302&isnumber=8401732

14. Desai, M. *et al.*, Effects of changing reliability on trust of robot systems. *2012 7th ACM/IEEE International Conference on Human-Robot Interaction (HRI)*, Boston, MA, USA, pp. 73–80, 2012, doi: 10.1145/2157689.2157702, URL: http://ieeexplore.ieee.org/stamp/stamp.jsp?tp=&arnumber=6249617&isnumber=6249474.
15. Richert, A., Shehadeh, M., Plumanns, L., Groß, K., Schuster, K., Jeschke, S., Educating engineers for industry 4.0: Virtual worlds and human-robot-teams: Empirical studies towards a new educational age. *2016 IEEE Global Engineering Education Conference (EDUCON)*, Abu Dhabi, United Arab Emirates, pp. 142–149, 2016, doi: 10.1109/EDUCON.2016.7474545, URL: http://ieeexplore.ieee.org/stamp/ stamp.jsp?tp=&arnumber=7474545&isnumber=7474513.
16. Meneweger, T., Wurhofer, D., Fuchsberger, V., Tscheligi, M., Working together with industrial robots: Experiencing robots in a production environment. *2015 24th IEEE International Symposium on Robot and Human Interactive Communication (RO-MAN)*, Kobe, Japan, pp. 833–838, 2015, doi: 10.1109/ROMAN.2015.7333641, URL: http://ieeexplore.ieee.org/stamp/stamp.jsp?tp=&arnumber=7333641&isnumber=7333553.
17. The welding Institute. What is industry 5.0? (top 5 things you need to know), TWI, Retrieved April 30, 2023, from https://www.twi-global.com/technical-knowledge/faqs/industry-5-0.
18. Bonini, L. and Bertone, N., The Use of Collaborative Robots (Cobots) in an Analog Modeling Laboratory, 2023, https://doi.org/10.5194/egusphere-egu23-6125.
19. Staff, C.T., Wessling, B., Crowe, S., Nicho, J., Khokhar, S., Demaitre, E., Hsu, C., Becky Ham | MIT Media Lab., Research archives, Collaborative Robotics Trends, https://www.cobottrends.com/category/research/.
20. Weiss, A., Wortmeier, A.-K., Kubicek, B., Cobots in industry 4.0: A roadmap for future practice studies on human–robot collaboration. *IEEE Trans. Hum.-Mach. Syst.*, 51, 4, 335–345, 2021, https://doi.org/10.1109/thms.2021.3092684.
21. What can we expect from cobots in 2023?, WiredWorkers, 2023, January 18, https://www.wiredworkers.io/blog/what-can-we-expect-from-cobots-in-2023/.

12

Aviation Bots for Decongesting Airports

Saptarshi Paul

Computer Science Department, Assam University, Silchar, India

Abstract

Developing interactive real-time aviation chatbots and robots has always been a challenge for the developers and the aviation industry as a whole. Collection of proper aviation resources determines how the chatbots and robots are going to perform and up to what accuracy. Translation analysis determines the accuracy of translations and transliterations, whereas other parameters determine how that particular aviation bots interact with passengers. Proper identifying, pre-processing, and then using of resources to get the desired results in real time are the ultimate challenges. In most countries, including India aviation resources like corpora and retrievable data are scarce. The pandemic had increased the challenge of resource collection, documentation, and classification manifolds. This chapter is aimed at discussing on how to deal successfully and efficiently the development of various bots that are related to aviation industry. While some simple bots for web check-in and airport check-in are making their presence felt in major airports, bots capable of complete passenger assistance are still a dream that can make their presence felt only in the distant future. From the passenger viewpoint, current bots can handle an array of jobs. These aviation bots are capable of generating baggage tags and boarding passes and in many cases automated identification of passengers, e.g., DigiYatra. They are still playing an important role in decongesting airport queues. From the viewpoint of preflight maintenance and check, ground engineers and pilots often use chatbots which are, nowadays, often AI- and NLP-enabled. These bots are not only capable of holding a huge amount of information but also capable of suggesting the probable or appropriate course of action.

***Keywords*:** Aviation bots, ML, NLP, DigiYatra, Timi

Email: paulsaptarshi@yahoo.co.in

V. Ramasamy, S. Balamurugan and Sheng-Lung Peng (eds.) Intelligent Robots and Cobots: Industry 5.0 Applications, (247–266) © 2025 Scrivener Publishing LLC

Abbreviations

ML	machine language
AI	artificial intelligence
NLP	natural language processing
UDAN	Ude Desh ka Aam Nagarik
PAX	passengers
MoCA	Ministry of Civil Aviation

12.1 Introduction

Being the fastest growing aviation market in the world, India is projected to be the largest by 2030 [1]. With the Ministry of Civil Aviation (MoCA) opening up advanced landing grounds and tens of new airports and routes, there is a flood of first time fliers in India. People are traveling more than ever before not only to avail facilities of UDAN but also as revenge flying to avenge the years of pandemic restrictions. The MoCA is upgrading the existing airports and handing over busier ones to provide players like the GMR/ADANI group to facilitate better on-ground services to PAX.

Being a country with the most population and an economy that's fifth largest, India has no dearth of passengers availing the areal route. Indian airports are facing PAX movement like never before. The Regular Passenger Information Systems and check-in methods are falling short to handle the huge PAX movements at each and every airport. Kiosks were introduced to facilitate automatic check-in with boarding and baggage pass printing, but these measures are making little difference to the huge congestions that are a regular feature in all Indian airports.

Such conditions have called for radical and drastic steps to be taken. This challenge has been met up by a combination of artificial intelligence (AI), natural language processing (NLP), robotics, and machine learning (ML). Chatbots, robots, and facial recognition systems coupled with data mining like DigiYatra are rising up to meet the demand to decongest Indian airports like never before.

The chapter first discusses features of aeronautics or aviation English, which is a unique one. The chapter then takes up the various sources for the resources that are used in aviation bots. It also discusses how various organizations have digitized aviation data, but a lot still remains to be digitized. The chapter then enters in a discussion on the current aviation bots, which are in use and the probable features of the future aviation bots.

The conclusion part discusses how various aviation bots are being used to decongest and streamline various airports around the world. The citations and references used in the chapter are all well established research work in journals and in use technologies by companies like Boeing and Airbus are listed at the end.

12.1.1 Aeronautic English

The English that is used in aeronautics encompasses civil aviation, maintenance, operation, and aero-space terms and terminologies. This unique version of English follows a structured pattern that is used in both spoken and written formats. Aeronautics demands that both the sentences be spoken and drafted in a uniform manner and it applies for both native and non-native speakers of English. To implement it, International Civil Aviation Organization has dictated a uniform and unique pattern that is to be used by Pilots, Air Traffic Controllers, maintenance staff, and ground operation employees. Each aircraft and airlines are allocated unique call-signs. These call signs are achieved by combination of alphanumeric characters. International Civil Aviation Organization has determined on how phonetic equivalents of every word and number are to be spelled and drafted. Table 12.1 shows a mix of the phonetic pronunciation of numbers in aviation, as well as alphabets equivalents used in aviation and aeronautics.

12.1.2 Perspective of NLP in Aviation

NLP applications generally encompass any tool or software that is related with development of E-Dictionary, parts-of-speech tagger, pre-processing tools, and post processing tools. Complex and complicated work may include developing bilingual translation models. NLP applications in aeronautics have huge scope as it is an unexplored branch of linguistics. Multiples of tools are used [1] around the world by companies but such tools and repositories are absent for Indian languages. This void also creates a huge pool of opportunities for researchers and technical project developers to contribute significantly.

Scope exists for creation of tools and models for English and Bengali/native Indian languages. Development of tools as such demands resources ranging from databases to parallel corpora. Such projects demand attention both in the programming aspect [2] as well as for creation of proper bilingual data repositories, which can be achieved through data collection, cleaning, preprocessing, and processing before being tested and implemented.

Table 12.1 Phonetic pronunciation in aviation.

English alphabet	Pronounced as	English alphabet	Pronounced as
A	*AL-FAH*	N	NO-VEM-BER
B	BRAH-VOH	O	OSS-CAR
C	CHAR-LEE	P	PAPA
D	DELLTAH	0	ZE-RO
E	ECK-OH	1	WAN
F	FOKS-TROT	2	TOO
G	GOLF	3	TREE
H	HOH-TEL	4	FOW-er
I	IN-DEE-AH	5	FIFE
J	JEW-LEE-ETT	6	SIX
K	KEY-LOH	7	SEV-en
L	LEE-MAH	8	AIT
M	MIKE	9	NIN-er

12.2 Aviation Resources

12.2.1 Collection of Resources Apart from Regular Sources

Gathering resources for the implementation of Smart airport bots/chatbots are very important. It not only helps us to understand the various types of NLP tools but also the various methods of translation. Each NLP tool or model requires corpora or data repository to act upon. Apart from that, testing the tools to achieve the desired output and higher translation accuracy is a must. The pandemic has made the collection of resources a painstaking affair.

12.2.2 Feedback from Passengers (PAX)

Feedbacks from passengers are the most unique, genuine, and best way to build the data repository for any sort of aviation resource. Online feedback forms, tweets, and feedback through SMS are the most convenient ones for airliners and companies to gather data for development and improvement of air bots. It helps them to develop terminal bots like "TIMI" as well as intelligent chatbots that can be accessed from cellular telephones and laptops alike.

12.3 Understanding Real-Time Results

Real-time bots/tools need to be executed and checked again and again before a satisfactory result can be obtained. Neural machine translation (NMT) models along with AI and pattern recognition needs to be preprocessed and post-processed to attain the requisite results in airports and in practical scenarios. No other theoretical metrics and scores can suffice the results obtained from practical implementation. All these are achieved through interacting with experts from linguistics, AI, and NLP. It is more of an inter-disciplinary research wing and less of a single integrated one. The pandemic has hastened us in attaining these interactions, and the results are quite praise worthy.

12.4 Resource Sharing and Sourcing

12.4.1 Accessing Online Corpora and Data Repositories

To gain access to the various out-of-vocabulary words used in aviation, the access to aviation data repositories are a must. Online resources are available that can be accessed in a multitude of ways. While some resources can be accessed only through proper permission, e.g., NPLT (https://nplt.in/demo/resources/text-corpus) and TDIL (http://www.tdil-dc.in/), there are many which can be accessed without any permission or charges. These free resources, especially for aeronautics are mainly restricted to reports of accidents, incidents and hence demands pre- and post-processing of the data. Maintenance manuals and procedures can be accessed only through proper channels and permissions. Resources for creation of corpora or data repository for aeronautics domain can be accessed from the following options:

1. Airport Authority of India (AAI)—manual of air traffic services [3].
2. Directorate General of Civil Aviation (DGCA)—manuals and accident/incident reports [4].

3. International Civil Aviation Organization (ICAO)—reports on safety/incidents [5].
4. National Transportation safety board—aviation accident data summary [6].
5. NASA-ASRS incident reports [7].
6. Aviation newsletters and aviation related blogs.

Apart from that a) phraseology manual of Chapter12-AAI [8], b) Pilot's handbook of Aeronautical knowledge 2016-2020 [9], and c) Dictionary of aviation English [10].

12.4.2 Online Tools and Applications

Various online tools and applications being used by airlines companies can provide an insight into passenger requirements. These all ultimately calls for work involving NLP, AI, and pattern recognition. Feedback obtained from twitter has been one of the major motivating factors in designing bots.

12.5 Current Aviation Bots and Their Limitations

12.5.1 Airport Terminal Bots

Gone are the days when passengers (PAX) were contented with kiosks printing out boarding pass and luggage tags to assist in the boarding process. Modern terminals come equipped with passenger information systems displaying latest flight numbers estimated time of departure, estimated time of arrival (ETA), gate number, and other information. Yet passengers prefer to check flight details like gate number and gate change information repeatedly. In such scenarios, bots can be of great help. As bots can be enabled with mobility, it can be a leveraging machine with internal navigation capability software that can be helpful with flight details. Bots such as "TEMI" can be seen helping out PAX at Bangalore Airport (KIAL) and other airports of India.

DigiYatra has born out of or emerged out of the need for fast check-in, thus relieving the airports of excess lines in check-in counters. It also helps in quick security check-ins, leading to hassle-free and quick check-ins toward security hold areas better known as SHA. DigiYatra is an artificial automation and facial recognition system or initiative that uses recognition of facial structure technology to facilitate entries to airport and clearance to security at the airport fast, free flowing, and automated process. Digiyatra is a platform that makes use of pattern recognition methodology to identify facial structures leading to automatic verification of passenger's identity. It is a system that makes use of distributed Cellular device–based information data repository that enables air passengers to safely store their photo- IDs and requisite documents needed during travel.

DigiYatra was first introduced by the Ministry of Civil Aviation better known as MOCA. DigiYatra officially known as DY-Foundation is the domain that leads to the next step toward creating a Digital India, i.e., better known as digitally empowered society. It is envisioned as the future of air travel. DigiYatra is making possible the automatic digital processing of passengers Identity. The DigiYatra facility is being offered by a series of airports starting at Delhi Airport, which provides PAX with the luxury of significant amount of lesser waiting time. It ultimately contributes to a faster boarding process, which is seamless and modern.

DigiYatra, the pride of Indian airports, now acts to facilitate paperless air travel. It is not only helping flyers "fly" through the check in and security process but also providing minimum physical contact that is a primary "Have-Not" for post-pandemic period. DigiYatra allows passengers to swiftly "Glide" through an array of intelligent cameras while using modern biometric security algorithms. DigiYatra is supposed to provide PAX with four facilities, concerned with connecting all passengers, seamless flying, modern digital airports, and ultimately under the umbrellas of a totally connected system. DigiYatra is the ultimately solution for a single window seamless check in system.

12.5.2 Maintenance Bots

Maintenance bots, as opposed to their "Physically Visible" Airport Terminal Bots Cousins are mainly restricted to laptops used for MRO (Maintenance, Repair, Overhaul) of aircrafts, airports, and related instruments. MROs are an integral and indispensible part of the aviation industry. All aircrafts have to undergo regular official maintenance from time to time.

Maintenance bots are being used in MRO for a long time, in aviation-related NLP systems, a credible translation system can be found in the form of the software known as TUAM AVIATION. This first of its kind work was attempted in 1976, in a Canadian Institute going by the name of CETADOL research centre. It was a work by University de Montreal Department of Computational Linguistics, which later went to be named as TAUM. The project was started with the aim to make a translation system for English to French language pair. The outcome was indented for the maintenance manuals of translation of aircrafts. The year 1979 saw the development of a preliminary design that was later developed to work on translation of manuals containing maintenance and hydraulic instructions. The TAUM AVIATION needed an environment to work on and the basic hardware consisted of CYBER 173. The O/S was Network-OS based on BE 1.4. The work needed development of dedicated dictionaries and grammars. These were later used to develop the object code. This code was transformed at run time. It considered the input text at all times. The source language was in English, whereas the destination language was in French. The dictionary, being the first of its kind consisted of vocabulary from core domain. It had the following features: 4054 one-of-its-kind entities. The English-to-French pair dictionary had a collection of 3280 one-of-its kind entities [11]. The MT used was a completely automated one. Pre-processing by humans was not required. In case of encounter with unidentified words, entry was made in the dictionary data repository [11].

After a few years with NLP and AI making way, a stream of tools made its way into the market. Examples include International Business Machine WATSON. It was a NLP tool that made use of NLP, machine translation, and artificial computing in aeronautics. International Business Machine has been working in multiple aspects of aviation. International Business Machine has introduced a fantastic query-reply capability to WATSON. It has used MT and NLP to help pilots tackle maintenance challenges in real time. International Business Machine has made sure to take one and all related to aeronautics and aviation to provide this service. WATSON can recognize human spoken language and can process it in NLP. It is capable of recommending suggestions based on previous instances in real time. NMT based on recurrent neural networks makes WATSON to estimate situations. Intelligent systems are being used in real time not only in aeronautics but also in medicine and agriculture [12].

For example, if a pilot encounter a problem with any of the aircrafts subsystems like hydraulics, she/he can ask the automated system like WATSON in his/her natural language. Systems like WATSON then consider the problem from technical viewpoint and then consider the probable solution from its data repository. The work of the NLP-AI system is then to then make an array of technical suggestions to the pilot on the future course of action to be performed on the aircraft to solve the issue. It can also suggest the safest approach to the nearest airport, if diversion is required. With use of NN systems like WATSON, they are able to take into consideration all possibility and suggest the best one [12].

Systems using of AI coupled with NLP and machine translation are forming the basic architecture of management, understanding and reporting machines of the aeronautics domain that involves airports to organizations like ICAO and DGCA.

Another intelligent system is AMRIT, a company named with location in Sydney, which is in the process of trying to help technicians easily understand documentations in technical writing. Their work domain includes helicopters, better known as helos manufactured by Airbus Group. They are making use of an intelligent tool known as AMRIT. AMRIT makes use of NLP to decode 120 million pages of documentation of various types.

Product maintenance manuals for aircrafts like ATR7, component documentations for helos like Eurocopter, service manuals for helicopters, and fitness directive and problem isolation documentations, for helos are all included. The underlying idea is to come up with a data repository linked to multiple tasks by thorough investigation and identification of sequence of words and in the array of documents mentioned. Statistics is used to derive from earlier positive cases of where problems were solved. If anyone comes up with same problem, the earlier positive cases are suggested for problem solving in real time. AMRIT can only be successful if proper training is imparted and also proper type of aircraft is chosen. Today AMRIT has developed into a fully working product, which is very useful in the helo/aviation market. Search engines are employed to go through all the required documents. European aircraft manufacturer AIRBUS uses AMRIT to find and isolate faults. People related to maintenance repair overhaul uses AMRIT found in NLP in AIRBUS [13] and NLP in MRO [14]. Parallel processing using multiple servers are used to achieve the desired results as in NLP in AIRBUS, 2016 [13]. AMRIT is now evolving to its next league where it is being readied to take on the big birds like A320NEO, A330NEO, A350NEO, and A380s. Proper field testing is to be performed along with training of MRO officials for successful implementation of the final product.

BLUE is another airline-related BOT that works on the basis of pipeline architecture. BLUE has many steps, each with a transformation one through which each sentence is processed. BLUE is capable of producing processed details through documentations of technical structure, but it may be required that humans have to go for post-processing or human-assisted machine translation to get proper in-order output [15].

PLUS is a safety-BOT by a company named SAFETY DATA. PLUS is used to using NLP to go through and analyzing information of flights. Natural language processing, which, in itself, is a part of AI, linguistics, and information technology, allows this BOT to assign weights to each and every data instead of binary Yes or No. These are assigned to each and every flight reports.

AToS, a company that has an international presence is an IT company with NLP-AI basis. AToS uses ML to perform better and provide better IT services. It is based on ML and has been able to undertake a multitude of projects whose primary work is to perform automatic BOT structural design of parts used in aircrafts. It uses NLP software as the work progresses. Implementation of these techniques brought about eye-catching positive notes in the aerospace industry.

These BOTs not only enhanced safety but also minimized environmental impact. The major disadvantage of SMT and NMT systems was that they suffered from data deficit for resource-scarce languages. As a result, it was not possible to get meaningful results in case of such languages, i.e., the input source language determined the output destination language quality.

Systems like ASRS are used to target multiple sources for dealing with data scarcity. Generally, regular ASRS reports contain of narrations, attributes, and entities of the event that is described. Depending on the report submitted, each submission has different narrative and entities. Used descriptors are capable of providing detailed information about the time, speed of wind, flight altitude, flight location, aircraft type used, and other factors such as cloud cover altitude. ASRS experts are capable enough to code them and give them technical values and encode them in the reports [16]. ASRS products and services are available in the following format [17].

A) Messages meant to alert safety data and suggestive correction courses. B) ASRS provides immediate response with very urgent basis and emergency information support. C) Data repositories of ASRS are made public for open source use by researchers. D) Monthly updates through newsletters and emails are available. E) ASRS latest updates for safety data are updated for pilots and others to update themselves. F) Regular research, conference, and meetings are conducted with various aviation organizations.

European Coordination Centre for Accident and Incident Reporting Systems better known as ECCAIRS is responsible for standardizing various incident reports. It collects information and performs exchange of data between the various European Union countries. Developed by the European Commission's Joint Research Center, the aim of ECCAIRS was to assist entities inside Europe in collection, sharing, and analyzing safety information in order to improve air safety. In order to do so, it made sure that information and data were freely available to researchers and others alike. It turned out to be a software umbrella using free flowing data and information from ICAO's data extraction reports. The European Coordination Centre for Accident and Incident Reporting Systems along with the Safety Recommendation Information System are functioning under the umbrella of european climate foundation (ECF). It is umbrella software developed by the JRCE. It is, in turn, responsible for reporting and sharing accident all information related to air incidents. It has been found to be useful by users of several countries such as DGAC of France. The reports collection mechanisms of India Directorate General of Civil Aviation are same as those of ASRS [16, 18, 19]. Table 12.2 shows the NLP tools commonly used in aviation.

Multiple systems spanning various countries and continents are used. Some of them can be summarized with their costs as follows: a) MLTS is a tool working for English French language pair with costing INR 32,240 only, English and Arabic language pair with cost of INR 38,400 only, and French and Arabic language pair costing INR 37,640 only; b) LogoMedia: English Spanish language pair with cost of INR 13,014 only; c) TranSmart: English and Finnish language pair with cost of INR 26,620; d) HonYaku: English to Japanese costing of around INR 63,760 only; e) SysTran: it works for French and Spanish language pair costing

Table 12.2 NLP tools used in aviation.

Tool name	Currently used by	Applicability
TUAM AVIATION	Canada	Transfer of maintenance manuals from French to English
IBM WATSON	USA	In-flight questioning answering capacity and maintenance
AMRIT	Airbus Helicopters	To provide guidance of use of digital manuals
BLUE	BOEING	Language understanding capability
PLUS	ATR, Air France, EASA, France	Generation of consolidated reliable information on aviation incidents
ATOS	France	Bezons speech-based NLP–partial automation of aircraft parts design
ASRS	NASA	North America–automated incident reporting database
ECCAIRS	European Union	Automated collection sharing and analyze safety data

around INR 26,604 only; f) LogoVista: capable of translation for French and Japanese language pair and Spanish and Japanese language pair costing around INR 48,800 only.

Price has always been playing a significant role. Prices are significantly higher with costs of around 8,000 Euros for server-based systems. For systems that are not server based, prices range from 88,000 to 160,000 INR. For server-based systems generally licensed for specific language pairs, tools can be used throughout updates of software, updates of TE, and updates of vocabulary data repository.

With huge orders from airlines like AIRINDIA, INDIGO, and AKASA coming up, the need for intelligent troubleshooting devices and software is rising day by day. Reference [18] gives us a detailed report that signifies that, after 2017, the market for AI, NLP, and MT together will be a multi-billion market not only in USA but also in countries like India and Sri Lanka where MRO has found a strong grip. The NLP parts discussed in the chapter are the ones that are forming the backbone of AI supported MRO and flying all throughout the world. Civilian, scientific, and military operators all are finding the use of this AI-based software more and more. The last few decades have seen MRO manuals that evolve systems like TUAM AVIATION to NLP-AI systems like AMRIT working in real time. These tools have been able to achieve not only smaller time frames but better performance. Domain for future research includes not only just increasing the number of documentations, work environment policies, and architectures for multiple organizations in the AI data repositories but also capable of translation from one language to other. Tools like International Business Machine WATSON have found its uses not only in aeronautics but also in law and agriculture. Uses of NLP instruments are now being slowly but surely introduced to domains like tourism and medicine. Domains of research now include auto-classifiers of documents. Such tools are seeing increase in demand on a day to day basis. Tools as such are improving with the use of huge data repositories including MRO, business use, and medical uses. Such tools employing MT-AI pairs can now be easily used to develop software and tools that will display aircraft positions in real time along with local weather situations from satellite mapping. These bots can easily thus predict the expected flight paths and routes. Social media being a huge source of information can also be used to develop aviation bots/tools capable of gathering information from platforms like twitter and displaying such information in a consolidated manner to the users. Unverified news can spread misinformation among PAX. Such information can be checked by bots using AI to verify the authentication and also employ sentiment analysis to find the status of various airlines/airports, etc.

12.5.3 Chatbots

In line with various intelligent chat-assistants used in banking systems, NLP- and AI-enabled chatbots are being created by individual airlines and airport service providers like the GMR/ADANI-group. Airlines chatbots are not only capable of providing information about flights but also able to provide information about ticket cancelation, rescheduling, and other features related to flights, like booking of wheelchairs, extra luggage, and onboard menu. While chatbots of airport operators provide all the information about services availability at airports, it also provides the weather information like delays due to fog, current congestion, etc. A chatbot was developed by the students of the Department of Computer Science, Assam University Silchar, that could provide information about the various rules for entry into different states related to COVID-19. The interface of the chatbot named as "airBot" was designed in Figure 12.1.

In addition, if you have any query regarding COVID guidelines in any northeast airports, then you have to input the data like "COVID rules in Silchar" and, as an output, the ChatBot will return a message that will contain a URL. The URL is the link that contains all the COVID-19–related entry-exit information and the various tests required for entering that particular state through that airport, e.g., as in Figure 12.2.

Figure 12.3 shows how the latest guidelines or SOP of COVID-19 guidelines by Airport Authority of India for Silchar Airport are being displayed here.

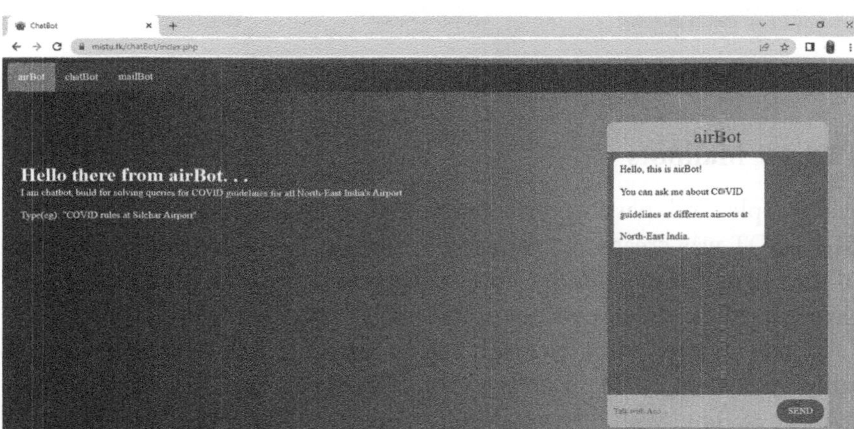

Figure 12.1 airBot interface.

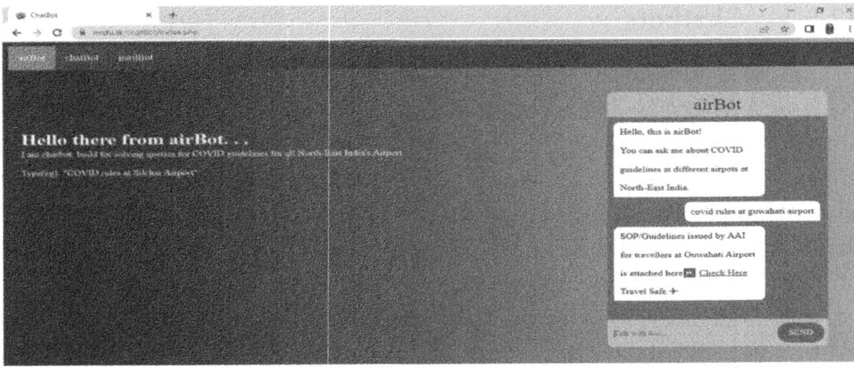

Figure 12.2 Interfacing with the airBot.

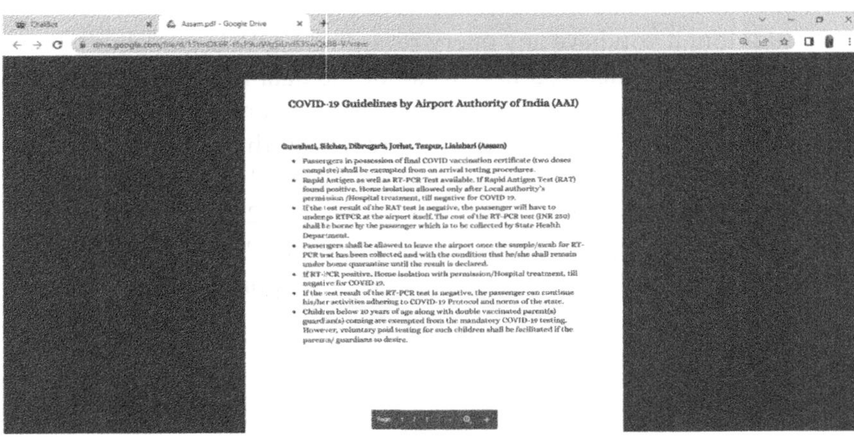

Figure 12.3 airBot showing SOP of Assam.

12.5.4 Mailbots

For people and passengers intending to communicate over email an MAILBOT was created by the same team that facilitated communication of the current states standard operating procedure, better known as SOP through email. The screenshot of the email is as given in Figure 12.4.

Where we need to fill the text field "To" with the mail ID of the person to whom you want the MailBot to response, then you have to type your query in place of "send your query to MailBot." After that, we need to click on the send button placed at the bottom (Figure 12.5).

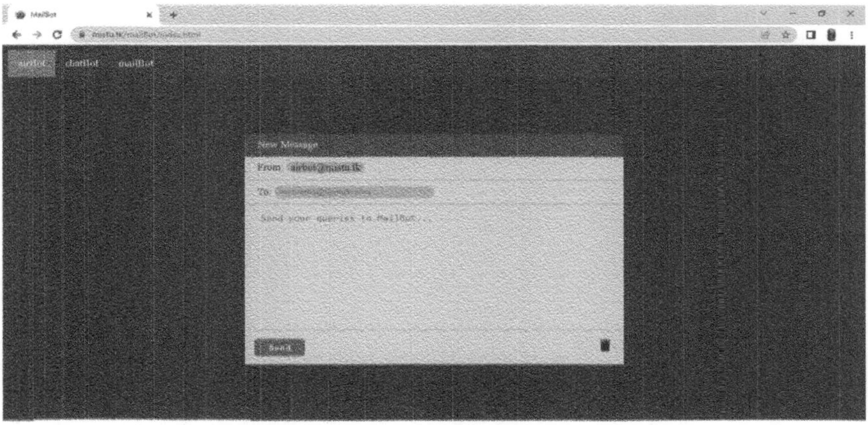

Figure 12.4 Mailbot interface.

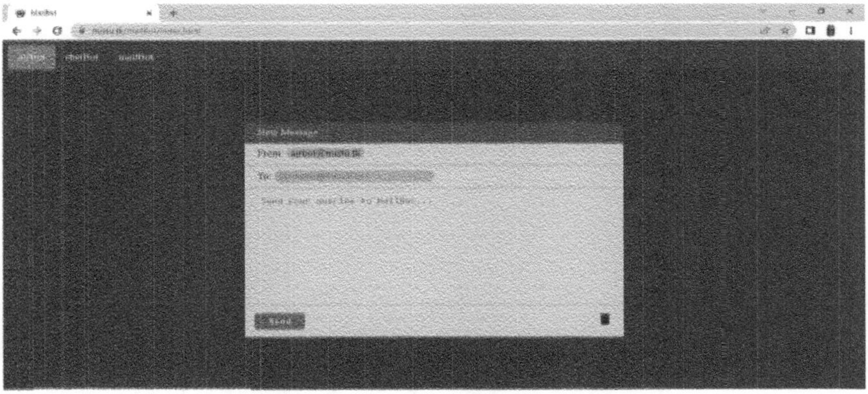

Figure 12.5 Working of MailBot.

Once we click on the send button, the query is send from the mailbot, and, as a result, a response to that query is send back to the mail ID that is provided by the user and to confirm that a confirmation message will be displayed to the screen (Figure 12.6).

Once the processing is over, the user will receive an email in the provided email ID (Figure 12.7).

After opening the mail, you can see that the mail contains an URL and a PDF. Now, after clicking on the URL or the PDF, you will be redirected to your expected outcome, i.e., the COVID guidelines of the desired state (Figure 12.8).

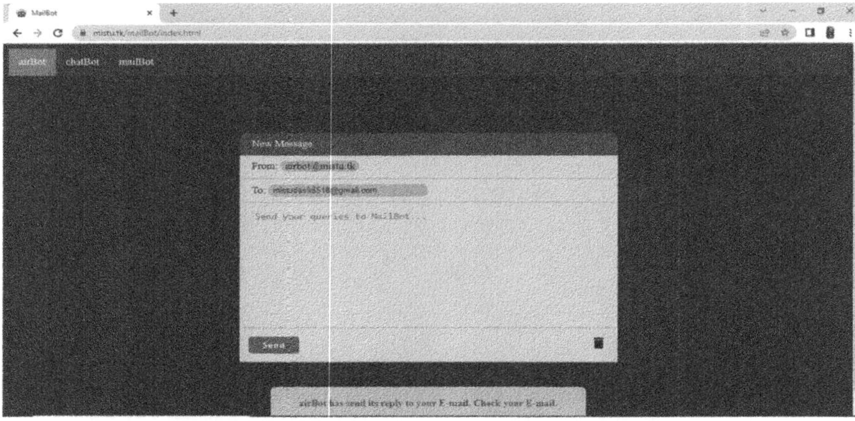

Figure 12.6 Response of MailBot.

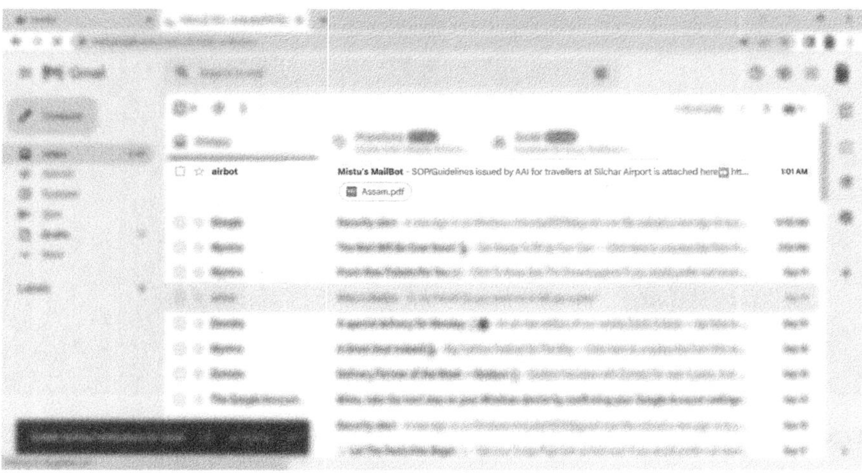

Figure 12.7 MailBot response to user's email.

12.5.5 Development of the Bots

During development of these two applications, the attempt was made to develop the Bots by using BrainShop api that helps us to train the brain or knowledgebase of the bot, so that we can get the desired output. In particular, during unknown pandemics like COVID-19, our work aims to provide AI information collecting chatbots and MailBots for travelers. During catastrophic epidemics, the described AI bot will have a significant influence on passenger safety.

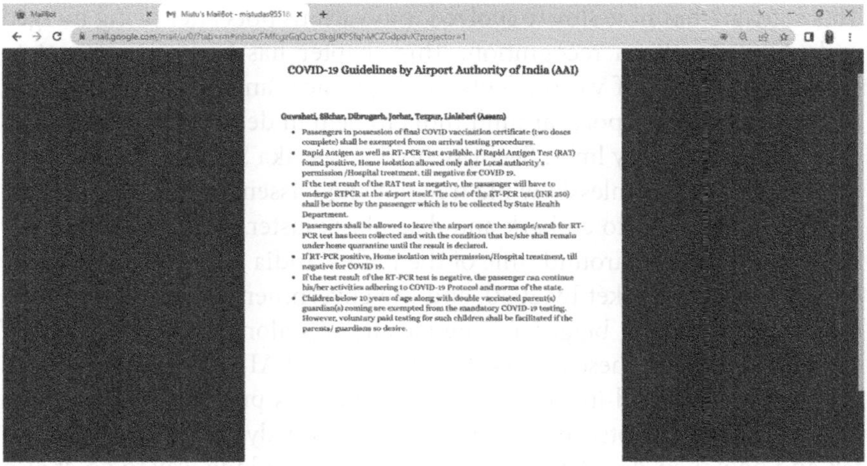

COVID-19 Guidelines by Airport Authority of India (AAI)

Guwahati, Silchar, Dibrugarh, Jorhat, Tezpur, Lialabari (Assam)

- Passengers in possession of final COVID vaccination certificate (two doses complete) shall be exempted from on arrival testing procedures.
- Rapid Antigen as well as RT-PCR Test available. If Rapid Antigen Test (RAT) found positive, Home isolation allowed only after Local authority's permission /Hospital treatment, till negative for COVID 19.
- If the test result of the RAT test is negative, the passenger will have to undergo RTPCR at the airport itself. The cost of the RT-PCR test (INR 250) shall be borne by the passenger which is to be collected by State Health Department.
- Passengers shall be allowed to leave the airport once the sample/swab for RT-PCR test has been collected and with the condition that he/she shall remain under home quarantine until the result is declared.
- If RT-PCR positive, Home isolation with permission/Hospital treatment, till negative for COVID 19.
- If the test result of the RT-PCR test is negative, the passenger can continue his/her activities adhering to COVID-19 Protocol and norms of the state.
- Children below 10 years of age along with double vaccinated parent(s) guardian(s) coming are exempted from the mandatory COVID-19 testing. However, voluntary paid testing for such children shall be facilitated if the parents/ guardians so desire.

Figure 12.8 COVID-19 guidelines of AAI, Assam.

12.6 Future Bots in Aviation

The current bots have given us an idea into what bots of the future will look like and what they will be capable off. Apron Bots that are capable of pushing the aircrafts from air-side aprons to the taxiways/runways are already making their approaches known and are creating huge savings of fuel and money. Current bots with scanning technologies are capable of isolating luggage with suspicious contents from those without. Future bots may take up the responsibility of not only safety and security but also fire and disaster management.

12.7 Conclusion

The first part of this chapter has highlighted the use of various software that were developed decades ago to the most recent ones. Software that is helping not only in machine translation but also in maintenance of aircrafts has been discussed here. Various companies around the globe are developing maintenance repair and overhaul software for helping the maintenance crew. The first part of the chapter has tried to explain on how the brain of the current bots like "TIMI" was developed and how they evolved. The brain was nothing else but the various software developed by researchers. These software and data repositories are both open source and commercially available. These span from machine translation to suggestive AI.

This software has now slowly evolved into various bots after being merged with AI and pattern recognition. This chapter has been an attempt to highlight the uses of various bots, both physical and software ones that are being used in airports around India to help in decongestion of airport check in and security lines. Security hold areas, aka SHA, are being made more accessible seamlessly while free flowing passenger lines are making it easier for airlines to embark and disembark faster. This step drastically takes down the turnaround time of the flights. India is projected to be the biggest aviation market by 2030, and the government is leaving no stone unturned in making bigger and better airports along with expansion of existing terminals. These are assisted by NLP- and AI-enabled bots in both physical domain and in the web. The chapter has projected how various web bots, security bots, and terminal bots are already making their impact felt and how their days have just started and have a long way to go. It also presents a golden opportunity to researchers and young minds working in the field of NLP, AI, and pattern recognition to build better and intelligent bots for the aviation Industry. In the foreseeable future, India will be leading the MRO and airlines industry, so it is a bus that should not be missed by researchers and technical minds alike.

References

1. Paul, S., Purkaystha, B.S., Das, P., NLP Tools Used In Civil Aviation: A survey-2018. *Int. J. Adv. Res. Comput. Sci.*, 9, 2, 110–114, March-April 2018, ISSN: 0976-5697.
2. OpenNMT, 2016, Available:https://opennmt.net/OpenNMT-py/main.html [Accessed: 2020/01/19].
3. AAI- Resources, 1995, Available: https://www.aai.aero/hi/system/files/resources/[Accessed: 2018/08/29].
4. DGCA reports, 1937, Available: http://dgca.nic.in/accident/reports/contents_acc_rep.html [Accessed: 2018/07/18].
5. I CAO Documents, 1947, Available: https://www.icao.int/safety/airnavigation/AIG/Documents /Forms /AllItems.aspx [Accessed: 2018/08/02].
6. NTSB, 1967, Available: https://www.ntsb.gov/_layouts/ntsb.aviation/index.aspx.[Accessed:2018-07-19].
7. ASRS-Reports, 1987, Available: https://asrs.arc.nasa.gov/search/reportsets.html[Accessed: 2018-01-28].
8. AAI MATS, 1995, Available: https://www.aai.aero/en/system/files/resources/1%20Commercial%20Manual%202019.pdf [Accessed: 2018/11/27].
9. Duncan, 2016, Available: https://www.faa.gov/regulations_policies/handbooks_manuals/aviation/phak [Accessed: 2021/12/07].

10. Dictionary-Aeronautical English, 2007, Available:https://www.theairlinepilots.com/forumarchive /aviation-regulations/aviationdictionary.pdf, eISBN-13: 978-1-4081-0226-8 [Accessed: 2019/03/27].
11. Isabelle, P. and Bourbeau, L., TAUM-AVIATION: its Technical Features and some experimental Results. *Comput. Linguist.*, 11, 1, 18–27, 1985.
12. IBM WATSON, 2010, Available: https: //www.ibm.com/watson/stories/aviation.html [Accessed: 2017/10/10].
13. NLP in AIRBUS, 2016, Available: http://www.intelligent aerospace.com/pt/2016/10/21/applying-natural-learning-processing-to-airbus-helicopters.html [Accessed: 2018/02/27].
14. NLP in MRO, 2015, Available: http://aviationweek.com/aviation-maintenance-andsupport-software/applying-natural-learning-processing-airbus helicopters [Accessed: 2017/12/02].
15. Clark, P. and Harrison, P., Boeing's NLP System and the Challenges of Semantic Representation. *Proceedings of the Conference on Semantics in Text Processing*, September 22 -24, WA 98124, ACL, STEP '08, Venice, Italy, pp. 263–2, 2008.
16. ECCAIRS, 2004, Available:http://eccairsportal.jrc.ec.europa.eu/index.php?id=2 [Accessed: 2017/12/21].
17. FS-Toolkits, 2012, Available: https://flightsafety.org/toolkits-resources/past-safetyinitiatives/global-aviation-safety-network-gain/global-information-sharingsystems/ [Accessed: 2017/12/15].
18. AI-Aviation Resources, 2018, Available: https://www.marketsandmarkets.com/Market-Reports/artificial-intelligencein-aviation-market-106037016.html [Accessed: 2018/11/20].
19. EU Accident Reports, 1989, Available: https://ec.europa.eu/jrc/en/scientific-tool/eccairs-european-central-repository-aviation-accident-and-incidentreports [Accessed: 2018/01/15].

Part 3
CHALLENGES

13

Cobot-Aided System for Hydroponically Grown Plants

Oliver Joel G.[1], B. Jafrin Rosary[2], Vallidevi Krishnamurthy[2*], Sujithra @ Kanmani[2] and Surendiran Balasubramanian[3]

[1]*School of Mechanical Engineering (SMEC), Vellore Institute of Technology, Chennai, India*
[2]*School of Computer Science and Engineering (SCOPE), Vellore Institute of Technology, Chennai, India*
[3]*Department of Computer Science and Engineering, National Institute of Technology, Puducherry, India*

Abstract

Plants are widely acknowledged as being a crucial component of a sustainable world, capable of sustaining humans through various challenges in life. They typically require sufficient amounts of sunlight, water, and nutrient-rich soil to flourish. However, can plants survive without these prerequisites? This chapter discusses "hydroponics," i.e., the subset of horticulture that deals with soilless growing methods, and how it can aid plant growth. Additionally, Cobots are utilized to simplify the process and provide further assistance to plant survival. Ultimately, hydroponics and Cobots work in conjunction to create an efficient and effective system for growing plants. Existing traditional farming systems can lead to soil degradation, reduced crop yields, and water scarcity due to high water usage for irrigation. Additionally, the use of chemicals such as fertilizers and pesticides can contribute to environmental pollution and harm wildlife. Moreover, traditional farming methods often include laborious working hours for the farmers, which might lead to poor health. Cobot-aided hydroponics farming offers numerous advantages that can improve the productivity, sustainability, and efficiency of farming while reducing labor costs and environmental impact.

Keywords: Cobot, hydroponics, hydroculture, IoT, automation, smart agriculture, sensors, collaborative robot

**Corresponding author*: vallidevi.k@vit.ac.in

V. Ramasamy, S. Balamurugan and Sheng-Lung Peng (eds.) Intelligent Robots and Cobots: Industry 5.0 Applications, (269–290) © 2025 Scrivener Publishing LLC

13.1 Introduction

Hydroponics is a subset of hydroculture. It is a different way of growing plants, predominantly crops, in a soil-free environment. There have been numerous methods that go with the traditional environment and use automation for watering the plants and taking care of them [12], but adding nutrients and changing the water levels of the plant based on its growth is a hard task in the traditional environment, and that is where hydroponic system plays a huge role. This method involves growing the required plants in nutrient-rich water. There are different ways to approach the system itself. However, every system consists of the important elements, namely:

1. **Water:** the water in a hydroponic system is allowed to flow in pipes with holes. This water is continuously monitored to check for the ideal pH levels and is adjusted accordingly.
2. **Root support:** the plant roots are anchored to the pipes using some material like coconut fiber and rock wool. These serve the functionality better than poor water-retaining compounds like gravel or sand.
3. **Nutrients:** every plant requires a different set of nutrients to reach optimal growth levels. These nutrients are fed to the plants through the flowing water and are absorbed by the roots.
4. **Oxygen:** different oxygenating compounds may be mixed with the water to fulfill the oxygen needs of the hydroponically grown crop.
5. **Light:** the sunlight required for photosynthesis is artificially replicated using special lighting techniques to facilitate the plants' growth.

Hydroponics as a method ensures that crops can be grown anywhere in the world, at any time of the year. It takes away from the uncertainty that conventional agriculture poses, by giving the farmer more control of the crop growth; it eliminates the unpredictability caused by weather conditions and animal damage.

13.1.1 Advantages of Using Hydroponics

- Space constraint is eliminated using this technique.

- The yield, when optimized, tends to show better results than conventional agricultural methods.
- The possibility of soil-borne plant diseases is eliminated.
- Water is saved using hydroponics.
- Hydroponically grown plants grow faster than soil-grown plants.

13.1.2 Use of IoT in Hydroponics

"Internet of Things," or IoT, can be described as an interconnection of sensors, both mechanical and digital, that can sense and transfer data over a network of devices. The hydroponically grown plants each require the flow of nutrients to be at a favorable pH and electrical conductivity (EC) to reach optimal growth. These parameters keep changing as the plants absorb the nutrients from the continuously flowing stream of water. Thus, it is imperative to constantly monitor these values. When done manually, this task is redundant and tedious. Therefore, using pH and EC sensors, levels can be recorded periodically and the data can be pushed into a cloud, which can send alerts on when the nutrients in the water need to be replenished.

13.1.3 Role of Cobots in an IoT-Based Hydroponic System

The utilization of Cobots in an IoT-based hydroponic system is paramount. The tireless and precise nature of a Cobot makes it an ideal assistant for various tasks such as loading plants, sensing pH and EC levels, and regulating nutrient levels. The efficiency of these tasks can be enhanced by training the Cobots through reinforcement learning, resulting in increased production output and elimination of human errors. Despite being previously limited to agricultural tasks such as milking cows, precision seeding, harvesting, fertilizing, and irrigating, this research demonstrates the potential benefits of using Cobots in hydroponics. Furthermore, it covers the process of data gathering from pH and EC sensors to regulate nutrient levels in the water.

13.2 Hydroponic System

The word "hydroponic" was coined by using the root words "hydro," meaning "water," and "ponos," meaning "labor," implying that no soil is to be used in this system. So, in a hydroponic system, nutrient solutions are used instead of soil-based environments.

13.2.1 Types of Hydroponic Systems

There are six types of hydroponic systems as shown in Figure 13.1.

i. **Nutrient Film Technique (NFT)**
In this technique, the nutrient solution will be collected and reused after it reaches past the bare roots of plants through a watertight channel.

ii. **Deep Water Culture (DWC)**
In this technique, the plant's roots are suspended in nutrient-rich, oxygenated water that is contained in a rectangular tank.

iii. **Wick System**
In this technique, a soft cotton or nylon string (often called a wick) absorbs the nutrient solution and supplies it to the plants in trays.

iv. **Ebb and Flow**
In this technique, the nutrient solution is pumped periodically into the plant tray. It is then absorbed by the plants, and the remaining solution is collected in the reservoir.

v. **Drip Systems**
It is also called the trickle system or the micro-irrigation system. In this technique, a drip is used to feed nutrients to the plants from the reservoir. The flow rate of the nutrients can

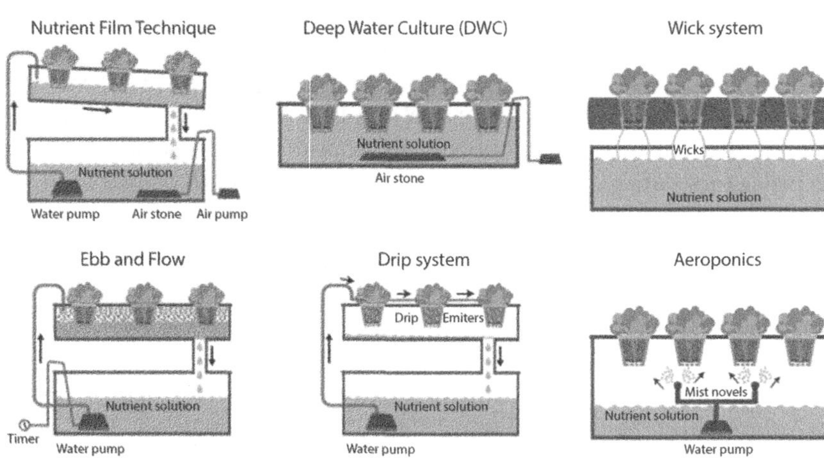

Figure 13.1 Types of hydroponic systems [1].

be adjusted with this technique; hence, it can be used in any type of plant.

vi. **Aeroponics**
   The primary benefit of this system is aeration. In this technique, a mist of the solution is sprayed at regular intervals. When compared to other systems, this one consumes a lesser amount of nutrients.

13.3 Cobot

A Cobot, short for collaborative robot, is a robot designed to work hand in hand with humans. These robots are used primarily for repetitive tasks such as inspection and material handling, to help humans focus on tasks that require problem-solving skills. Because they operate in close proximity to humans, they work at a comparatively lower speed than traditional robots and have inbuilt safety features.

13.3.1 Types of Cobots

In production, collaborative robots are becoming more and more common. Let us look at the four different ways that collaborative robots can operate as shown in Figure 13.2.

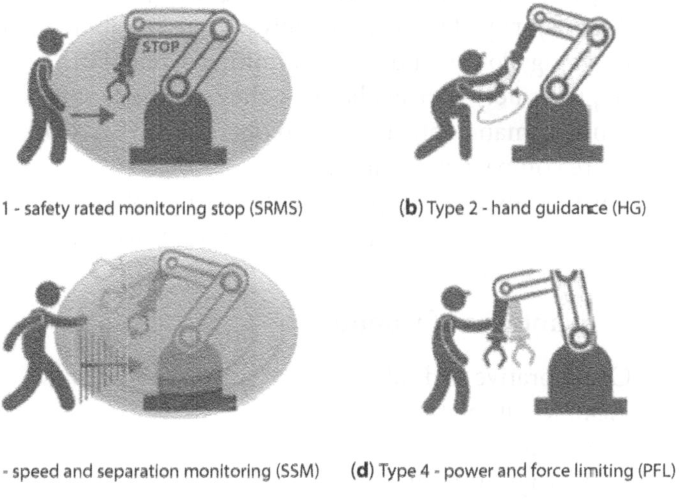

(**a**) Type 1 - safety rated monitoring stop (SRMS) (**b**) Type 2 - hand guidance (HG)

(**c**) Type 3 - speed and separation monitoring (SSM) (**d**) Type 4 - power and force limiting (PFL)

Figure 13.2 Types of Cobots [2].

(a) **Type 1: Safety-Rated Monitoring Stop (SRMS)**
SRMS Cobots recognize when a person enters a specified area and halt operation until the person departs that communal workspace, doing away with the requirement to manually restart a robot. When the person leaves, the robot resumes its work on its own. These are typically employed when human–robot contact is less intense.

(b) **Type 2: Hand Guidance (HG)**
To allow a person to manually direct or move the robot, hand-guiding Cobots include a safety-rated device attached to the end of their arm. Robot courses and positions can be readily and quickly programmed by hand using hand direction. It is perfect for mobile applications because Cobots must constantly change jobs and require reprogramming because they move between stations and have to learn new duties.

(c) **Type 3: Speed and Separation Monitoring (SSM)**
When there is little interaction with a person yet a person needs easy access to the robot, Speed and Separation Monitoring Cobots, sometimes known as "fenceless" Cobots, operate effectively. To create zones close to the robot's workspace, safety laser area scanners are frequently employed. The robot's movements slow down when a person approaches and eventually cease if the person is within reach of the workstation. The robot automatically resumes its usual speed as soon as the operator leaves the workspace.

(d) **Type 4: Power and Force Limiting (PFL)**
The majority of people probably picture collaborative robots as being Power and Force Limiting Cobots. They do not require safety barriers because they are made specifically to sense human contact and stop operating. They work well in more compact applications.
For the next three types of Cobots, any robot can be made into a Cobot by integrating safety devices with it.

13.3.2 Advantages of Cobots

- Collaborative and safe
- Flexible automation
- Fast setup
- Easy programming
- Consistency and accuracy

13.4 Cobot-Aided Hydroponic System

13.4.1 Workflow of the Proposed Cobot-Aided Hydroponic System

The basic components that make up a Hydroponic Farm consist of a light source, nutrient solution (different nutrient-rich solutions for different plants), and the plants interconnected as shown in Figure 13.3. Every plant type has a pH solution that provides maximum output (in this case, plant growth).

Sensors will be connected to the Hydroponic Farm using IoT concepts. The role of the sensors is to monitor the growth of the plant and if the pH level and EC of the nutrient solution need alteration. These data are then collected and stored in a database through a software application. Based on this, the data are analyzed.

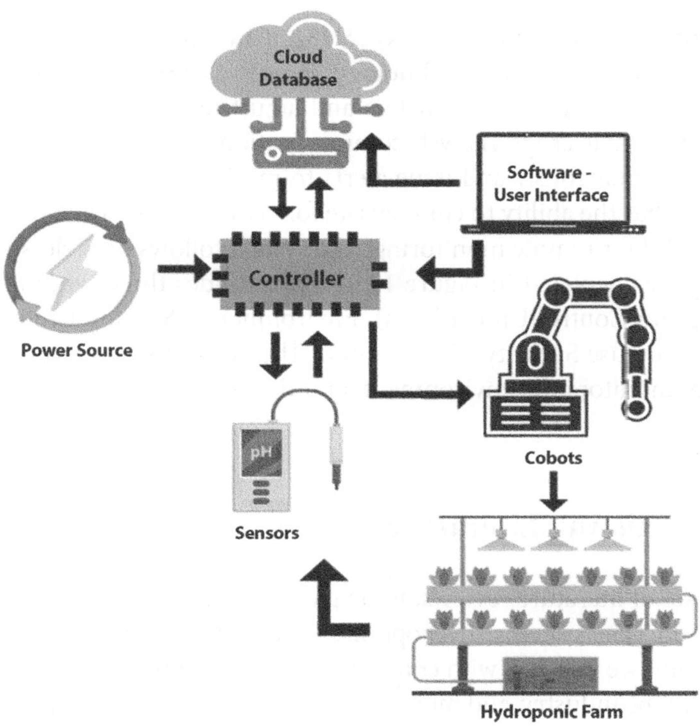

Figure 13.3 Workflow of the proposed Cobot-aided hydroponic system.

The analyzed data are then received by the Cobot. So, every time a plant needs to be loaded/re-loaded into another channel, the Cobots will get the corresponding commands to do the same. This process continues.

13.4.2 Novelty and Contributions

A Cobot-aided hydroponic system is a novel approach that combines Cobots with hydroponics to automate and optimize the process of growing plants. The system offers several contributions and benefits, including the reduction of water and nutrient waste, and increased food security. The concept of combing the repetitive nature of Cobots and the soilless farming concept of hydroponics has not been dealt with in such depth before. This system can contribute a lot to the farming and food industries, like consistency in quality and increased crop yields.

13.5 iPONICS: IoT-Based Monitoring and Control System for Hydroponics Greenhouses

A recent study introduces iPONICS [3], a hydroponic monitoring and control system using IoT technology. It uses a wireless sensor network to monitor hydroponic parameters and adjust pumps. Growers can easily monitor their crops via web communication, monitor multiple greenhouses simultaneously and issue alerts to minimize human intervention. iPONICS has the ability to compensate for errors to a high standard, making it useful for remote monitoring. The system follows a wireless network topology (as depicted in Figure 13.4) that includes three types of nodes: Sensing and Control Units (SCUs), Environmental Sensing Units (ESNs), and Greenhouse Security Units (GSNs). The SCU acts as the central core, the ESN monitors the environment, and the GSN provides basic security measures.

13.6 Existing Literature

In this set of literature reviews, we explore various topics related to modern production systems, hydroponics, and human–robot collaboration. Specifically, we will deal with concepts such as smart hydroponics, usage of Cobots in the industry, and monitoring and control of hydroponic systems.

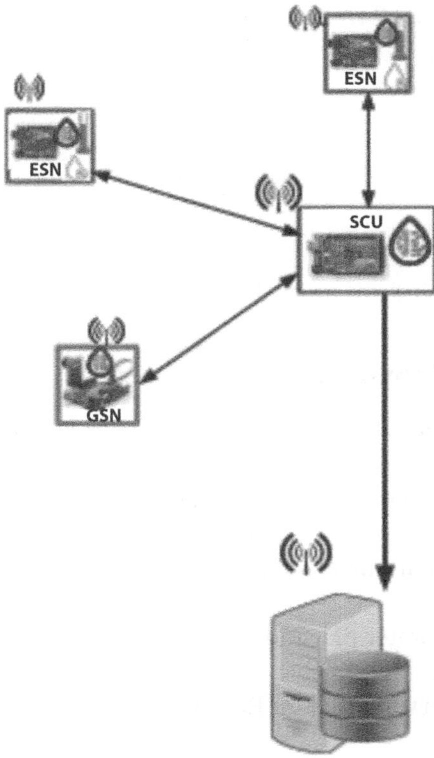

Figure 13.4 The topology of iPONICS WSN [3].

13.6.1 Smart Hydroponics

Automated hydroponic systems have long been studied in the literature. This is because automated systems offer high throughput and efficiency and have the advantage of not requiring specialized labor. Furthermore, the tasks performed in hydroponic systems can be easily automated, so the automation of hydroponics is not infeasible. Computers can do this better because they are more available compared to humans.

Smart hydroponics is divided into four main categories as given in Figure 13.5.

Automation of nutrient supply is mainly controlled by proportional control, which is notorious for vibration. The existing research work that has been done using AI showed that AI is a good technique but is costly and requires a lot of computing power. IoT has proven to be one of the most attractive forms of automation due to the flexibility it offers when designing systems.

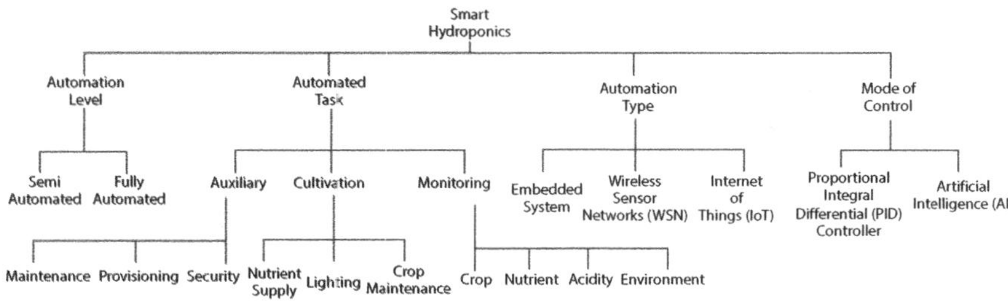

Figure 13.5 Taxonomy for smart hydroponic systems [4].

13.6.2 IoT in Hydroponic Systems

The temperature, pH, and conductivity of the water should be constantly monitored to determine what action to take. For this purpose, a Raspberry Pi, a credit card–sized mini-computer, and an Arduino-based IoT device with multiple sensor modules were used. Internet of Things technology could be used to monitor key parameters of all hydroponic systems.

a. Temperature sensor, water pH, and conductivity.
b. Arduino IoT module.
c. Database and web server based on Raspberry Pi Pocket PC [5].
d. Camera module helps detect plant illness and well-being [6].
e. Low-cost optoelectronic sensor that senses the radial growth of crops.

Figure 13.6 shows the architecture of the iPONICS hydroponic system.

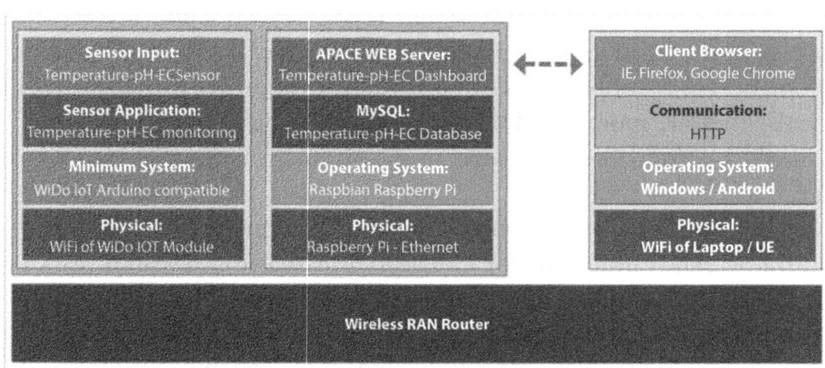

Figure 13.6 Hydroponic IoT system architecture [5].

This sensor detects alternating black and white narrow bars printed on reflective tapes which are installed encircling the fruit (or some other part of the plant). The movement of reflective tapes gets detected by the infrared sensor on the optoelectronic. This helps measure continuously and monitor plant growth with minimum maintenance [7].

13.6.3 Cobots in Industry 4.0

As part of the progressive future of robotics, Cobots have become safer and more efficient. It can perform various tasks. The configuration and programming process can be done in real-time on-site. The global demand for collaborative robots is increasing year by year. In the future, Cobots will become a key tool in industrial production processes. This optimizes processes, reduces production costs, and increases productivity [8]. Only transient collisions can occur, in the case of a human-hand collision, pressure, and force values are below the threshold [9]. Humans and Cobots should be seen less as a dichotomy but rather as human and nonhuman actors in a sociotechnical network that needs to be designed as such [10].

13.6.4 Human–Robot Collaboration (HRC) Applications

The adoption of state-of-the-art HRC (human–robot collaboration) applications has been observed across various industries, including automotive, metal, and plastic sectors. These industries possess significant expertise in integrating industrial robots and other automation solutions within their manufacturing facilities. Notably, there is a clear inclination toward utilizing HRC systems for assembly processes in all sectors. Among such processes, screwdriver activities have emerged as the dominant operation in these industries. Additionally, instances of material handling manipulation have been noted across industries, albeit with limited occurrences reported in HRC literature [11].

As anticipated, both research institutes and manufacturing companies are swiftly embracing the fenceless safety collision capabilities offered by collaborative robots (Cobots) for assembly and material handling tasks.

13.6.5 Literature Survey

Table 13.1 given below is a summary of a literature survey conducted for this field.

Table 13.1 Literature survey.

Paper title	Main topic	Methodology	Key findings
[2] Human factors in cobot era: a review of modern production systems features	Human factors in Cobot era	Literature review	Identified the main features of modern production systems and the challenges and opportunities for human–robot collaboration
[3] Reliable IoT-Based Monitoring and Control of Hydroponic Systems	IoT-based monitoring and control of hydroponic systems	Case study	Developed an IoT-based system for monitoring and controlling hydroponic systems and demonstrated its reliability and effectiveness
[4] A survey of smart hydroponic systems	Survey of smart hydroponic systems	Literature review	Summarized the different types of smart hydroponic systems, their features, advantages, and limitations
[5] Development of IoT at hydroponic system using raspberry Pi	IoT at hydroponic system using Raspberry Pi	Case study	Developed an IoT-based system using Raspberry Pi for monitoring and controlling a hydroponic system

(*Continued*)

Table 13.1 Literature survey. (*Continued*)

Paper title	Main topic	Methodology	Key findings
[6] IoT based growth monitoring system of guava (Psidium guajava L.) Fruits	IoT-based growth monitoring system of guava fruits	Case study	Developed an IoT-based system for monitoring the growth of guava fruits and demonstrated its effectiveness
[7] An IoT-Based Smart Plant Monitoring System	IoT-based smart plant monitoring system	Case study	Developed an IoT-based system for monitoring plant growth and health and demonstrated its effectiveness
[8] Cobots and the benefits of their implementation in intelligent manufacturing	Benefits of Cobot implementation in intelligent manufacturing	Literature review	Reviewed the benefits of Cobots in different stages of the manufacturing process
[9] Cobots in industry 4.0: A roadmap for future practice studies on human–robot collaboration	Cobots in Industry 4.0	Literature review	Reviewed the state-of-the-art research on Cobots in Industry 4.0 and proposed a roadmap for future practice studies
[10] Complementary Manipulator Tool Development for Safe Cobot-Assisted Hydroponics	Cobot-assisted hydroponics	Case study	Developed a complementary manipulator tool for safe Cobot-assisted hydroponics

(*Continued*)

Table 13.1 Literature survey. (*Continued*)

Paper title	Main topic	Methodology	Key findings
[11] Human–Cobot interaction fluency and Cobot operators' job performance. The mediating role of work engagement: A survey	Human–robot interaction fluency and job performance	Survey	Investigated the relationship between human–robot interaction fluency and Cobot operators' job performance
[12] Design for an IoT based automated plant watering system	IoT-based automated plant watering system	Design	Proposed an object-oriented design for an IoT-based automated plant watering system
[13] The digitization of agricultural industry–a systematic literature review on agriculture 4.0	Digitization of agricultural industry	Literature review	Reviewed the state-of-the-art research on Agriculture 4.0 and its implications for the digitization of the agricultural industry
[14] Comparative Assessment of Environmental/Energy Performance under Conventional Labor and Collaborative Robot Scenarios in Greek Viticulture	Collaborative robot scenarios in Greek viticulture	Case study	Compared the environmental/energy performance of conventional labor and collaborative robot scenarios in Greek viticulture

(*Continued*)

Table 13.1 Literature survey. (*Continued*)

Paper title	**Main topic**	**Methodology**	**Key findings**
[15] A Review of Hydroponics and Conventional Agriculture Based on Energy and Water Consumption, Environmental Impact, and Land Use	Hydroponics and conventional agriculture	Literature review	Compared hydroponics and conventional agriculture based on energy and water consumption, environmental impact, and land use
[16] Industry 5.0: Analysis, Applications and Prognosis	Industry 5.0	Literature review	Reviewed the state-of-the-art research on Industry 5.0, its applications, and its prognosis

13.7 IoT-Cobot Integrated System Architecture

The IoT-Cobot Integrated System can be mainly divided into three sub-categories as shown in Figure 13.7, namely:

(a) **Automation System**
The Automation System consists of all the IoT sensors, namely, pH, light, temperature, humidity, and the app "Ponix," which retrieves data from both the reservoir and the farm and helps to convey the information to the worker present and also initiate the work of Cobots when and where it is needed.

(b) **Hydroponic Farm**
The Hydroponic Farm consists of the plants being kept in different water levels based on their growth, type of plant, and other requirements as shown in Figure 13.1.
The nutrient-rich water supply is connected directly to the roots of the plants on the farm.
The Cobot plays an important role in unloading and re-loading the plants in different water levels once they reach the required growth level in their present position.

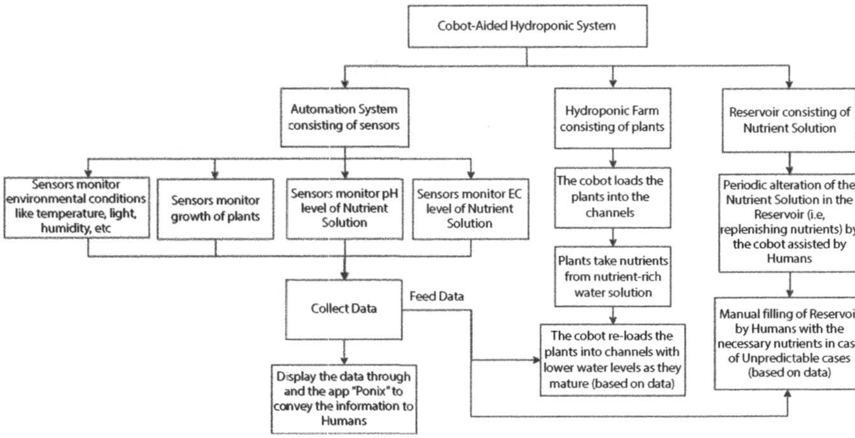

Figure 13.7 Architecture of the integrated system.

(c) Reservoir

The reservoir consists of a nutrient-rich water solution, which is monitored by the sensors and the nutrients are added periodically by the Cobot so that the amount of nutrients do not get depleted.

13.7.1 Data Collection Using IoT Sensors

The IoT is a system of physical objects that contain various technologies like sensors and are interconnected through the internet. Hence, they can exchange information with each other and other systems over the internet.

- Now, what is Data Collection in IoT and how does it work? Sensors that are a part of the IoT network collect real-time data regarding the target's performance and the updates in the IoT network.
- There are three types of Data Collection using IoT sensors:
  - ➢ Equipment Data Collection: This is mainly used to monitor the status of the objects connected through the IoT network. The data are used in Predictive Maintenance.
  - ➢ Submeter Data Collection: This is mainly used to automate the process of computing the usage of a single utility in a system with multiple users.
  - ➢ Environmental Data Collection: This is the monitoring and collection of environmental data like heat, temperature, and humidity.

In this Cobot-aided hydroponic system, the IoT sensors will monitor the pH level, the EC, and the growth of the plants in the channel.

Every plant will have pH level and EC conditions where it grows and performs best. So, the job of the IoT sensors is to monitor and check if the preferred conditions are met and pass on the information to the humans via an application/UI.

In some hydroponic systems, once the plant has reached a certain level of growth, i.e., maturity, it will have to be removed from its current channel and reloaded into a different channel with lesser water levels. For this purpose, the IoT sensors will have to monitor the growth of the plants, and, in the same manner, as stated above, they will pass on the information to the humans via an application/UI.

13.7.2 Data Processing and Algorithm to Determine the Point of Regulation

Data processing is the method by which the raw data (which is not useful by itself) is processed and analyzed and is thus made into useful data. Using this data, in this case, the data can be collected from the Hydroponic Farm. After analyzing the different attributes and values of the plant, the growth factor can be determined, i.e., if the plant has matured enough to be reloaded into another channel. Based on this, the Cobot will reload the matured plants into different channels with lower water levels

If the resources in the reservoir are being used rapidly, then humans will have to re-fill/replenish the resources in the reservoir so that the Cobot can use it to again fill the channels with the nutrient solution. The Cobot can also concentrate the solution with any nutrient, if required.

All of this information has to be made visible to humans in a readable manner using a mobile application (or any user-friendly user interface).

The process flowchart has been clearly detailed in Figure 13.8.

13.7.3 Receipt of the Trigger

"Ponix" is an app that will act as the storage unit for all the data collected by the sensors connected through the IoT network. When the demands for the preferred conditions for the favorable environment of a plant are not met, an E-mail (essentially acting as a receipt of the trigger) will be sent to the owner of the app with an alert message consisting of all the details (like "pH value lesser than it should be"). Normally, Ponix will display the status of the Hydroponic Farm with all the details about the data collected by the sensors, but no E-mail will be sent.

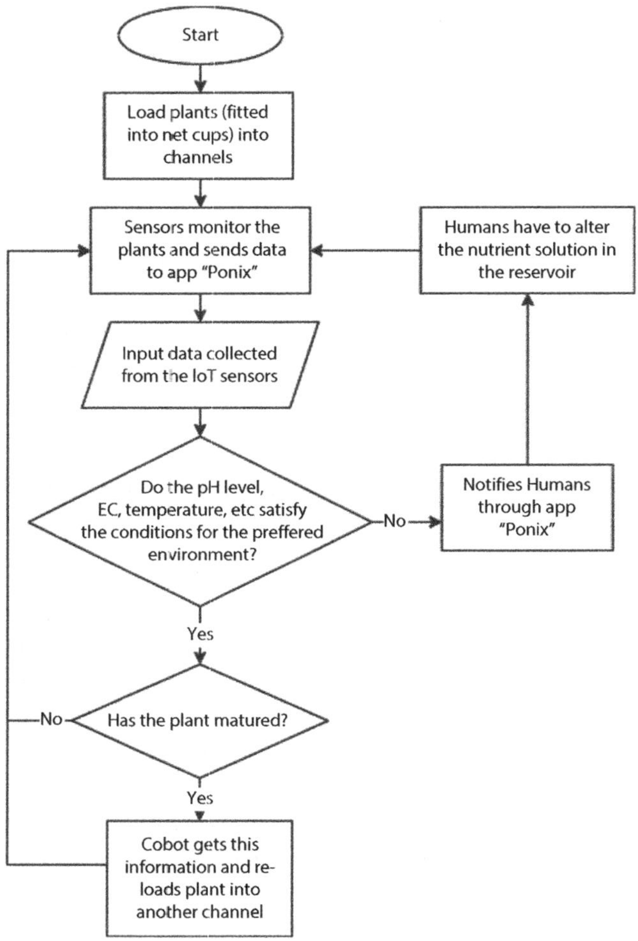

Figure 13.8 Flowchart of the algorithm used in the proposed system.

13.7.4 Regulation by the Cobot

As explained previously, when the Cobot gets the information that the plants in a certain channel have matured, it unloads the plant and reloads into another channel with a lower water level in the same system. Then, it will methodically and periodically reload the matured into different channels with lower water levels. This is done in some cases so that the previously vacant spots can be re-loaded with new saplings. Cobots will be used for loading the new saplings too, making the process of loading, unloading, and reloading the plants the main regulation point of the Cobot.

When the nutrient solution starts losing the nutrients little by little, the Cobots have to carry out the task of concentrating the nutrient solution with the required nutrients, so that the nutrient-rich solution can flow from the reservoir to the channels and the plants can now take in the nutrients that were depleted.

In the case of nutrient solution getting depleted, humans will have to work collaboratively with the Cobot to get the work done. Humans will have to re-fill/replenish the resources so that they can be used again to fill the channels with the nutrient solution.

13.8 Results

A short survey was conducted, where the respondees were given a questionnaire on the concept of the Cobot-aided hydroponic system. The majority of the responses received were positive. Integrating automated systems into the proposed system received many positive responses, saying that it could increase the efficiency of the proposed system. Many of the respondees found this to be a very futuristic and helpful approach.

The advantages of the proposed system over the existing system have been summarized in the Table 13.2 given below.

Thus, the Cobot-aided hydroponic system offers numerous advantages over the traditional system, making it a very helpful concept for hydroponic workers to utilize in their farms. Figure 13.9 shows a pie chart of the survey takers' suggested improvements of the proposed system.

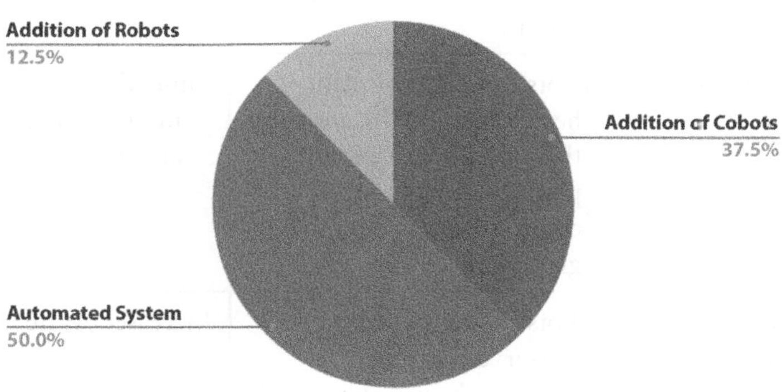

Figure 13.9 A pie chart of the responses to the question "How can this system be improved?".

Table 13.2 The advantages of the proposed Cobot-aided hydroponic system over the existing system.

Advantages	Cobot-aided hydroponic system	Existing system
Efficiency	Cobots can perform repetitive tasks with high accuracy and speed, allowing for faster and more efficient harvesting, planting, and maintenance.	Limited efficiency due to the reliance on manual labor.
Safety	Cobots can perform tasks that are hazardous or require heavy lifting, protecting human workers from injuries and strain.	Increased risk of injuries and strain for human workers due to manual labor.
Cost-effectiveness	Cobots can be programmed to perform multiple tasks, reducing the need for specialized equipment and labor, ultimately lowering costs and increasing profitability.	Higher costs due to the need for specialized equipment and labor.
Flexibility	Cobots can be easily reprogrammed to perform different tasks, allowing for greater flexibility in the operation of the hydroponic system.	Limited flexibility due to the reliance on manual labor.
Plant Growth	Cobots can monitor plant health and growth, ensuring that plants receive the proper nutrients and environment for optimal growth.	Limited ability to monitor plant health and growth.
Environmental Impact	Cobots can reduce waste, conserve water, and decrease the use of harmful chemicals, resulting in a more environmentally sustainable operation.	Increased waste, water usage, and harmful chemicals can have a negative environmental impact.

13.9 Conclusion

The Cobot-aided hydroponic system offers numerous advantages over traditional hydroponic systems, like increased efficiency, accuracy, and safety. The integration of collaborative robots into hydroponic systems allows for better monitoring and control of plant growth, leading to higher yields and crops of better quality. Additionally, Cobots can perform tedious and repetitive tasks, reducing labor costs and freeing up workers for other tasks. Future works regarding this system may include the integration of extra safety measures into the Cobot's mechanism, research about the different types of crops that can be benefitted from the system, etc. But an interesting addition to the proposed system can be the integration of automated systems, which will take care of the menial jobs, and will do almost the work for the workers. This will leave room for only a little human intervention, making this a very futuristic approach. Overall, Cobot-aided hydroponic systems provide a promising solution for sustainable agriculture and food production in the future.

References

1. Hydroponic system, Byjus, 2020, https://byjus.com/neet/hydroponic-system/.
2. Faccio, M., Granata, I., Menini, A., Milanese, M., Rossato, C., Bottin, M., Rosati, G., Human factors in cobot era: a review of modern production systems features. *J. Intell. Manuf.*, 34, 1–22, 2022.
3. Tatas, K., Al-Zoubi, A., Christofides, N., Zannettis, C., Chrysostomou, M., Panteli, S., Antoniou, A., Reliable IoT-Based Monitoring and Control of Hydroponic Systems. *Technologies*, 10, 1, 26, 2022.
4. Modu, F., Adam, A., Aliyu, F., Mabu, A., Musa, M., A survey of smart hydroponic systems. *Adv. Sci. Technol. Eng. Syst. J.*, 5, 1, 233–248, 2020.
5. Lukito, R.B. and Lukito, C., Development of IoT at hydroponic system using raspberry Pi. *TELKOMNIKA (Telecommunication Comput. Electron. Control)*, *17*, 2, 897–906, 2019.
6. Slamet, W., Irham, N.M., Sutan, M.S.A., IoT based growth monitoring system of guava (Psidium guajava L.) Fruits. *IOP Conf. Ser.: Earth Environ. Sci.*, 147, 1, 012048, IOP Publishing, 2018, May.
7. Athawale, S.V., Solanki, M., Sapkal, A., Gawande, A., Chaudhari, S., An IoT-Based Smart Plant Monitoring System, in: *Smart Computing Paradigms: New Progresses and Challenges*, pp. 303–310, Springer, Singapore, 2020.
8. Galin, R., Meshcheryakov, R., Kamesheva, S., Samoshina, A., Cobots and the benefits of their implementation in intelligent manufacturing. *IOP Conf. Ser.: Mater. Sci. Eng.*, 862, 3, 032075, IOP Publishing, 2020, May.

9. Weiss, A., Wortmeier, A.K., Kubicek, B., Cobots in industry 4.0: A road-map for future practice studies on human–robot collaboration. *IEEE Trans. Hum.-Mach. Syst.*, *51*, 4, 335–345, 2021.
10. Paniti, I., Nacsa, J., Szűr, D., Rácz, S., Tóth, J., Complementary Manipulator Tool Development for Safe Cobot-Assisted Hydroponics. *Hung. J. Ind. Chem.*, *49*, 2, 85–89, 2021.
11. Paliga, M., Human–Cobot interaction fluency and Cobot operators' job performance. The mediating role of work engagement: A survey. *Rob. Auton. Syst.*, *155*, 104191, 2022.
12. Rajagopal, S. and Krishnamurthy, V., OO design for an IoT based automated plant watering system, in: *2017 International Conference on Computer, Communication and Signal Processing (ICCCSP)*, 2017, January, IEEE, pp. 1–5.
13. Abbasi, R., Martinez, P., Ahmad, R., The digitization of agricultural industry–a systematic literature review on agriculture 4.0. *Smart Agric. Technol.*, 2, 1–24, 2022 Feb 24.
14. Tziolas, E., Karapatzak, E., Kalathas, I., Lytridis, C., Mamalis, S., Koundouras, S., Pachidis, T., Kaburlasos, V.G., Comparative Assessment of Environmental/Energy Performance under Conventional Labor and Collaborative Robot Scenarios in Greek Viticulture. *Sustainability*, 5, 1–21, 2023 Jan.
15. Pomoni, D.I., Koukou, M.K., Vrachopoulos, M.G., Vasiliadis, L., A Review of Hydroponics and Conventional Agriculture Based on Energy and Water Consumption, Environmental Impact, and Land Use. *Energies*, 16, 1–26, 2023 Feb 8.
16. Joglekar, S., Kadam, S., Dharmadhikari, S., Industry 5.0: Analysis, Applications and Prognosis. *Online J. Distance Educ. e-Learn.*, 11, 257–264, 2023 Jan.

14

Data Analytics and Collaborative Robots in Smart Territory: Research Methodology, Applications, and Open Challenges

Bipradeep Biswas[1], Sushruta Mishra[1]* and S. Balamurugan[2]

[1]*Kalinga Institute of Industrial Technology, Deemed to be University, Bhubaneswar, Odisha, India*
[2]*Albert Einstein Engineering and Research Labs (AEER Labs), Vice Chairman-Renewable Energy Society of India (RESI), Tamil Nadu, India*

Abstract
The integration of collaborative robots, known as cobots, with the Internet of Things (IoT), marks a new era where everyday items seamlessly connect to the web. This convergence envisions a future where previously isolated objects gain an online presence, resulting in a significant increase in connected devices. This surge in connections generates vast amounts of data, which aligns with the expansion of big data. Data analytics for IoT within the realm of cobots involves collecting, processing, and analyzing the copious data generated by IoT devices to extract insights and inform decision-making. Cobots and other IoT devices produce various data types, ranging from structured to unstructured, originating from sources like sensors and wearables. The integration of data analytics and these intelligent devices offers numerous opportunities for advancing big data and analytics within IoT systems. This paper explores the latest advancements in big data analytics for cobots, addressing the challenges researchers that face in both IoT settings and cobots. The discussion encompasses gaps in the analytics, research methodologies, and techniques for data management. Additionally, it outlines the IoT architecture for big data analytics, taxonomy, use cases, and applications of the devices. Lastly, it concludes with an examination of opportunities and prospects in the field of IoT data analytics for cobots.

***Keywords*:** Data analytics, Internet of Things, data, cobots, machine learning

**Corresponding author*: sushruta.mishrafcs@kiit.ac.in

V. Ramasamy, S. Balamurugan and Sheng-Lung Peng (eds.) Intelligent Robots and Cobots: Industry 5.0 Applications, (291–316) © 2025 Scrivener Publishing LLC

14.1 Introduction

The Internet of Things (IoT) has been growing rapidly, leading to a significant increase in the amount of data being generated from various sources. This enormous amount of data has resulted in an urgent need for efficient data analytics techniques to extract valuable insights and trends from the huge amount of data collected by IoT devices, and this can be achieved by implementing the correct methods of big data analytics over the original data. The data that is collected by the IoT is heterogeneous in nature as it originates from multiple sources such as e-commercial sites, transactional details from online booking of flight tickets, movie tickets, railway tickets, hotel booking, data of the patients from hospitals, data from smart devices, from smart homes, and so on. In this context, the integration of cobots introduces an intriguing perspective. These cobots, equipped with advanced sensors and communication capabilities, contribute to the influx of data into the IoT ecosystem. However, by employing appropriate big data analytics methods, one can identify the emerging trends, concealed patterns, and all other sorts of valuable information from this unstructured data.

Cobots, also known as collaborative robots, a lightweight robot arms that can safely share a workspace with humans and are mainly used in manufacturing and assembly tasks. The cost, size, and programming can limit the use of traditional robots in such industries. This is where the cobots come into play. The network formed by the IoT, combined with the capabilities of cobots, creates a dynamic ecosystem that not only connects people and objects but also enhances the quality and quantity of data generated.

Embedded sensors in physical objects allow for monitoring, tracking, and communication remotely for various purposes. The embedded sensors have excellent capabilities to sense and collect vast amounts of data, which they send to various data collection points usually via the cloud. Once the data has been gathered, it must be analyzed to extract valuable insights so that better decisions can be made and the objective of comprehending users or the subject matter under investigation can be achieved. Ahmed *et al.* [1] discussed that the number of devices connected to the internet will spike to more than 50 billion, and, therefore, efficient data analysis is required to extract valuable insights to achieve greater awareness of the environment.

Main highlights of the work are as follows:

- Recent advancements in the field of collaborative robots are discussed.

- The challenges encountered along with research gaps present on the IoT data analytics are addressed.
- The research methodology, tools, and techniques used for factors affecting and evaluation metrics used in an IoT environment are discussed.
- The relevant literature is categorized based on significant parameters in the context of collaborative robots.
- The practical role of data analytics in IoT applications is examined.
- Potential opportunities in big data processing and analytics in the smart computing domain and future research directions for big data analytics on the IoT are highlighted.

14.2 Advancements in Data Analytics for Collaborative Robots

A framework has been proposed by Bashir *et al.* [2] for the IoT and big-data analytics to tackle the significant challenges of inspecting and storing big data generated from cobots and other sources. It consists of three main components data analytics, big-data management, and IoT sensors. Real-time analytics are utilized to regulate the various sections of an automated building such as the light intensity, the air conditioning or heating, the oxygen levels, or for moving any machinery or circuit parts. The suggested framework is carried out using the Cloud Era-Hadoop distribution, and pySpark executes the analytics. The findings demonstrate that this framework is apt for big data analytics in an IoT-enabled environment.

- Edge Analytics for Robots: Jayashree *et al.* [3] said that edge analytics has been identified as a potential solution to address the difficulties associated with managing and processing the large volume of data that the cobots generate. This approach involves analyzing the data at the edge, i.e., closer to the data source. This helps to reduce the latency linked with sending data to the cloud and thus also helps to enhance the privacy and security of the system. Although there are such advantages, several obstacles prevail which must be resolved such as the demand for more effective algorithms and better alignment with cloud-based platforms.

- Machine Learning for Smart Setting: The use of machine learning has become critical for data analytics. It enables the extraction of valuable insights from huge amounts of data produced by IoT devices as well as cobots by applying various machine learning algorithms. These insights help in predictive maintenance, anomaly detection, and fault detection. Nevertheless, there are still challenges to overcome such as the need for more efficient algorithms that can manage the huge volumes of data produced by IoT devices.
- Blockchain for the Collaborative Robots: The utilization of blockchain technology on the IoT has the potential to transform the security and privacy of data. Implementing a distributed ledger can enable secure and decentralized access to IoT data while also ensuring privacy and ownership.
- Cloud Computing for Sensory Units: Cloud computing continues to play a significant role in IoT data analytics. Cloud-based platforms offer scalable and cost-effective storage and processing capabilities, which can be critical for handling huge amounts of data produced by the cobots. With cloud computing, they can send their data to the cloud for storage and processing, which authorized users can access at any time. However, cloud computing has its limitations such as latency, network, bandwidth, and reliability issues which must be addressed for the smooth functioning of the bots.
- Fog Computing and IoT: Hussain *et al.* [4] said that fog computing has emerged as a promising solution to overcome the challenges associated with processing massive volumes of data produced by IoT devices. This approach involves conducting data analysis closer to the data source at the network's edge, like edge analytics. However, fog computing differs from edge analytics by providing more computational resources and storage capabilities at the network edge.
- Deep Reinforcement Learning (DRL) for Robots: Yu *et al.* [5] discussed that DRL has great potential for improving data analytics in cobots. DRL can be used to optimize the systems by training an agent to learn from its actions, interact with the environment, and make decisions that maximize a reward signal. It can also be applied to a variety of use cases, including energy management, resource allocation, and anomaly detection, to improve overall system performance.

- Stream Data Mining Techniques for IoT: Stream data mining has become a crucial element in managing the uninterrupted flow of data generated by IoT devices. This data can include sensor readings, device logs, and other forms of structured or unstructured data. This technique can help in identifying the patterns, trends, and anomalies in the data, which can provide valuable insights for various IoT applications.
- Artificial Intelligence (AI) Techniques for Cobots: Ahanger *et al.* [6] discussed that AI techniques have great potential for improving data analytics in cobots used in manufacturing, automobile, food industries, and many more. However, it encounters several obstacles, which include ensuring data quality, security concerns, and the requirement for efficient and scalable algorithms.
- An Overview of Time-Series Analysis Techniques for Data Generated From Cobots: Cook *et al.* [7] said that time-series analysis has become a critical tool for extracting insights from the time-dependent data generated by the cobots. In various research papers, time-series analysis has been highlighted as an essential tool for forecasting, anomaly detection, and pattern recognition.
- Privacy-Preserving Data Analytics for IoT: Dwivedi *et al.* [8] discussed that privacy-preserving data analytics has become a critical issue in IoT applications, where sensitive data is often generated by devices. This is a field that ensures the confidentiality, integrity, and privacy of sensitive data to be extracted from the data. Internet-of-Things devices produce a massive amount of data; however, this data can contain sensitive information, making it vulnerable to privacy breaches.

Figure 14.1 shows the rise in the number of devices being connected to the internet over the recent years.

14.3 Challenges in Analysis of the Sensory Data

The main challenges faced by researchers in data analytics for IoT devices such as cobots include data heterogeneity, data quality, data privacy, security, and the scalability of data analytics algorithms. Acharjya *et al.* [9] said

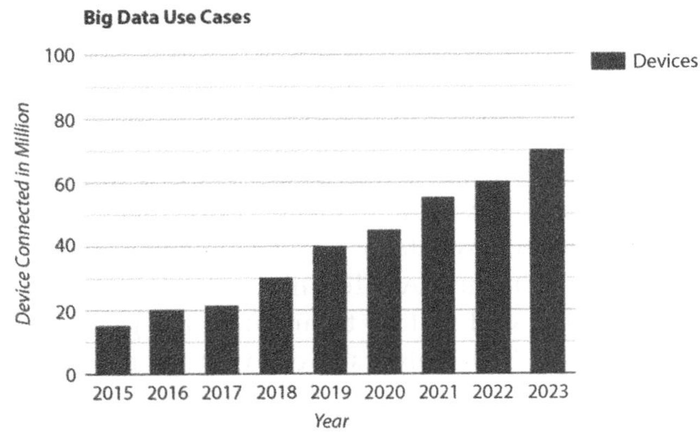

Figure 14.1 Big data use cases over the past years.

that other challenges include the lack of standardized data formats and protocols and the need for real-time analytics. To overcome these challenges, researchers need to develop new algorithms and techniques that can handle the unique characteristics of data generated from cobots and collaborate with industry partners to develop standardized protocols and frameworks for the analysis of such data.

The security and privacy of IoT data is a major challenge for data analytics researchers. IoT devices often collect sensitive personal information, and this data needs to be protected from unauthorized access or use. Furthermore, IoT devices themselves are vulnerable to cyber-attacks, which can compromise the integrity and availability of data.

Another challenge on the IoT data analytics is the need for real-time or near–real-time processing. IoT devices often generate data that requires immediate attention, such as monitoring critical infrastructure or detecting anomalies in manufacturing processes. Therefore, developing analytics techniques that can provide real-time insights is essential.

The field of IoT data analytics faces lots of challenges, particularly in terms of scalability and interoperability. The ability to effectively analyze the data which is produced by various IoT devices becomes more and more difficult as there is a significant increase in the number of IoT devices. The lack of standardization and interoperability between IoT devices and platforms presents a significant obstacle to the integration and analysis of data from various sources. Another major issue faced by researchers in this field is the energy consumption and sustainability of IoT systems, given that these devices rely on limited power sources and must operate for extended periods of time.

Deploying and managing the cobots and their infrastructure can be a complicated and expensive process. The installation of sensors and gateways in remote or dangerous areas may necessitate the use of specialized equipment and expertise. Therefore, data analytics researchers must consider the expenses and logistical challenges associated with implementing and maintaining IoT systems. The complexity of cobot data, including heterogenous data types and real-time processing requirements, presents difficulties in designing effective analytics frameworks.

Internet-of-Things data analytics also requires addressing issues related to data quality and reliability. Because IoT devices operate in dynamic and unpredictable environments, the data generated by these devices may be noisy, incomplete, or inconsistent. Therefore, data cleaning, preprocessing, and validation are critical steps in the data analytics pipeline.

By incorporating IoT data from diverse data sources such as weather data, social media data, and financial data, organizations and businesses can gain valuable insights. However, merging and analyzing heterogeneous data sources requires very sophisticated data analytics and integration techniques. Thus, researchers in IoT data analytics must devise techniques for integrating and analyzing a variety of data sources.

Finally, the ethical and legal implications of IoT-enable cobots need to be considered. IoT data may contain sensitive personal information, and the use of this data needs to follow privacy regulations and ethical standards. Therefore, researchers in IoT data analytics need to develop methods for ensuring data privacy, security, and transparency. Below in Figure 14.2, the various challenges prevailing in cobot-based data analytics are shown.

14.4 Research Gaps in Cobot-Enabled Data Analytics

- One of the major gaps in IoT data analytics is the lack of standardized data formats and communication protocols. This makes it difficult to share and integrate data from different sources, which hinders the development of advanced analytics techniques. Al-Qaseemi *et al.* [10] said that standardization efforts are needed to promote interoperability and data sharing across IoT systems. This problem can be resolved by developing and promoting standardized formats and protocols for IoT data exchange, as for example, the Message Queuing Telemetry Transport (MQTT) protocol and the JavaScript Object Notation for Linked Data (JSON-LD) format.

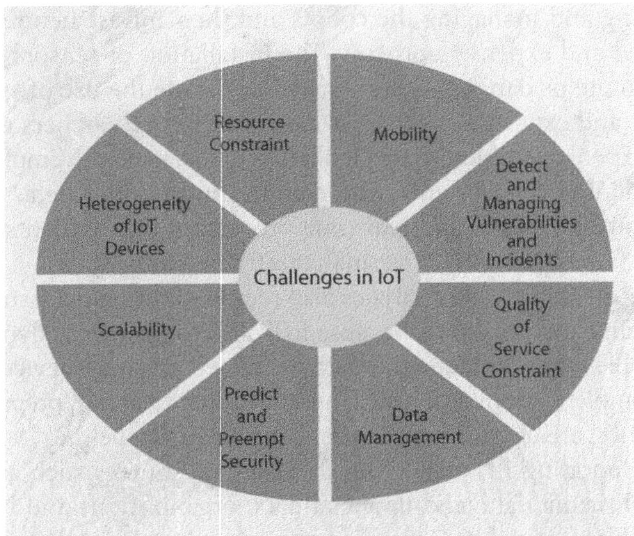

Figure 14.2 Challenges in Internet-of-Things data analytics.

- Another gap in data analytics for cobots is the limited availability of high-quality, labeled data for training and validation. The data collected from these systems is often unstructured and noisy, making it challenging to build accurate models. The development of benchmark datasets and standard evaluation metrics can help researchers compare and validate their analytics techniques. Method to resolve: Develop and promote techniques for unsupervised and semi-supervised learning that can leverage unlabeled data to improve the accuracy and robustness of machine learning models in those machineries.
- Another gap that needs to be addressed is the lack of transparency and interpretability of IoT data analytics models. Complex models, such as deep learning models, can be difficult to interpret and explain, which can limit their usefulness in practical applications. Mineraud *et al.* [11] discussed that developing more interpretable and transparent models can help build trust and confidence in IoT data analytics. Method to resolve: Develop and promote interpretability and transparency techniques for machine learning models, such as explainable AI methods, to improve the trustworthiness of the analytics results.

- Limited scalability of existing data analytics techniques for IoT data: It can be resolved by developing new distributed and parallel computing techniques for processing and analyzing large-scale IoT data, such as MapReduce and Hadoop frameworks.
- Another gap in IoT data analytics is the need for sophisticated visualization techniques and user interfaces. The data collected from the collaborative robots is often intricate and challenging to interpret, and presenting the data in a user-friendly manner can assist end users in making informed decisions. Developing interactive visualization tools and intuitive interfaces can bridge the gap between IoT data analytics and end users.
- The lack of robustness and adaptability of IoT data analytics models is another gap that needs to be addressed. Internet-of-Things data can be highly dynamic and nonlinear, and models that are developed for one application may not generalize well to other applications. Developing models that can adapt to changing data environments and be transferred to new applications can help address this gap.
- Finally, another gap that needs to be addressed is the lack of collaboration and interdisciplinary research in the collaborative robots. Internet-of-Things data analytics requires expertise in data science, computer science, engineering, and domain-specific knowledge. Developing interdisciplinary collaborations can help address complex challenges in IoT data analytics and promote innovation. Additionally, provide training and support for end users to ensure that they can effectively use and maintain these systems.

14.5 Research Methodology in Cobot-Driven Data Analytics

There are various techniques in which data can be collected in collaborative robots. These techniques can be categorized based on various characteristics. Some of these classifications are discussed below.

- Granat *et al.* [12] discussed that data collection techniques can be classified into two categories: passive and active.

Passive data collection techniques are used when the data is generated automatically without any user interaction, whereas active data collection techniques are used when the data is generated through user interaction. In IoT applications, passive data collection techniques are widely used due to their ease of implementation and lower cost.

- The data collection techniques can also be divided into the following three groups, namely, direct, indirect, and hybrid. Direct techniques use sensors or other devices to collect data directly from the environment. Indirect techniques gather data through prediction or inference and hybrid techniques are a combination of both direct and indirect techniques.
- Alturki *et al.* [13] said that the data collection techniques can also be classified into two groups: centralized and decentralized. In centralized techniques, data is collected from all IoT devices by a single server or data center. In decentralized techniques, multiple data centers or servers collect data from different IoT devices. Decentralized techniques are considered more reliable and robust than centralized techniques.
- Pouryazdan *et al.* [14] said that the data collection for smart cities can be categorized into three main techniques: wireless sensor networks, mobile crowdsensing, and big data. Wireless sensor networks utilize sensors placed throughout the city to gather data, whereas mobile crowdsensing involves collecting data via mobile devices carried by citizens. Big data is the collection of data from multiple sources, including social media and traffic cameras. The selection of a particular technique depends on the specific application's requirements as each technique has its own advantages and disadvantages.

14.6 Deployment of Tools and Techniques in Context-Aware Cobots

The integration of IoT technologies with cobots has enabled real-time data collection and analysis, offering unprecedented opportunities for process improvement, predictive maintenance, and enhanced decision-making. Husamaldin *et al.* [15] discussed that the analytical techniques can be classified into four groups: descriptive analytics, diagnostic analytics, predictive analytics, and prescriptive analytics. Descriptive analytics employs statistical techniques to describe the data. Diagnostic analytics uses machine learning algorithms to identify issues. Predictive analytics applies machine learning algorithms to predict future events. Prescriptive analytics utilizes optimization techniques to recommend actions.

The analytical techniques for IoT data analytics can again be classified into two types: supervised and unsupervised. In supervised analytics, labeled data is utilized to train machine learning models, which make predictions. Unsupervised analytics makes use of unlabeled data to uncover patterns and relationships in the data. Supervised analytics is more appropriate for applications that require accurate predictions, whereas unsupervised analytics is more suitable for applications that require data exploration and discovery.

- Apache Hadoop: Apache Hadoop is used for storing and the distributed processing of big data, and it is an open-source framework. It is capable of handling huge amounts of data and consists of two main components namely Hadoop Distributed File System (HDFS) and MapReduce. HDFS is a distributed file system that facilitates the storage of large files across several nodes in a cluster. Gulia *et al.* [16] discussed that MapReduce is a programming paradigm that facilitates the parallel processing of enormous datasets over a cluster of nodes.
- Apache Spark: Apache Spark is a publicly available platform used for processing substantial amounts of data. It is composed of several components, such as Spark Core, Spark SQL, Spark Streaming, and MLlib. Spark Core is the primary processing engine that allows for distributed data processing across a cluster of nodes. Spark SQL is a module that enables support for SQL queries on Spark data. Spark Streaming is a module for processing data in real time, making it possible to handle streaming data, whereas MLlib is a library for machine learning that offers assistance for various types of machine learning algorithms.
- TensorFlow: TensorFlow is a machine learning library that is used for deep-learning applications and is open-source. It provides a framework for the development of various deep neural networks, such as convolutional and recurrent neural networks. TensorFlow comprises several components, including

TensorFlow Core, Keras, and Estimators. The TensorFlow Core is a fundamental library used to develop machine learning models. Keras is a high-level application programming interface (API) that facilitates the quick development of machine-learning models. Estimators are also high-level API that enables the creation of intricate machine-learning models.

- Mahout: Mahout is an open-source machine-learning library that supports a range of machine-learning algorithms, such as clustering, classification, and collaborative filtering. It is optimized to operate with large-scale datasets and has distributed processing support through Apache Hadoop. Mahout provides a framework for machine learning model development, including custom algorithm development.
- Cloudera Data Hub: Cloudera Data Hub is a platform widely used for analytics on big data. It offers a toolkit for data management, processing, and analysis. The platform includes multiple components, such as Cloudera Manager, HDFS, Apache Spark, and Apache Hive. Cloudera Manager is a web-based console used for managing and monitoring Cloudera clusters. HDFS is a distributed file system that enables the storage of large datasets across multiple nodes in a cluster. Apache Spark is a distributed computing framework that processes massive datasets across a group of nodes. Apache Hive is a data warehouse system that allows SQL queries on Hadoop data.
- 1010data: 1010data is a cloud-based platform that enables users to analyze data, manage it, and process it. The platform consists of multiple components, including a high-performance database, a powerful query engine, and a suite of analytics tools. The 1010data database is optimized for handling vast quantities of data, helps with intricate queries, and supports various data analysis techniques, including machine learning and predictive analytics. The analytics tools in the platform are designed to support a wide range of data visualization techniques, including graphs, charts, and dashboards.
- Hortonworks: Hortonworks is a big data analytics platform designed to handle and process large-scale data. The platform includes multiple components such as, for instance, the HortonWorks Data Platform (HDP), Apache Hadoop, Apache Spark, and Apache Hive. HDP provides support

for data management, ingestion, processing, and analysis. Apache Hadoop and Spark allow for the storage and processing of large datasets across clusters of nodes. Apache Hive is a data warehousing system, and it supports SQL queries on Hadoop data. Together, all these components enable efficient and effective analysis of big data.

- HP-HAVEn: HP-HAVEn is a big data analytics platform that is designed to handle large-scale data processing and analysis by combining multiple components into an integral system. It consists of various components, namely, HP Vertica, HP IDOL, and HP HAVEn OnDemand. HP Vertica is a columnar database management system that supports real-time analytics on large datasets. HP IDOL is a text analytics engine that helps in natural language processing and sentiment analysis. HP HAVEn OnDemand is a cloud-based analytics platform that supports various data analytics techniques, including machine learning, predictive analytics, and data visualization.

Figure 14.3 shows some of the different tools that are used to analyze big data in the Internet of Things.

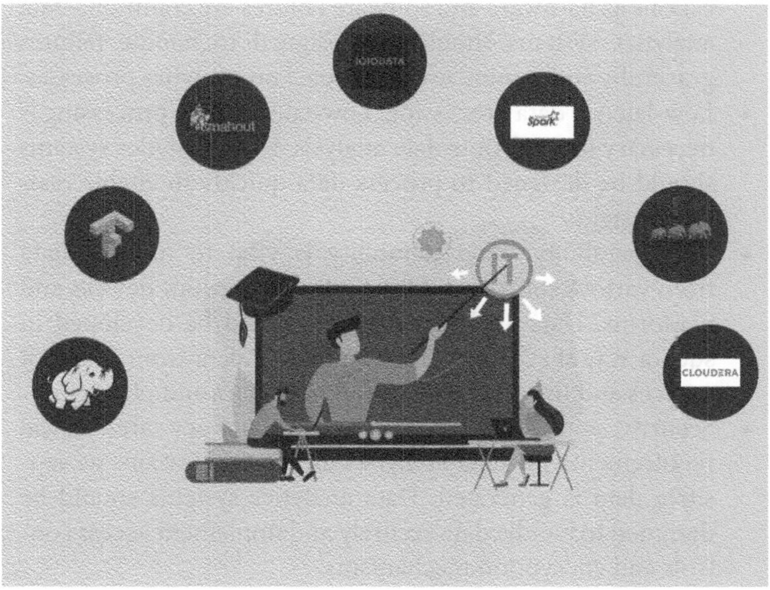

Figure 14.3 Data analytical tools in cobot-driven units.

14.7 Provisions of Smart Architecture for Data Analytics in Collaborative Robots

Internet-of-Things data analytics has certain requirements to handle large columns of data generated by the cobots. These requirements include scalability, interoperability, security, privacy, and real-time processing. Scalability is necessary to accommodate the large amounts of data produced by such IoT devices. Interoperability ensures that data can be exchanged and processed across different devices and systems. Security and privacy are essential to protect data from unauthorized access and misuse. Yaqoob *et al.* [17] discussed that real-time processing is crucial to making timely decisions based on the data produced by IoT devices.

- Scalability: Scalability is a major requirement for data analytics in IoT applications due to the large volumes of data generated. To ensure scalability, data analytics systems should be designed to handle data in a distributed manner.
- Real-Time Processing: This is critical to generate insights and alerts as soon as data is generated. Data analytics systems should be designed to analyze data in real-time to enable real-time processing.
- Fault tolerance: It is essential for data analytics in IoT applications as hardware and software failures are common. Data analytics systems should be designed to handle failures gracefully and minimize their impact on analytics processes.
- Low-latency data processing: Low-latency data processing is necessary for real-time data analysis. Data analytics systems should be designed to process data quickly to enable real-time insights.
- Interoperability: Interoperability is vital to manage data from various sources and integrate with various devices and platforms. Data analytics systems should be developed in such a way that allows them to handle data from different sources and integrate with different devices and platforms.
- Security and privacy: Security and privacy are critical requirements for data analytics in IoT applications as sensitive data is generated. Data analytics systems should be designed to handle data securely and implement access controls and encryption mechanisms.

14.8 Advanced Sensory-Based Framework for Big Data Analytics

The architecture for big data analytics on the IoT-based Cobots is composed of four layers, namely, the sensing layer, network layer, data processing layer, and application layer. Dey *et al.* [18] said that the sensing layer involves the IoT devices that create data, whereas the network layer is responsible for transferring data from these devices to the data processing layer. The data processing layer handles the storing and processing of large volumes of data, whereas the application layer delivers analytics services and insights to end users. The sample framework is shown in Figure 14.4.

Several evaluation metrics have been proposed to measure the effectiveness of the analytics approach in the working of such robots that are able to work alongside humans. Such metrics include accuracy, precision, recall, F-measure, and receiver operating characteristic (ROC) curves. Accuracy refers to the proportion of data points that are correctly classified. On the other hand, precision measures the proportion of true-positive classifications relative to all positive classes. Recall assesses the percentage of

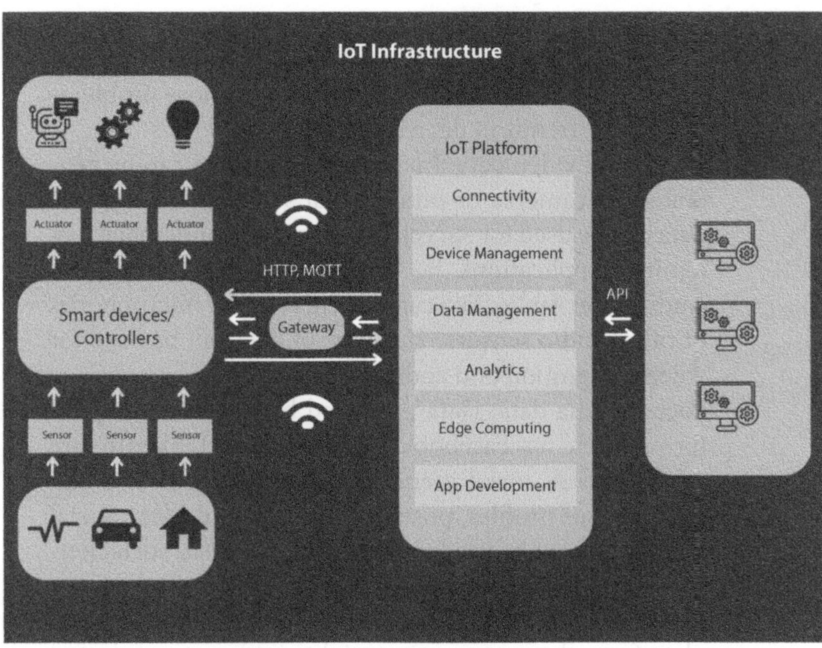

Figure 14.4 Internet-of-Things architecture for big data analytics.

true-positive classifications relative to all actual positive cases. Mahmudul *et al.* [19] said that F-measure is the harmonic means of precision and recall. ROC curves are used to measure the balance between the true-positive rate and the false-positive rate.

14.9 Taxonomical Categorization of Collaborative Cobots

Based on context-specific features and taxonomy, collaborative robots are classified into various categories.

- Type of data: Siow *et al.* [20] discussed that the types of data produced by IoT-based Cobots can be highly diverse and may include sensor data, log data, image data, video data, audio data, social media data, and many more.
  - ➢ Sensor data: This is the most common type of data generated by IoT devices. Sensors are used to quantify physical quantities such as temperature, humidity, pressure, motion, and light.
  - ➢ Log data: Log data captures events that occur on a device or system, such as error messages, status updates, or user interactions. Log data can be used to diagnose problems or identify trends in device usage.
  - ➢ Image and video data: Image and video data can be generated by cameras or other visual sensors. This type of data can be used for surveillance, security, and image recognition applications.
  - ➢ Audio data: Audio data can be generated by microphones or other audio sensors. This type of data can be used for voice recognition and audio surveillance.
  - ➢ Social media data: Data produced by social media platforms such as Facebook, Twitter, and LinkedIn are referred to as social media data. This type of data can include user profiles, posts, comments, and likes. Social media data can be used for sentiment analysis, social network analysis, and targeted advertising.
- Level of Analysis: Descriptive analytics involves analyzing data from the past to gain insights into past trends and events. Diagnostic analytics focuses on understanding why past events occurred. Predictive analytics utilizes statistical

and machine learning techniques to forecast future events. Prescriptive analytics involves using optimization and simulation techniques to recommend courses of action.

- Analytics Techniques: Machine learning algorithms such as decision trees, neural networks, and support vector machines have been widely used for IoT data analytics. Statistical analysis techniques such as regression analysis and clustering analysis have also been applied in IoT data analysis. Data visualization tools such as dashboards, charts, and graphs are used to present the insights derived from the data.
- Application Domains: Healthcare is an important application domain for IoT data analytics, where it is used for patient monitoring, disease management, and drug discovery. In manufacturing, IoT data analytics is used to optimize production processes and reduce downtime. Transportation is another area where IoT data analytics is applied to optimize logistics and improve safety. Smart cities use IoT data analytics to optimize energy consumption and improve urban planning.

Figure 14.5 demonstrates various search interests analyses on smart-driven collaborative robots in a real-time setting.

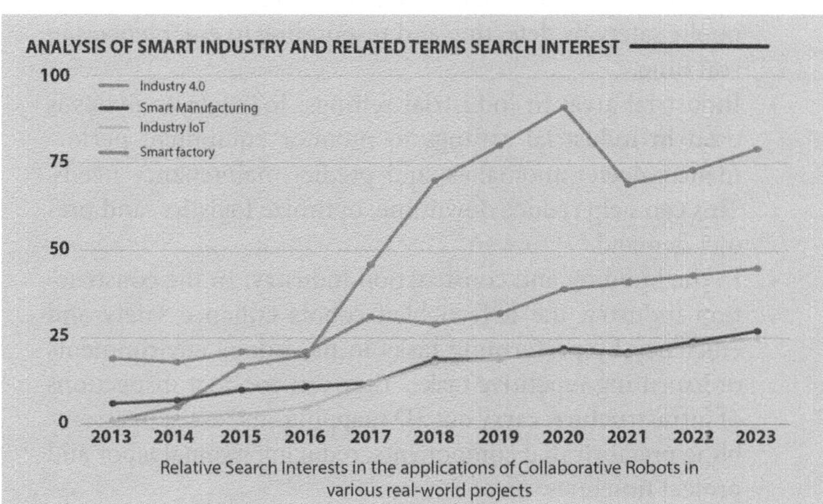

Figure 14.5 Comparison of search interests on Internet of Things–based Cobots in real-world applications.

14.10 Use Cases of Cobots and Data Analytics

- Warehouse and Logistics: In the logistics sector, IoT-based cobots aid in automated material handling, inventory management, and order fulfillment. These cobots can navigate through warehouses using radio frequency identification (RFID) or barcode systems, locate and retrieve items, and optimize storage space. The real-time tracking of goods and automated inventory updates enhance accuracy, reduce errors, and improve overall operational efficiency.
- Agriculture: Cobots equipped with IoT sensors and cameras contribute to precision agriculture by monitoring crop health, detecting pests, and automating planting and harvesting. These cobots can be programmed to navigate fields automatically, collect data about soil conditions, and apply targeted treatments, leading to increased yields and reduced resource wastage.
- Smart Cities: In smart cities, IoT data analytics can improve the efficiency and effectiveness of various systems such as transportation, environmental monitoring, and public safety. Internet-of-Things sensors can collect traffic data to optimize traffic flow and reduce congestion, monitor air quality to detect pollutants, and optimize energy usage, and reduce waste. Additionally, IoT data analytics can improve public safety by detecting and responding to emergencies in real time.
- Industrial area: In industrial settings, IoT data analytics is used in industrial settings to monitor equipment performance, detect anomalies, and predict maintenance needs. This can help reduce downtime, optimize logistics, and predict demand.
- In the building and construction industry: In the construction industry, the IoT-enabled cobots enhance safety and efficiency by performing tasks in hazardous environments or executing repetitive tasks. They can conduct inspections of infrastructure, carry out 3D mapping, and assist in assembling prefabricated components, reducing manual labor and project timelines.
- Energy sector: IoT data analytics is used to improve the efficiency and effectiveness of the energy sector. Energy consumption data can be collected from IoT devices to identify

patterns and optimize energy usage. Additionally, it can predict and prevent equipment failures, improve asset management, and reduce downtime.

- Retail industry: Retail businesses utilize IoT-based cobots to enhance customer service through interactive kiosks, shelf monitoring, and inventory management. Cobots can guide shoppers to desired products, answer frequently asked questions, and ensure that shelves are stocked appropriately, improving the shopping experience.
- Transportation: Ta-Shma *et al.* [21] discussed that the IoT data analytics can be used to improve the performance and reliability of transportation systems, such as trains and airplanes. For example, IoT sensors can be used to monitor equipment health, predict maintenance needs, and prevent breakdowns. Additionally, IoT data analytics can be used to optimize routes, reduce fuel consumption, and improve safety.
- Smart Manufacturing Industry: IoT-enabled cobots play a crucial role in revolutionizing manufacturing processes by facilitating real-time monitoring and control of production lines. These cobots can collaborate with human workers in tasks like assembly, quality control, and material handling. Sensors and cameras integrated into cobots gather data about product quality, machine performance, and many more.
- Security: The application of IoT data analytics can enhance the protection and reliability of vital infrastructure, including transportation systems, water treatment facilities, and power plants. For example, data collected from IoT sensors can be utilized to monitor equipment health, detect anomalies, and prevent accidents. Additionally, IoT data analytics can be used to analyze security threats, detect intruders, and respond to emergencies.
- Healthcare: IoT-integrated cobots find applications in healthcare settings by assisting medical professionals in various tasks. They can transport supplies, deliver medications, and even perform routine tasks under supervision. IoT connectivity enables remote monitoring of patient vital signs, ensuring timely interventions and personalized care.

Figure 14.6 shows various real-life applications of the IoT-based cobots and its relative importance.

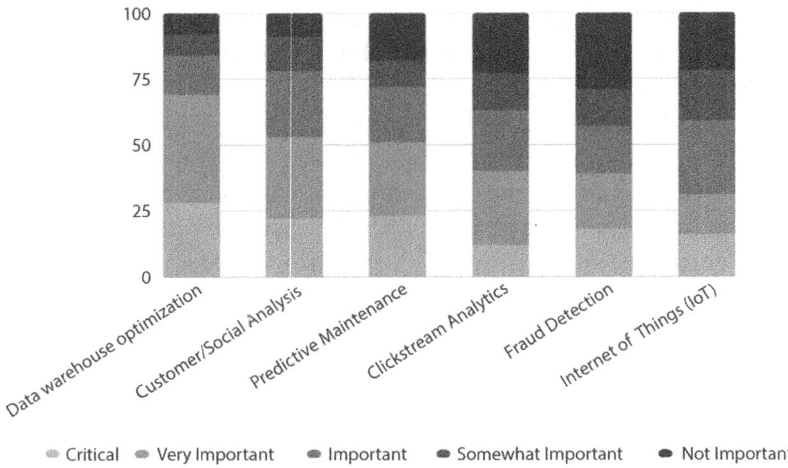

Figure 14.6 Various practical uses of data analytics for the Internet of Things.

14.11 Futuristic Opportunities in Data Analytics for Collaborative Robots

The integration of IoT-based devices such as the collaborative robots and data analytics is opening new opportunities for businesses to create innovative business models and generate additional revenue streams. Utilizing behavior can help businesses develop new products and services that are tailored to their customers' needs. Furthermore, by optimizing their operations through IoT data, businesses can reduce their costs and boost their profitability.

The potential impact of data analytics for the IoT is far-reaching, with numerous industries such as healthcare, manufacturing, energy, and transportation poised to benefit from this technology. Ejaz *et al.* [1] said that, by tapping into IoT data, businesses can gain a deeper understanding of their operations and their customers, leading to better decision-making, greater efficiency, and a stronger competitive advantage. Some critical open challenges and directions are cited below.

- Data privacy and Security: Balaji *et al.* [22] said that the security and privacy of IoT data analytics is an essential aspect that requires special attention. Analyzing IoT data can offer valuable insights and enable numerous applications, but it also poses significant privacy and security risks. To guarantee the privacy and security of data produced by these cobots, it is crucial to protect it through encryption, access controls, and other security measures. Furthermore, safeguarding user privacy necessitates obtaining informed consent for the collection, storage, and processing of their data. The complexity of the IoT data ownership arises from its generation by multiple devices and its sharing across diverse organizations and domains. Research on developing privacy-preserving data analytics techniques that permit IoT data analysis while safeguarding user privacy is imperative. Future research should also prioritize the creation of secure and reliable bots and systems that can withstand cyber-attacks and other security threats, as well as addressing legal and regulatory challenges related to privacy and security in IoT data analytics.

- Scalability and Real-Time Processing: Vyas *et al.* [23] said that, in data analytics of such devices, the issues of scalability and real-time processing are of paramount importance. As IoT devices generate an immense volume of data, this data is expected to increase rapidly in the years to come. To ensure the efficient analysis of such data, IoT data analytics systems must be scalable and capable of handling large volumes of data and processing it promptly. Real-time processing is fundamental to numerous applications, including industrial automation, traffic management, and smart homes. However, real-time processing poses a challenging task that requires both low latency and high throughput. To tackle the issues of scalability and real-time processing in IoT data analytics, the use of distributed and parallel computing techniques, such as MapReduce, Hadoop, and Spark, is necessary. These techniques can facilitate the efficient processing of large volumes of data on clusters of commodity hardware. Nevertheless, the implementation of distributed systems in IoT environments is challenging due to the limited resources and unreliable connectivity of IoT devices. Future research should explore energy-efficient data analytics techniques that minimize energy consumption while processing IoT data efficiently. Developing new hardware and software architectures that can support the efficient processing of IoT data is another important area of research. By addressing these challenges, researchers can develop scalable and real-time IoT data analytics systems that can handle large data volumes, deliver timely results, and operate efficiently in resource-constrained IoT environments.

- Data Interoperability and Integration: Data interoperability and integration are important challenges in IoT data analytics, as the cobots often generate data in different formats and standards. Future research should focus on developing data integration and fusion techniques that can handle different data types and formats and enable seamless data exchange and interoperability among IoT devices and systems.
- Human-in-the-Loop Analytics: Human-in-the-loop analytics is a burgeoning field in IoT data analytics that seeks to combine human expertise with machine intelligence. The goal is to develop methods that can integrate human input and feedback into data analytics workflows and allow humans to interact with IoT data in a more natural and intuitive way. Human-in-the-loop analytics can be applied to many IoT applications, including healthcare, transportation, and smart cities. It can help to create individualized healthcare plans for patients who have specific health needs optimize traffic flows and reduce congestion in smart cities. To advance this field, future research should focus on developing advanced techniques for human-in-the-loop analytics in IoT data analytics and exploring the ethical and social implications of this approach.
- Explainable AI and Trust: Explainable AI is an important concern in IoT data analytics, as machine learning algorithms used in IoT devices often make decisions that are not easily understandable. To ensure that decisions are fair, ethical, and unbiased, future research should focus on developing transparent and interpretable AI techniques that can provide clear explanations of machine learning models and decisions. Trust is a crucial factor in IoT data analytics because users must be confident in the accuracy and reliability of the analytics results to make informed decisions based on the data. Building trust requires the use of not only explainable AI techniques but also robust security and privacy measures to safeguard sensitive data and guarantee the integrity of the analytics results.
- Integration of Multiple Data Sources: Integrating data from multiple sources is a significant challenge in IoT data analytics because these many IoT devices generate data from

diverse domains and sources. Integrating data from various sources in the IoT data analytics can result in more accurate insights and predictions, as well as the development of more advanced models and algorithms. For instance, in smart city applications, integrating data from traffic sensors, weather sensors, and social media can provide a better understanding of traffic patterns and congestion, leading to the development of improved techniques of traffic management. Future research should focus on developing techniques for integrating data from multiple sources and leveraging complementary information from different data domains and types to improve the outcomes of data analytics.

- Domain-Specific Data Analytics: A crucial area of focus in IoT data analytics is domain-specific data analytics, which involves creating data analytics techniques customized for specific domains and applications. This requires understanding the distinct demands and difficulties of the domain and devising analytics models and algorithms optimized for those challenges. It may involve integrating specialized knowledge and expertise into the analytics models, as well as adapting pre-existing analytics techniques to suit the unique characteristics of the domain. To advance the field, future research should concentrate on developing domain-specific data analytics methods that can effectively capture the unique qualities and needs of diverse domains and can adapt to the changing needs and contexts of IoT applications.
- Standardization and Interoperability: Kumar *et al.* [24] discussed that, in the data analytics for such devices, standardization and interoperability play a critical role in ensuring seamless data exchange and interoperability among various devices, systems, and domains. By establishing standardization and interoperability, it is possible to ensure that IoT data analytics systems are capable of functioning effectively and efficiently, as well as to facilitate the creation of more sophisticated and advanced analytics models and algorithms. For example, standardized protocols for data exchange and communication can aid in developing more sophisticated predictive maintenance models for industrial IoT applications, resulting in improved equipment reliability and reduced maintenance costs. To advance the field of IoT data analytics, future research must concentrate on

developing standardization and interoperability frameworks that allow for cross-domain data exchange, data sharing, and data fusion.

14.12 Conclusion

To summarize, data analytics is a crucial element of IoT systems that enables the collection, processing, and analysis of large volumes of data produced by IoT devices. To fully harness the power of IoT data analytics, it is necessary to develop more advanced and sophisticated analytics models and algorithms that can support a variety of applications and use cases in areas like healthcare, smart cities, and industrial IoT.

Internet of Things–based cobots are transforming various industries by providing intelligent, flexible, and collaborative automation solutions. These practical use cases illustrate the tangible benefits of incorporating IoT technology into cobots, ranging from increased productivity and efficiency to improved safety and customer satisfaction. As technology continues to advance, the potential for the IoT-based cobots to drive innovation and reshape industries remains substantial.

However, there are many challenges and unanswered research questions that need to be addressed to unlock the full potential of IoT data analytics. These include ensuring data privacy and security, addressing scalability and real-time processing concerns, enabling human-in-the-loop analytics, developing explainable AI and trust mechanisms, integrating multiple data sources, and ensuring standardization and interoperability across different devices and systems.

To address these challenges, future research must focus on developing innovative approaches to IoT data analytics that involve collaboration across various fields like data science, machine learning, computer science, and engineering. By addressing the challenges and open research questions outlined in this paper, a more efficient, effective, and sustainable future can be created for the people.

In conclusion, the potential benefits of IoT data analytics are vast and diverse, and the development of more sophisticated and advanced analytics models and algorithms will be crucial to realizing this potential. By overcoming the challenges and unanswered research questions, a better future can be created that improves the quality of life and drives innovation.

References

1. Ahmed, E., Yaqoob, I., Hashem, I.A.T., Khan, I., Ahmed, A.I.A., Imran, M., Vasilakos, A.V., The role of big data analytics in Internet of Things. *Comput. Networks*, 129, part 2, 459–471, 2017.
2. Bashir, M.R. and Gill, A.Q., Towards an Internet of Things big data analytics framework: Smart buildings systems, in: *High Performance Computing and Communications; IEEE 14th International Conference on Smart City; IEEE 2nd International Conference on Data Science and Systems (HPCC/SmartCity/DSS), 2016 IEEE 18th International Conference on*, IEEE, pp. 1325–1332, 2016.
3. Jayashree, N., Sathish Babu, B., Raj, P., Saini, K., Surianarayanan, C., Chapter Seven - Edge data analytics technologies and tools, in: *Advances in Computers*, Elsevier, Cambridge, United States, 2022.
4. Hussain, M.M. and Beg, M.M.S., Fog Computing for Internet of Things (Internet of Things)-Aided Smart Grid Architectures. *Big Data Cogn. Comput.*, 3, 1, 8, 2019, https://doi.org/10.3390/bdcc3010008.
5. Yu, L., Xie, W., Xie, D., Zou, Y., Zhang, D., Sun, Z., Zhang, L., Zhang, Y., Jiang, T., Deep Reinforcement Learning for Smart Home Energy Management. *IEEE Internet Things J.*, 7, 4, 2751–2762, April 2020, doi: 10.1109/JInternet of Things.2019.2957289.
6. Ahanger, T.A., Aljumah, A., Atiquzzaman, M., State-of-the-art survey of artificial intelligent techniques for Internet of Things security. *Comput. Networks*, 206, 108771, 2022, ISSN 1389-1286, https://doi.org/10.1016/j.comnet.2022.108771.(https://www.sciencedirect.com/science/article/pii/S138912862200007X).
7. Cook, A.A., Misirli, G., Fan, Z., Anomaly Detection for Internet of Things Time-Series Data: A Survey. *IEEE Internet Things J.*, 7, 7 6481–6494, 2020, doi: 10.1109/jInternet of Things.2019.2958185 10.1109/JInternet of Things.2019.2958185.
8. Dwivedi, A.D., Srivastava, G., Dhar, S., Singh, R., A Decentralized Privacy-Preserving Healthcare Blockchain for Internet of Things. *Sensors*, 19, 2, 326, 2019, https://doi.org/10.3390/s19020326.
9. Acharjya, D.P. and Ahmed, K., A survey on big data analytics: challenges, open research issues and tools. *Int. J. Adv. Comput. Sci. Appl.*, 7, 2, 511–518, 2016.
10. Al-Qaseemi, S.A., Almulhim, H.A., Almulhim, M.F., Chaudhry, S.R., Internet of Things architecture challenges and issues: Lack of ization, in: *2016 Future Technologies Conference (FTC)*, San Francisco, CA, USA, pp. 731–738, 2016, doi: 10.1109/FTC.2016.7821686.
11. Mineraud, J., Mazhelis, O., Su, X., Tarkoma, S., A gap analysis of Internet-of-Things platforms. *Comput. Commun.*, 89–90, 5–16, 2016, ISSN 0140-3664, https://doi.org/10.1016/j.comcom.2016.03.015. (https://www.sciencedirect.com/science/article/pii/S0140366416300731).

12. Granat, J., Batalla, J.M., Mavromoustakis, C.X., Mastorakis, G., Big Data Analytics for Event Detection on the Internet of Things-Multicriteria Approach. *IEEE Internet Things J.*, 7, 5, 4418–4430, May 2020, doi: 10.1109/JInternet of Things.2019.2957320.
13. Alturki, B., Reiff-Marganiec, S., Perera, C., A hybrid approach for data analytics for internet of things, in: *Proceedings of the Seventh International Conference on the Internet of Things (Internet of Things '17)*, Association for Computing Machinery, New York, NY, USA, Article 7, pp. 1–8, 2017, https://doi.org/10.1145/3131542.3131558.
14. Pouryazdan, M., Fiandrino, C., Kantarci, B., Soyata, T., Kliazovich, D., Bouvry, P., Intelligent Gaming for Mobile Crowd-Sensing Participants to Acquire Trustworthy Big Data in the Internet of Things. *IEEE Access*, 5, 22209–22223, 2017, doi: 10.1109/ACCESS.2017.2762238.
15. Husamaldin, L. and Saeed, N., Big Data Analytics Correlation Taxonomy. *Information*, 11, 1, 17, 2020, https://doi.org/10.3390/info11010017.
16. Gulia, P. and Chahal, A., Big Data Analytics for Internet of Things (2020). *Int. J. Adv. Res. Eng. Technol. (IJARET)*, 11, 6, 593–603, 2020, Available at SSRN: https://ssrn.com/abstract=3657061.
17. Yaqoob, I., Ahmed, E., Hashem, I.A.T., Ahmed, A.I.A., Gani, A., Imran, M., Guizani, M., Internet of Things Architecture: Recent Advances, Taxonomy, Requirements, and Open Challenges. *IEEE Wireless Commun.*, 24, 3, 10–16, June 2017, doi: 10.1109/MWC.2017.1600421.
18. Dey, N., Hassanien, A.E., Bhatt, C., Ashour, A.S., Satapathy, S.C. (Eds.), *Internet of Things and Big Data Analytics Toward Next-Generation Intelligence*. Studies in Big Data, Springer Cham, Berlin, 2018, doi:10.1007/978-3-319-60435-0.
19. Hasan, M., Islam, M.Md., Zarif, Md.I., Hashem, M.M.A., Attack and anomaly detection in Internet of Things sensors in Internet of Things sites using machine learning approaches. *Internet Things*, 7, 100059, 2019, ISSN 2542-6605, https://doi.org/10.1016/j.InternetofThings.2019.100059. (https://www.sciencedirect.com/science/article/pii/S2542660519300241).
20. Siow, E., Tiropanis, T., Hall, W., Analytics for the Internet of Things. *ACM Comput. Surv.*, 51, 4, 1–36, 2018, doi: 10.1145/3204947.
21. Ta-Shma, P., Akbar, A., Gerson-Golan, G., Hadash, G., Carrez, F., Moessner, K., An Ingestion and Analytics Architecture for Internet of Things Applied to Smart City Use Cases. *IEEE Internet Things J.*, 5, 2, 765–774, 2018, doi: 10.1109/jInternet of Things.2017.2722378 10.1109/JInternet of Things.2017.2722378.
22. Balaji, S., Nathani, K., Santhakumar, R., Internet of Things Technology, Applications and Challenges: A Contemporary Survey. *Wirel. Pers. Commun.*, 108, 363–388, 2019, https://doi.org/10.1007/s11277-019-06407-w.
23. Vyas, D.A., Bhatt, D., Jha, D., Internet of Things: trends, challenges and future scope. *IJCSC*, 7, 1, 186–197, 2015.
24. Kumar, S., Tiwari, P., Zymbler, M., Internet of Things is a revolutionary approach for future technology enhancement: a review. *J. Big Data*, 6, 111, 2019, https://doi.org/10.1186/s40537-019-0268-2.

15

Comprehensive Analysis on Design, Working, and Manufacturing of Soft Robots

Mithun N. S.*, Saravanagowri S. P., Sashankh V. R., Rithish Hari G., Nithish Kumar S., Saravanan B. K. and Ramesh Babu S.

Department of Mechanical Engineering, KPR Institute of Engineering and Technology, Coimbatore, TN, India

Abstract

Robotics is an emerging field of technology that is based on the design, fabrication, construction, and development of robots that are used in broad-spectrum applications in today's world. The robots of the day must be industrially, economically, as well as commercially viable to cater the needs of humanity. Conventionally, it is seen that robot bodies are constructed using stiff materials like aluminum, titanium, steel, polymers like acrylonitrile butadiene styrene (ABS), and other materials. In the context of improving efficiency not just in terms of performance but also in terms of operation and fabrication, light substitutes replace the body-make of the robots. With the shortcomings of conventionally operated robots with respect to natural environments and human interaction, soft robots are engineered in order to create a proximity between the same. While having tremendous abilities that normal robots lack, such as the ability to morph and regenerate, the soft robots have a leading edge solely because of their key interaction ability.

Soft robots are well-known for their ability to control the range of force in which they act such as the ones used in material prehension. Also, soft robots are smart robots. This is because they depend on smart materials such as composites for fabrication, which serve a multitude to their functioning. This book chapter primarily deals with design and control of a soft robot, few widely used materials in the manufacturing with a review on their characteristics. The robots encompass a lot of day-to-day applications that include wearables, mechanical motion, surgeries, and others.

**Corresponding author*: mailtomithunns@gmail.com

V. Ramasamy, S. Balamurugan and Sheng-Lung Peng (eds.) Intelligent Robots and Cobots: Industry 5.0 Applications, (317–336) © 2025 Scrivener Publishing LLC

Keywords: Soft robots, shape regeneration, composite, graphene oxide, sustainability

15.1 Introduction

This review talks about the potential for soft robots, which are created in a way that mimic living human cells, in the field of robotics and automation. These soft robots' design anticipates a number of upcoming technological advancements for a better portrayal of the many different natural-looking morphologies [1]. It is explored how soft robots can be optimized with the least amount of human involvement [2]. Many embedded sensors are employed to estimate the robot states in order to activate soft robotics [3, 4]. Genetic algorithms and PhysXI are the two main software solutions that use the lumped modeling technique to optimize soft robots for control and operation. Even composites are frequently utilized for the construction of soft robot structures that improve their capability to mimic mammalian cells.

There is no denying that nature is a never-ending source that serves as an inspiration for the creative designing of contemporary AI-automated systems and equipment. Thus, to create soft robots, bioinspired materials are used. One such kind of soft robot is one made of graphene oxide (GO). Graphene oxide (GO), which has a large number of oxygen functional groups, holds greater potential for its use in moisture-responsive sensors and actuators because of the strong GO–water interaction and rapid movement of water molecules within the piled-up GO sheets. The mass manufacture of GO using graphite as the raw material is possible, and, by the chemical exfoliation process, its size, flaws, and oxidation degree can be well controlled [5, 6].

New, less invasive surgical techniques have been developed as a result of recent breakthroughs in robotics technology. The DaVinci research kit (DVRK) surgical robotic platform was successfully achieved by the integration of software and hardware of the proposed soft manipulator [7]. Although soft robots are a promising development in the realm of futuristic robots, controlling, operating, manufacturing, etc., do present certain difficulties. The creation of sophisticated actuators and algorithms that can successfully control the motion and behavior of soft robots is thought to be the most difficult task of all [8]. There is no question that soft robots are the future of robotics, and, by stating that, we distinctly imply that it has a vast range of potential applications in the future. Since kt presents the possibility of being implemented in a variety of practical domains, such as biomedical engineering, we may even predict the development of 4D soft robots. Technical and implementable development of 4D soft robotics is extremely important.

15.2 Design of Soft Robots

The methods of designing are either getting better or are still being studied in light of current or fresh advancements in the burgeoning field of soft robots. Technical developments in soft materials and subsystems let the soft robot function autonomously, which is not possible for rigidly built robots. Innovative designs for soft robots have been created with the help of specialized software in finite element analysis, such as VoxCAD, that supports structures with a wide modulus range and is then integrated with design optimization utilizing evolving algorithms [1]. Also, modern designs offer greater room for automation tools, which aid in more effective operating space exploration. Moreover, design solutions are improved. From the beginning to the conclusion, morphology is improved to produce a variety of forms that look natural in order to ensure a good fit in the environment. A network that may be used to describe robot morphologies employing morphology utilizing spatial coordinates as a function can be created using a compositional pattern [2]. A complete new variety of multi-material additive manufacturing techniques are developed to design new amorphous or soft robots as a result of the much increased design space [3].

Olson's linear "Materials by Design" theory was used in optimization studies to create soft robots by taking advantage of material–structure and property–performance relationships. In this, manifolds of an object can be mapped onto a 2D rectangular domain using a unified computational framework built for generative designs [4]. Simulators based on ChainQueen are used to study the physics of robot design. This scenario makes use of the differentiable Moving Least Squares Material Point Method simulator. Robotics design and controller parameter trajectory simulation both use analytical gradients [5]. In order to demonstrate the accomplishment of gripping task and the way in which performance is impacted by changing the system's parameters, a hypothetical model, usually rod-based for fingers that act as the hand of the soft robot was created. Guidelines were created using a thorough analysis of soft robots theoretically along with the observations that followed [6].

To produce technical wonders in the area of soft robot design, a bio-inspired design has been introduced as a successful one. For instance, design elements were taken from grasshoppers and other leaping insects, and further research was conducted using an AI paradigm [7]. For the design and control of soft robotics, mathematical formulation is applied in a flexible framework in the area of differential simulation. Although the development is still in its early stages, fully automated design and pipeline manufacture of soft robots were demonstrated using the relations. The automated pipelines are ideal for integrating with design projects that are simulation-driven. These designs themselves are automatable and can specify the requirements for working robots [8]. Octopus-inspired OctArm soft robot series of manipulators, which demonstrated an ability of adaptive and unique manipulation in difficult or hostile situations, were created to add enhancements. Its construction and design are based on field testing and performance in internal laboratories. Moreover, ideas for using soft robots in military settings are examined [9]. In the near future, soft robots may benefit greatly from robotics, artificial intelligence, and flexible materials. The DOF (degrees of freedom) in the deformation process and the accessible geometrical parameters are increased by the design of this structure [10]. Figure 15.1 shows a brief analogy of the various available design principles of soft robots in the recent days. One can identify the various steps to design the Soft robot systems by referring Figure 15.2.

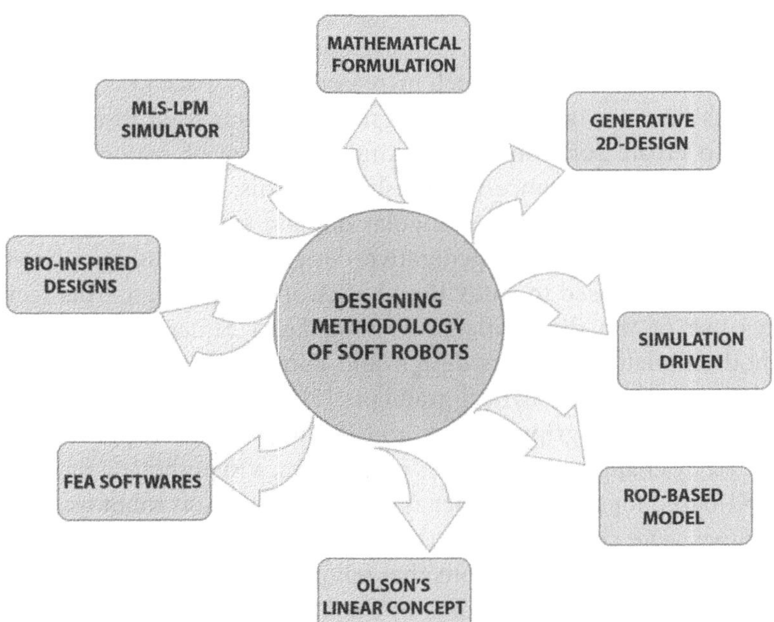

Figure 15.1 Some design principles of soft robots.

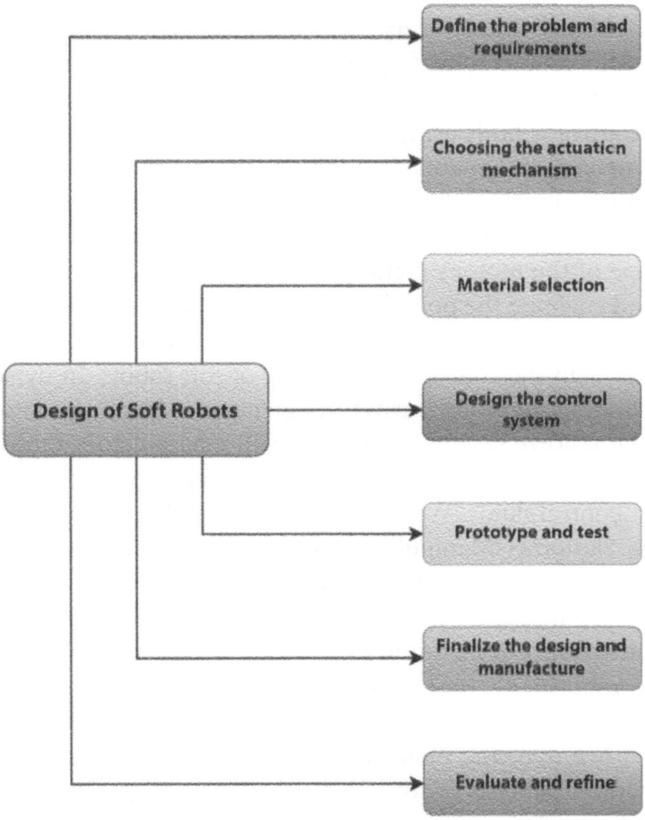

Figure 15.2 Steps under designing of soft robots.

15.3 Working and Control of Soft Robots

Soft robotics modeling and control pose significant challenges due to the absence of a universal paradigm for managing dynamic systems with infinite dimensionality. Soft robots rely on embedded sensors to estimate their states and operate their soft actuators or active materials [7, 11]. Three types of control approaches are discussed for soft robots: compliance control for lightweight arms, joint torque control shaping the actuator kinetic energy, and impedance control for complex kinematic chains. The composite control of lightweight arms framework is utilized to enable active real-time compliance control, which is usually based upon joint torque sensing, resulting in more responsive soft robots. To achieve a lightweight design, gear production ratios are relatively high, which leads to joints that have moderate intrinsic compliance and are barely back-drivable. As a result, the robot is modeled as a flexible joint system, and torque alerts are utilized to reduce joint friction and quiet vibrations related to joint compliance while implementing compliant management rules. The framework for control is built using a passive control perspective by using the distinct state vector elements. In addition, the use of motor position feedback imposes complementary behavior. By controlling joining torque and modeling the actuator kinetic energy, the joint torque feedback, in relation to the shape of the motor's meniscus, acts a vital role in streamlining the analysis and enabling the point level method to be generalized to Cartesian coordinates [12, 13]. Figure 15.3 depicts the sequences and the corresponding flow in the control and working of soft robots. In dynamic control, the planning algorithm learns through iterative learning control, how to identify all unknown system parameters (soft manipulators, fluid actuators, and continuously actuated cylinders or valves), and new dynamic models of spatial soft control fluids. We describe computing locally optimal dynamic operations. Elastomer-Manipulator: In this method, high compliance of the manipulator, actuation limitations, and self-loading effects due to gravity are presented as optimization constraints. Examples of different control strategies for soft robots can be found in dynamic models of hard and semi-soft continuum robots [14–16]. The evolution of morphology and control is also an important parameter of discussion, and the two main software tools that utilize intensive dynamic modeling approaches to optimize the morphology and control patterns of soft robots are genetic algorithms and PhysXI.

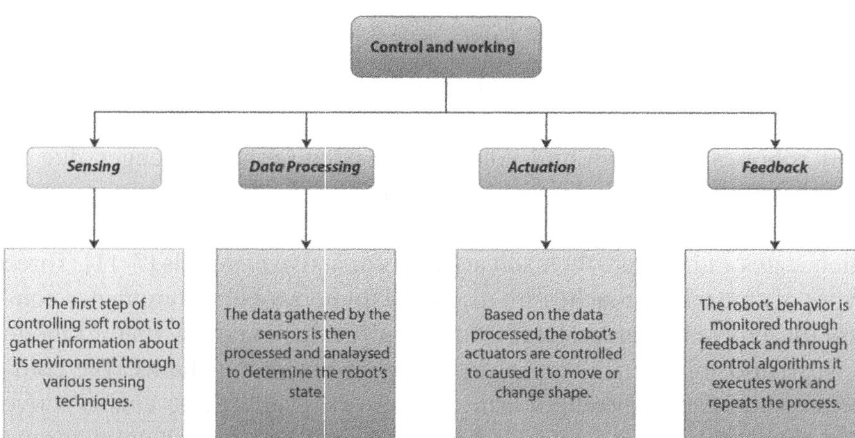

Figure 15.3 Working and control of soft robots: an illustration.

In this approach, the genotype is defined in terms of physical and control parameters, and PhysXI is used to simulate the robot's movements.

15.4 Composite Soft Robots

As soft robots are gaining progression in terms of structure, mechanical strength and durability along with enhanced properties assimilated by new methods of manufacturing, composites come into the play of the same. Some designs of soft robots along with its material make are discussed. Some materials like Rayon and Spandex, which come under the stretchable fabric composites, are used to demonstrate maintenance of low elastic modulus when under a large deforming load. It also helps to improve tear, puncture, and wear resistance along with protection from sudden failure. They can be used in some new-end applications like robotic skins, soft elastomer sensors in robotic hands, instrument fingers, and soft-fluid actuators. Composites like cotton have low-load supporting tendencies, increased chances of failure in multiple instances, and slow stress-relieving properties before a major failure. Though on the nemesis, these properties add to the enhancement of the production of soft robotic skins. Some new silicon-based composites, as compared against elastomers, can offer improved properties of materials that are incorporated within a large range of elastic moduli when combined with materials such as embedded fabric and rubber. This combination certainly provides improved expansion strength, resistance against tear, residual properties against puncture, and resistance against external factors like acid action, oil, temperature, and base actions [17]. Another class of composites, called the IMPC is very widely used in the same. They are abbreviated as ionic metal polymer composites. The difference among these and others is mainly on the composition of the polymer used [19, 20]. Here, a polyelectrolyte polymer that is ion-conducting is used. If the voltage in the range of <5V is applied, then the cations of the electrolyte get displaced toward the cathode while the anions move toward the anode. This shifting of ions leads to the whole structure getting bent on the side of the cathode. The mode of operation is hence very simple [18]. Zhou *et al* described a dynamic model framework of a soft robotic gripper, which is combined with several fluid composite actuators that were prestressed. An actuator, in specific a flexible printed circuit, makes use of a prestressed elastomer surface in order to form an already curved shape that fits in an equilibrium, and airbase pressure, operating using pneumatics, is used to flatten the same. A soft gripper is made using actuators that were already curved or precurved and requires no manual effort to grip a structure, aiding in energy efficiency. In dynamic control, an iterative learning control (ILC) algorithm is employed to progressively identify and adapt to unknown system parameters, including components like soft manipulators, fluid actuators, and continuously actuated cylinders or valves. ILC aims to improve system performance over time by learning from previous iterations and refining control actions. Dynamic operations are computed to be locally optimal, tailored to achieve the best performance under given constraints, such as high compliance of soft manipulators, actuation limitations, and self-loading effects due to gravity. Various control strategies have been developed for hard, semi-soft, and soft continuum robots. Semi-soft continuum robots blend rigid and flexible components, providing a balance between strength and adaptability, and require unique control strategies to manage their dual characteristics [14–16]. The evolution of both morphology and control mechanisms is critical, with genetic algorithms and PhysXI being key software tools used for optimizing robot design and control patterns. Genetic algorithms evolve physical and control parameters, while PhysXI simulates the robot's movements based on these parameters, enabling researchers to refine designs virtually before real-world implementation. The field of dynamic control in soft robotics is advancing through these iterative learning and optimization methods, leading to more versatile and efficient robotic systems. Figure 15.4 depicts

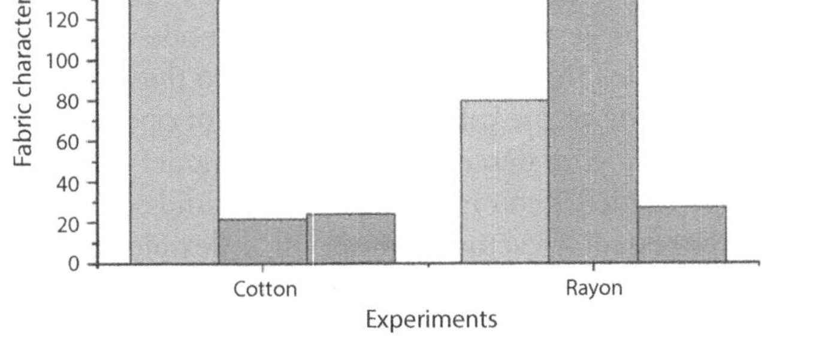

Figure 15.4 Comparison between various mechanical parameters of Cotton and Rayon composite materials used to fabricate soft robots.

the comparison between various mechanical parameters of two different materials used in the fabrication of Soft Robots.

15.5 Graphene Oxide Soft Robots

There is no denying that nature is a never-ending source of inspiration for the creative designing of contemporary AI-automated systems and equipment. The creation of GO-based soft robots using moisture responsive actuators mimics the behavior of living beings. Stimuli-responsive actuators (SRAs), which immediately translate environmental cues into mechanical work, are the essential elements in the development of microelectromechanical systems. SRAs, which are said to be sensitive to environmental cues including temperature, magnetic field, humidity, and chemical reagents, among others, allow for quick and dynamic actuation in plant tendrils and the snapping of Venus. The SRAs that are built on graphene are thought to be special tools that transform environmental cues into autonomic activity. Graphene oxide (GO), a graphene-based actuator with many hydrophilic oxygen-containing-groups (OCGs), is an emerging and intriguing contender for moisture-responsive actuators. Graphene oxide may be produced in large quantities using graphite as the raw material, and, through the chemical exfoliation process, its size, flaws, and degree of oxidation can all be tightly controlled. Graphene oxide is easily mouldable into morphologies including continuous films, fibers, and foams owing to their explicable easy solution-processing property. The number and distribution of OCGs on the GO can be managed using a variety of chemical or physical treatments. When hydroxyl, epoxy, and carboxyl groups are present on GO sheets, GO can establish hydrogen bonds with the molecules of water.

When exposed to moisture, GO's strong water absorption capacity causes a moisture-responsive swelling effect. Due to the existence of OCGs, GO appears to interact with water molecules more strongly than graphene, making it an attractive choice for moisture-responsive smart devices. Due to the many hydrophilic OCGs produced by the chemical oxidation of natural graphite, the conjugation of graphene sheets is broken. The highly oxidized graphite that becomes a single layer of GO as a result of ultrasonication can simply be converted into an aqueous solution of GO. Because we are creating a soft robot, as the name suggests, single GO sheets can be stacked to create strong, flexible GO films. Because to the water molecules' rapid velocity of diffusion among the GO sheets, GO not only exhibits a strong water interaction but is also extremely sensitive to moisture. Because we are creating a soft robot, as the name suggests, single GO sheets can be stacked to create strong, flexible GO films. Because of the water molecules' rapid velocity of diffusion among the GO sheets, GO not only exhibits a strong water interaction but is also extremely sensitive to moisture [20, 23]. Three primary mechanisms—transport over in-plane defects, transport via sheet edges, and interlayer transport—are involved in the movement of water molecules within GO sheets. When GO films come into contact with moisture, the moisture actuates the GO films, causing a high number of molecules to be adsorb, which causes the GO films to rapidly sweep away. The adsorption of the water molecules causes changes in the GO's structural and physical characteristics [20–22]. Occasionally soft robots can also be made to do some grasping functions, and, to improve gripping function control, the GO actuators are integrated with a plastic tube that links to a moisture source. GO and its hybrid are frequently used as scaffolds in the growth of human mesenchymal stem cells.

15.6 Application in Minimally Invasive Surgery

Recent advancements in robotics technology have produced novel, less invasive surgical methods. In this paradigm, less invasive procedures like laparoscopy frequently take the role of open surgery along the direction of surgeries that are minimally invasive (MIS). The patient has benefited greatly from MIS, which includes: (a) reduced trauma, (b) lower chance of infection, (c) quicker recuperation, and (d) better cosmetics. In laparoscopic surgery, several small incisions (ranging in diameter from 1 mm to 15 mm) are used instead of a single, big one to introduce lengthy instruments and carry out the procedure. The laparoscopic technique has been embraced in many medical specialties, including urological, thoracic, and paediatric surgery, because of its benefits in decreasing trauma caused due to surgery, post-operative discomfort, and issues involving cosmetic products. In this study, the authors discovered the procedure for the complementary usage of traditional rigid tools and soft robotics technologies to enhance the overall capabilities of the robot's system without compromising the degree of usefulness of the robotic environment. The end result shows that the suggested manipulator involving soft robots and the DVRK operational robot platform's software and hardware integration were effectively accomplished. Another developer used soft materials to create grippers operated using robots for brittle, mild, and secure working combined with the external interactions. The influence of surgical tools' lack of sensorization, which continues to be a constraint based on the feedback generated due to the force in MIS operations, is reduced and the proposed tool, which is identified using a particular mechanism which is self-regulated, is used here instead. A novel method called scale jamming is used in a soft robotic manipulator to regulate the stiffness factor during surgeries. In order to manage the extent of stiffness of manipulators, particularly continuum manipulators, the frictional force generated per the Coulomb's principle in the centre portion of rigid scales, a scale jamming theory is proposed, which draws inspiration from fish and snake scales. In order to reduce weight, volume, and complexity, the scale jamming depends on managing the torsion generated by shear force in the cross section of a spring, particularly helical spring, using minute sized stiff scales [24].

A soft robot could be used to identify the position of internal organs or spaces between the internal organs using a highly expert two-module soft robot during a minimally invasive surgery. This research aims to create an inherently safe robot with remarkable proficiency in the limited MIS operating space while maintaining the geometrical dimensions of common MIS devices. It offers all directional motion, which may not be achieved using conventional tools. Design improvement not only complies with the requirements of the operating room but also improves the robot's performance [25]. Preclinical assessments can be made in animals using a soft robot that is cable-driven, especially in the field of cardiothoracic endoscopy and its related surgeries. The soft robotic systems adaptability, controllability, and reachability in surgical space have all been verified. Additionally, it has been demonstrated that this device can be employed for a variety of surgeries in the pericardial and thoracic cavities. For minimally invasive surgery, the soft robot manipulator has a greater diameter. Our upcoming effort includes minimizing the diameter of manipulator of the soft robot along with the feeding station's dimensions [26]. Pop-up soft robot, used primarily for surgeries that are minimally invasive, is the first invasive surgical robot, to the authors' knowledge, whose frame and actuators were built utilizing laser welding manufacturing. As a result, the design is low-profile and affordable. The soft hydraulic actuators described in are used in place of the lengthy force transmission cables used in the inflatable Cyclops robot in this study. As a result, flexible plastic film is used to create the robot's structure as well as its actuators. A key advantage of this architecture is the ease with which the robot may be folded, rolled, and/or compressed before being deployed [7, 27]. It is important to ponder over the advancements and futuristic looks on the minimally invasive surgeries achieved using soft robots. This article gives a thorough summary of recent developments and lists the obstacles still to be overcome during the creation of soft robotics employed in minimally invasive surgery. While many surgical procedures are now made simpler by robot-assisted MIS techniques, laparoscopic surgery has not been able to be entirely replaced by them. This may seem counterintuitive, provided the stiffness and dynamic function restrictions of linear-operating tools and the relatively simple action with which soft robots would be able to avoid barriers and pass further into the limited spaces of the cavity of the abdomen (in addition to their intrinsic qualities involving safety) [28].

15.7 Challenges of Soft Robots

Soft robotics is a rapidly developing area of study that offers maximum potential for a wide range of applications; however, it also shows a list of challenges that are needed to be addressed. A typical challenge is the development of new-fangled control algorithms and actuators that can adequately control the motion and behavior of soft robots [29]. Another major challenge is the creation of cosmopolitan sensors that can meticulously measure and interpret the deformation and movement of soft materials [30]. The selection and development of appropriate materials for soft robots is also crucial, as the integration of different materials can improve the performance and accomplishment of these systems [31]. Additionally, soft robots are often unified with other technologies such as electronics and artificial intelligence, which presents additional challenges in terms of control, sensing, and materials [7]. Another main challenge in soft robotics is the development of competent and efficient power and energy systems for soft robots [32]. Another considerable challenge is the development of methods for strengthening soft robots to larger sizes and more complex designs [9]. Figure 15.5 briefly outlines the various challenges faced with soft robot systems, starting from Modeling to Stability control. The durability of soft robots is also a concern, as soft robots are often constituted by compliant materials that can be subject to repeated wear issues and tear in excess of a set limit [33]. Assuring the safety of soft robots, particularly when they are interacting with humans, is also an important challenge. The development of self-healing and self-repair capabilities and biomimetics and bio-inspired design can also improve the efficiency of soft robots [34]. Real-time control of soft robots is a challenging issue because of the intricacy and nonlinearity of soft materials, and the interaction of soft robots with chaotic environments requires advanced sensing and control algorithms [35]. Autonomous behavior is a challenging task in soft robotics, as it requires advanced control algorithms, advanced sensors, and efficient

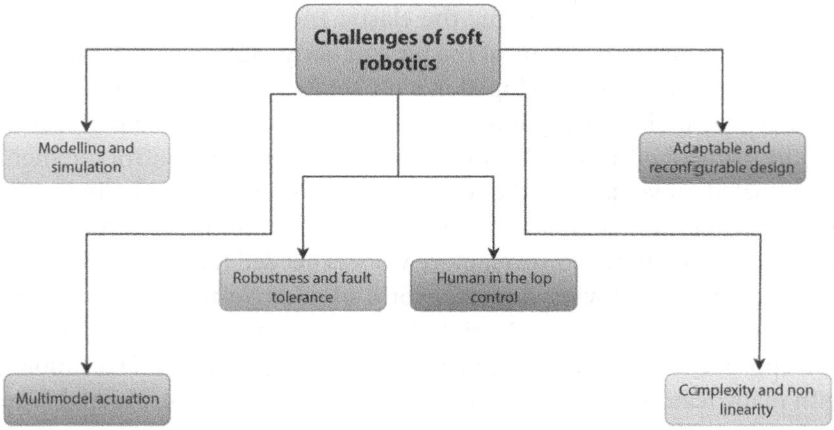

Figure 15.5 Challenges to soft robotics.

power and energy systems [36]. Modeling and simulating the behavior of soft robots is a challenging task due to the entanglement and intermittence of soft materials. This is a critical challenge as it helps in understanding and predicting the behavior of soft robots [37]. The development of adaptive and fault tolerant soft robots is another challenging task that can improve the performance and capabilities of these systems [38]. Soft robots are subject to a variety of environmental and operational conditions, and the development of robust and fault-tolerant systems is a challenging task that can improve the dependability of these systems [39]. Human-in-the-loop control is also a challenging task in soft robotics, as it requires advanced control algorithms, advanced sensors, and efficient power and energy systems [40]. Lastly, the integration of different actuation modalities in the design of soft robots is a challenging task that can improve the performance and capabilities of these systems by allowing them to perform multiple tasks [41].

15.8 Future Scope and Development

As the future era of robotics accelerates, it is important to achieve maximum access to the extended production and working principles of robots in today's world from the point of view of the research aspects related to the more effective and efficient work of machines. Although the field of soft robotics blends seamlessly into the environment in terms of flexibility and ease of movement, it is important to understand the future requirements of the same category. Several new biomedical concepts can be applied in the development field to mimic the elastic and rheological properties of biological tissues and organs of the human body [42]. Elastically soft, versatile, and biologically developed machines can represent a very multidisciplinary technological paradigm that can certainly revolutionize robotics also in the field of health care [43]. Controllers used to control the highly coordinated movements of the soft robot can also help. Morphological features of complex manipulators can be helpful for very precise, robust, and skilled manipulation. However, more work is needed than implementation and coding [44]. Moving into the world of additive manufacturing, it is now possible to produce 3D printed soft robots with them. In addition, new approaches are used to design or build soft robots using 4D technology with many features such as geometric, functional, and control capabilities. The process is done using machine learning and robot algorithm control. The construction of 4D-printed soft robots uses sensors printed with a combination of 3D-printed soft features, combined with actuators and artificial intelligence and machine learning (AIML) algorithms and finite element method (FEM) showing a feasible manufacturing process [45]. As the future era of robotics accelerates, it is important to achieve maximum access to the extended production and working principles of robots in today's world from the point of view of the research aspects related to the more effective and efficient work of machines. Although the field of soft robotics blends seamlessly into the environment in terms of flexibility and ease of movement, it is important to understand the future requirements of the same category. Several new biomedical concepts can be applied in the development field to mimic the elastic and rheological properties of biological tissues and organs of the human body [42]. Elastically soft, versatile, and biologically developed machines can represent a very multidisciplinary technological paradigm that can certainly revolutionize robotics also in the field of health care [43]. Controllers used to control the highly coordinated movements of the soft robot can also help. Morphological features of complex manipulators can be helpful for very precise, robust, and skilled manipulation. However, more work is needed than implementation and coding [44]. Moving into the world of additive manufacturing, it is now possible to produce 3D printed soft robots with them. In addition, new approaches are used to design or build soft robots using 4D technology with many features such as geometric, functional, and control capabilities. The process is done using machine learning and robot algorithm control. The construction of 4D-printed soft robots uses sensors printed with a combination of 3D-printed soft features, combined with actuators and AIML algorithms and FEM showing a feasible manufacturing process [45]. As 4D soft robots are getting closer to being used in various practical fields such as biomedical engineering, the technical and practical development of 4D soft robotics is very important. Intensive research is required to obtain reliable data to assess the same. A wide range of versatile techniques and materials combining two modern technologies, 4D printing and the use of soft robots in nanoscale biomedicine, are being developed to create advances [46]. A new series of soft robots with arm configurations are then combined with biotensegrity and thin artificial muscles. The analysis analyzed methods for its rapid production using a unit called DEMES and several cells with two robot systems. The scope of the dielectric elastomer minimum energy structure (DEMES) study should be extended to more than one area. This extended study [47] uses a dynamic particle spring network model to identify energy states with minimum energy and maximum stability. Additional aspects using robot production systems (SRP) technology also include extensive improvements in fabrication, motion, characterization, force modeling, and a brief introduction to one particular application of SRP to carefully equip it for future applications in the given technology. Patient trials may be conducted to develop the scope of research to improve SWAP designs in the future, such as attaching SWAP to a person's arm. All of this is possible with SRP, which has proven to be an important contribution to the soft robotics community. This expands and nurtures the design diversity of soft robots by expanding their scope [48]. Microfluidic systems are widely used in procedures such as capillary electrophoresis, isoelectric focusing, and other areas involving fluids in biomedical applications. Figure 15.6 introduces the futuristic possibilities that lie ahead in the deployment of Soft Robots in everyday life - industrially and commercially. To work effectively with microfluidics, soft robots are synergized with them as a step toward the future. The synergy between microfluidics and soft robotics is introduced as a potential research topic, and future research can reveal the advantages of these two fields. That field serves creativity in implementation because it focuses on the miniaturization of both the control system and the fluid actuators without changing the functional aspects whenever possible [49]. Soft robotics can also be extended as future perspectives in minimally invasive marine biology. Gesture controls, including haptics, 3D imaging, dark imaging, microscopy, and advanced genomics, can be added to advance the field of soft robots [50]. Soft robotic arms with tactile sensors in underwater robotics enhance the ability to interact delicately and precisely with fragile marine environments, providing improved adaptability and sensitivity for tasks such as exploration and sample collection. The history, development, and research projects in the field of soft robotics, combined with the functioning of the sense of touch in underwater environments, highlight significant

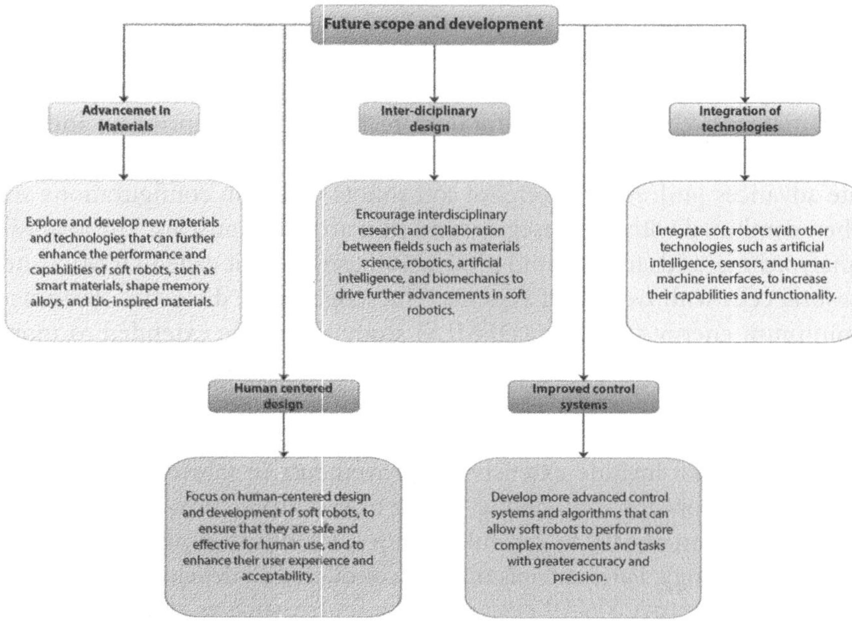

Figure 15.6 Methodologies in future development of soft robots.

advancements in creating adaptable, flexible robots capable of delicate interactions with marine ecosystems. This interdisciplinary approach has led to the design of soft robotic arms equipped with tactile sensors, enhancing their ability to perform precise tasks such as exploration and sample collection while minimizing damage to fragile underwater habitats. The integration of tactile sensing in soft robotics continues to evolve, promising further improvements in underwater operations and broader applications across various fields. The further development of soft robotic systems offers great opportunities to change and develop the field of ocean research, paying special attention to the interaction of conservatives with various foreign bodies without damaging the systems [51]. AI systems can also provide autonomous vehicle research using soft robots as well. This is done by combining soft robotic mechanical systems and artificial intelligence algorithms [52].

15.9 Conclusion

With the benefit of the usage of principles of artificial intelligence, embodied deeply into the robot dynamics, soft robots permit us to develop artificial intelligence, incorporating biological principles, in methods that are not reliable or possible with robots that are rigid bodied [53]. It is essential to measure torque after gears to enforce high performance soft robots. PhysXI is used to effectively analyze the performance of a designated soft robot and represent how a system responds to actuation. With the increased usage of composites in the making of soft robots more eco-friendly and bio-inspired soft robots can be developed.

Graphene oxide (GO) can be made with graphene as the raw material [54], which is present in abundance. The phenomenal water adsorption ability of GO when contact with moisture results in a swelling effect [54, 55]. Design improvement not only complies with the requirements of the operating room but also improves the robot's performance [56]. Even though the process of developing soft robots proved to have many challenges, with keen research ways to overcome it, identifying unique and creative solutions that are not only sustainable but also futuristic can be obtained along with its many applications. Because soft robots are designed in a particular way to imitate the animal cells, it finds its future scope in various practical fields that could bring about a promising good technological improvement.

References

1. Hiller, J. and Lipson, H., Automatic design and manufacture of soft robots. *IEEE Trans. Rob.*, 28, 457–466, 2012.
2. Auerbach, J.E. and Bongard, J.C., Evolving CPPNs to grow three dimensional physical structures. *Proceedings of the Genetic& Evolutionary Computation Conference*, ACM, New York, pp. 627–634, 2010.
3. Bendsoe, M.P. and Kikuchi, N., Generating optimal topologies in structural design using a homogenization method. *Comput. Methods Appl. Mech. Eng.*, 71, 2, 197–224, 1988.s.
4. Laschi, C., Cianchetti, M., Mazzolai, B., Margheri, L., Follador, M., Dario, P., Soft robot arm inspired by the octopus. *Adv. Robot.*, 26, 7, 709–727, 2012, Doi: 10.1163/156855312X626343.
5. Barbic, J. and Popovič, J., Real-time control of physically based simulations using gentle forces. *ACM Trans. Graphics (TOG)*, 27, 5, 163, 2008.
6. Miriyev, A., Stack, K., Lipson, H., Soft material for soft actuators. *Nat. Commun.*, 8, 596–605, 2017.
7. Rus, D. and Tolley, M., Design, fabrication and control of soft robots. *Nature*, 521, 467–475, 2015.
8. Truby, R.L. *et al.*, Soft Somatosensitive Actuators via Embedded 3D Printing. *Adv. Mater.*, 30, 34, 1801360, 2018.

9. Polygerinos, P. *et al.*, Soft Robotics: Review of Fluid-Driven Intrinsically Soft Devices; Manufacturing, Sensing, Control, and Applications in Human-Robot Interaction. *Adv. Eng. Mater.*, 17, 4, 476–494, 2015.
10. Laschi, C. *et al.*, Soft Robotics: Technologies and Systems Pushing the Boundaries of Robot Abilities. *Sci. Rob.*, 1, 1, eaah3690, 2016.
11. Luo, Y., Zhao, Z., Kim, K., Yi, J., Shen, Y., Inchworm locomotion mechanism inspired self-deformable capsule-like robot: Design, modeling, and experimental validation. *Proceedings of the IEEE International Conference on Robotics and Automation*, Brisbane, Australia, pp. 6800–6805, 2018.
12. Albu-Sch€ffer, A., Ott, C., Hirzinger, G., A unified passivity based control framework for position, torque and impedance control of flexible joint robots. *Int. J. Rob. Res.*, 26, 1, 23–39, 2007.
13. Ott, C., Albu-Sch€ffer, A., Kugi, A., Hirzinger, G., On the passivity based impedance control of flexible joint robots. *IEEE Trans. Robot. Autom.*, 24, 2, 416–429, 2008.
14. Brown, E. *et al.*, Universal robotic gripper based on the jamming of granular material. *Proc. Natl. Acad. Sci. U.S.A.*, 107, 18809–18814, 2010.
15. Gravagne, I.A., Rahn, C.D., Walker, I.D., Large deflection dynamics and control for planar continuum robots. *IEEE/ASME Trans. Mechatron.*, 8, 299–307, 2003.
16. Chirikjian, G.S., Hyper-redundant manipulator dynamics: a continuum approximation. *Adv. Rob.*, 9, 217–243, 1994.
17. Qi, H. and Boyce, M., Stress–strain behavior of thermoplastic polyurethanes. *Mech. Mater.*, 37, 817–839, 2005.
18. Lumia, R. and Shahinpoor, M., IPMC microgripper research and development. *J. Phys. Conf. Ser.*, 127, 0–15, 2008, doi: 10.1088/1742-6596/127/1/012002.
19. Zhou, Y., Headings, L.M., Dapino, M.J., Modeling of Fluidic Prestressed Composite Actuators with Application to Soft Robotic Grippers. *IEEE Trans. Rob.*, 38, 2166–2178, 2021.
20. Chen, Z., Shatara, S., Tan, X., Modeling of biomimetic robotic fish propelled by an ionic polymermetal composite caudal fin. *IEEE/ASME Trans. Mechatron.*, 15, 448–459, 2010.
21. Brunetto, P., Fortuna, L., Graziani, S., Strazzeri, S., A model of ionic polymer-metal composite actuators in underwater operations. *Smart Mater. Struct.*, 17, 2, 025029, 2008.
22. Aureli, M., Kopman, V., Porfiri, M., Free-locomotion of underwater vehicles actuated by ionic polymer metal composites. *IEEE/ASME Trans. Mechatron.*, 15, 603–614, 2010.
23. Ul Haq, M., Gang, Z., Ahad, F.E., Hussain, M., Aftab, S., Dynamic analysis of IPMC actuated fin of a micro fish like device. *J. Biomim. Biomater. Biomed. Eng.*, 24, 82–96, 2015.
24. Stassen, H.G., Grimbergen, C.A., Dankelman, J., Introduction to minimally invasive surgery. *Engineering for Patient Safety: Issues in Minimally Invasive Procedure*, pp. 2–18, 2004.

25. Kazanzides, P., Chen, Z., Deguet, A., Fischer, G.S., Taylor, R.H., DiMaio, S.P., An open-source research kit for the da Vinci® Surgical System. Paper presented at: *Robotics and Automation (ICRA), 2014 IEEE International Conference*, Hong Kong, China, May 31 to June 7, 2017.
26. Weber, B. and Schneider, S., The effects of force feedbackon surgical task performance: a meta-analytical integration, in: *Haptics: neuroscience, devices, modeling, and applications*, pp. 150–157, Springer, Berlin, 2014.
27. Michalczyk, K., Analysis of helical compression spring support influence on its deformation. *Arch. Mech. Eng.*, LVI, 25–28, 2009.
28. Wang, H., Wang, C., Chen, W., Liang, X., Liu, Y., Three dimensional dynamics for cabledriven soft manipulator. *IEEE/ASME Trans. Mechatron.*, 2, 8–28, 2016, doi: 10.1109/TMECH.2016.2606547.
29. Runciman, M., Avery, J., Zhao, M., Darzi, A., Mylonas, G.P., Deployable, Variable Stiffness, Cable Driven Robot for Minimally Invasive Surgery. *Front. Rob. AI*, 6, 2020, doi: 10.3389/frobt.2019.00141.
30. Runciman, M., Avery, J., Darzi, A., Mylonas, G., Open Loop Position Control of Soft Hydraulic Actuators for Minimally Invasive Surgery. *Appl. Sci.*, 11, 16, 2021, doi: 10.3390/app11167391.
31. Fras, J., Arezzo, A., Shiva, A., Althoefer, K., Soft robotics solutions for minimally invasive surgery: The need for stiffness controllability, in: *Soft Matter for Biomedical Applications*, pp. 684–719, The Royal Society Of Chemistry, London, U.K., 2021.
32. Wang, Z. and Galloway, K.C., Soft robotics: a review of recent advances. *Annu. Rev. Biomed. Eng.*, 20, 275–302, 2018.
33. Ilievski, F., Mazzeo, A.D., Shepherd, R.F., Chen, X., Whitesides, G.M., Soft robotics for chemists. *Angew. Chem. Int. Ed.*, 50, 8, 1890–1895, 2011.
34. Laschi, C. and Cianchetti, M., Soft robotics: new perspectives for robot bodyware and control. *Front. Bioeng. Biotechnol.*, 2, 3, 2014.
35. Shepherd, R.F., Ilievski, F., Choi, W., Whitesides, G.M., Multigait soft robot. *Proc. Natl. Acad. Sci.*, 108, 51, 20400–20403, 2011.
36. Cianchetti, M., Ranzani, T., Gerboni, G., De Canio, G., Petroni, G., Menciassi, A., Soft robotics technologies to address shortcomings in today's minimally invasive surgery: the STIFF-FLOP approach. *Soft Rob.*, 1, 2, 122–131, 2014.
37. Hines, L., Petersen, K., Lum, G.Z., Sitti, M., Diller, E., Soft actuators for smallscale robotics. *Adv. Mater. Technol.*, 2, 1, 1600176, 2017.
38. Yuk, H., Lin, S., Ma, C., Zhao, X., Hydrogel bioelectronics. *Chem. Soc. Rev.*, 46, 15, 5959–5984, 2017.
39. Keplinger, C., Sun, J.Y., Foo, C.C., Rothemund, P., Whitesides, G.M., Suo, Z., Stretchable, transparent, ionic conductors. *Science*, 341, 6149, 984–987, 2013.
40. Martinez, R.V., Branch, J.L., Fish, C.R., Jin, L., Shepherd, R.F., Nunes, R.M., Whitesides, G.M., Robotic tentacles with three-dimensional mobility based on flexible elastomers. *Adv. Mater.*, 25, 2, 205–212, 2013.

41. Kim, S., Laschi, C., Trimmer, B., Soft robotics: a bioinspired evolution in robotics. *Trends Biotechnol.*, 31, 5, 287–294, 2013.
42. Chen, Y. and Majidi, C., Soft robotics for locomotion. *Proc. IEEE*, 107, 3, 464–474, 2019.
43. Trivedi, D., Rahn, C.D., Kier, W.M., Walker, I.D., Soft robotics: biological inspiration, state of the art, and future. *Appl. Bionics Biomech.*, WILEY, 5, 3, 99–117, September 2008.
44. Polygerinos, P., Lyne, S., Wang, Z., Nicolini, L.F., Mosadegh, B., Whitesides, G.M., Towards a soft pneumatic glove for hand rehabilitation. *Rob. Auton. Syst.*, 93, 1–10, 2017.
45. Spielberg, S., Gopaluni, R., Loewen, P., Deep reinforcement learning approaches for process control, in: *Advanced Control of Industrial Processes (AdCONIP), 2017 6th International Symposium on*, IEEE, 2017.
46. Takaoka, M., Suzumori, K., Wakimoto, S., Iijima, S., Tokumiya, T., Fabrication of thin McKibben artificial muscle with various design parameters and their experimental evaluations, in: *Proceedings of International Conference on Manufacturing, Machine Design and Tribology*, p. 82, 2013.
47. Kofod, G., Wirges, W., Paajanen, M., Bauer, S., Energy minimization for self-organized structure formation and actuation. *Appl. Phys. Lett.*, 90, 081916, 2007.
48. Robertson, M.A., Sadeghi, H., Florez, J.M., Paik, J., Soft pneumatic actuator fascicles for high force and reliability. *Soft Robot*, 4, 23–32, 2017.
49. Liu, M., Zentgraf, T., Liu, Y., Bartal, G., Zhang, X., Light-driven nanoscale plasmonic motors. *Nat. Nanotechnol.*, 5, 8, 570–573, 2010.
50. Gorissen, B. *et al.*, Hardware sequencing of inflatable nonlinear actuators for autonomous soft robots. *Adv. Mater.*, 31, 3, 1804598, 2019.
51. Abondance, S., Teeple, C.B., Wood, R.J., A dexterous soft robotic hand for delicate in-hand manipulation. *IEEE Robot. Autom. Lett.*, 5, 5502–5509, 2020.
52. Laschi, C., Octobot-A robot octopus points the way to soft robotics. *IEEE Spectr.*, 54, 38–43, 2017. [CrossRef].
53. Smith, J. and Johnson, R., Soft Robotics: Pushing the Boundaries of Automation, in: *Advances in Robotics: Exploring New Frontiers*, R. Thompson (Ed.), pp. 205–220, Wiley, Springer, 2023.
54 Gupta, A. and Patel, S., Soft Robotics: A Pathway to Human-Robot Collaboration, in: *Advances in Robotics and Automation: Trends and Future Directions*, K. Williams (Ed.), pp. 175–190, Wiley, 2023.
55. Anderson, M. and Lee, H., Soft Robotics: From Inspiration to Application, in: *Advances in Robotics: Bridging the Gap Between Theory and Practice*, S. Kim (Ed.), p. 135150, CRC Press, 2023.
56. Chen, L. and Wang, C., Soft Robotics in Industrial Automation: Challenges and Opportunities, in: *Advances in Robotics and Automation: Applications and Trends*, R. Gupta (Ed.), pp. 220–235, Springer, 2023.

16

Investigation on Deployment of Microservices for Swarm Intelligence of Collaborative Robots

Abhijit Bora[1]*, Jonti Deori[1] and Uzzal Sharma[2]

[1]*Department of Computer Applications, Assam Don Bosco University, Assam, India*
[2]*Department of Computer Science, Birangana Sati Sadhani Rajyik Vishwavidyalaya, Assam, India*

Abstract

Collaborative robots, microservices, as well as swarm intelligence can work as a component to support different functional execution of industrial aspects. The swarm intelligence can be developed for logical analysis where the microservice can work as a broker among the collaborative robots and the knowledge hub. Each can work in collaborative manner to support human being while solving functional task of different challenges in the industry. Evaluating the deployment of the collaborative robot service is necessary in the industrial sectors as well as for researchers. So, in this work, a novel model will be proposed where the collaborative robots can utilize the features of microservices and swarm intelligence while serving the human being in the common workspace of industry. The evaluation process of such deployment will be proposed which can be followed while utilizing such collaborative service in different aspects of industrial units. The work will also highlight the deployment constraint and performance aspects of service while utilizing the swarm intelligence through microservices.

***Keywords*:** COBOT, swarm intelligence, microservice, artificial intelligence markup language

**Corresponding author*: abhijit.bora0099@gmail.com

V. Ramasamy, S. Balamurugan and Sheng-Lung Peng (eds.) Intelligent Robots and Cobots: Industry 5.0 Applications, (337–354) © 2025 Scrivener Publishing LLC

16.1 Introduction

The intelligence of the service can be developed and generated through Artificial Intelligence Markup Language (AIML). The microservice (MS) is an approach of developing software applications where different business logic is divided into multiple independent services that communicate through Application Programming Interface. It supports loosely coupled services that communicate through lightweight protocols such as Simple Object Access Protocol. The principle of Service-Oriented Architecture can be deployed through MS [1, 2]. The MS can be handled independently by independent team of developers [3]. As such, the additional complexity occurs while deploying can be reduced.

Throughout the past few decades, the field of robotics has rapidly advanced, and robotics systems are now an essential component of many different industries, including manufacturing, healthcare, logistics, and agriculture. Swarm robotics is one of the most exciting advances in this field; it involves groups of robots working together to do jobs quickly and adjust to changing conditions.

The concept of swarm robotics is inspired by the ideas of swarm intelligence (SI) found in nature, wherein simple creatures like fish, insects, or birds may collaborate to accomplish complicated tasks through decentralized coordination and interaction. In this context, deploying MS appears to be a novel and interesting way to improve the capabilities of collaborative robots.

With the advancement of technology in last few decades, there is a tremendous changes are evolving in the data transfer, internet, and computing world. The monolithic client server scenario is being replaced by cloud computing paradigm. Cloud computing is a system of accessing online resources on-demand by the users globally.

The resources available in the cloud are applications, servers, tools for mobile application development, storage space, networking tools, etc. The resources and services are made available by vendors that any end user can access those functions by purchasing the features for a specific time period. The vendors hosted the applications or services in some remote data centres and virtualized and managed in the cloud by Cloud Service Providers.

MS have become more and more popular across a range of software engineering areas due to their scalability, modularity, and ease of deployment. Their use in robotics offers a chance to get around many of the problems that conventional monolithic robotic systems have, especially when it comes to swarm intelligence. Collaborative robots can interact, adapt, and cooperate more effectively by dividing their functions into smaller, self-contained MS. This allows the robots to leverage the potential of collective intelligence.

In order to investigate the potential synergies between these two paradigm-shifting approaches, this study explores the deployment of MS for SI in collaborative robots. Through our investigation on the incorporation of MS into swarm robotics, we hope to reveal opportunities for improved scalability, coordination, and adaptability in robotic systems. In order to lay the groundwork for future developments in the fields of collaborative robots and swarm intelligence, this research aims to deliver insightful information about the opportunities and difficulties posed by this innovative method. We hope to show the practical significance of this novel method and its potential to redefine the capabilities of collaborative robotic systems through experiments, case studies, and real-world applications.

The public interface of the service can be designed and treated as per requirement of the Collaborative Robots (COBOTS) that used for. Multiple interface of the same service can be developed without disrupting the existence of COBOTS for the task. The sizes of MS can be of small and message enabled that can be developed autonomously [4]. It can be released based on the requirement of the COBOTS. The MS can work beyond monolithic applications that with its business interface can support in implementing layered architecture. The MS can work in independent mode where a change in part of the code requires only redeploying the particular part of the MS [5]. It supports fine-grained interfaces and domain-driven design.

The COBOTS are robotic devices that are deployed for collaborative work with human operators. The operator handles the basic working strategies, while the COBOTS use its software service and constraints for the functional execution of the task. The constraint of the COBOTS is safe as it follows the software-defined protocols. It can generate the surface that is feasible for the workspace of human being.

The COBOTS can utilize the remote MS over Hyper Text Transfer Protocol (HTTP) to perform the objective. The SI through AIML for specific operation can be developed that support the COBOTS and MS while working with human operator. The interface of MS can be developed separately for the specific COBOTS of respective industry. The industry specific operational intelligence and logic can be populated through MS, which, in turn, the COBOTS utilize for the specific task.

However, a proper mechanism is required to be developed while using the components through COBOTS, MS, and SI in the industry.

16.2 Related Work

The study of swarm robotics, in which numerous robots cooperate to finish tasks through decentralized coordination and interaction, is one of the most promising fields of robotics research. Swarm robotics has witnessed tremendous breakthroughs recently. The core of this method is swarm intelligence, which draws inspiration from natural phenomena such as ant colonies and bird flocking.

Because MS design is flexible and scalable, it has become more popular in software engineering at the same time. The deployment of MS for SI in collaborative robots is one area where these two domains overlap, and this literature review examines this junction by assessing the state of the field and its implications for robotics in the future. A potential approach to improve the flexibility, scalability, and efficiency of robotic systems is to combine SI with MS architecture in the context of collaborative robots. An overview of the research done in this area is given in this review of the literature, along with a focus on the major discoveries and contributions.

Jiang *et al.* (2020) proposed a novel monitoring mechanism for identification of fault in the particular MS among multiple set of MS [6]. The importance of intelligent alarm for fault detection of failures in MS was highlighted. Wan *et al.* (2020) discussed the importance of latency for cloud-based software services [7]. The utilization of resources in the MS and balancing the load during service execution can be evaluated through their mechanism. They proposed a technique to reduce the overall latency of the system.

Aktas *et al.* (2022) developed a mechanism to identify the fault in telecommunication system through deployment of MS [8]. A novel monitoring mechanism was introduced that can quickly observe the error and ensures smooth operability of the service. Ma *et al.* (2022) discussed the importance of MS architecture over monolithic conventional architecture [5]. They proposed a software project called as microservices identification using analysis for database access (MIADA) for identification of MS through analysis of database access. The efficiency and applicability of the MIADA was evaluated using comparative analysis.

Bicchi *et al.* (2008) stated that Human Right Council (HRC) refers to application scenarios where a robot, usually, COBOT, and a human occupy the same workspace and interact to accomplish collaborative tasks [9]. Following the introduction of the Universal Robot (UR), a COBOT was produced by Universal Robotics in 2008 [10], industrial interest in applying HRC and COBOTS to factory floors has escalated. Many other robotics manufacturers, such as KUKA, ABB, and Rethink Robotics, have also developed their own COBOTS, each tackling a particular niche.

An overview of COBOTS on the market is discussed [11], comparing their costs, payload capabilities, and safety features. Because COBOTS are built for close-proximity interaction with humans, they must adhere to stringent safety requirements, such as power and speed limiting, soft padding, and the absence of trap points (i.e., points that can trap body parts or clothing) as per ISO-TS 15066 [12].

Considering the distinguishing characteristics of COBOTS over regular robots, they are envisioned to pave the way for mass customization, decrease required floor space, increase product quality and production efficiency, and improve working conditions for humans [13, 14].

Yin *et al.* (2021) proposed a delivery robot cloud platform based on MS that enabled the rapid development and deployment of applications and reduced the development time [15].

An approach called cooperation offloading was proposed by Yuanyzhao *et al.* (2021) that offloads the intensive communication among robots along with compute-intensive and data-intensive tasks [16]. Based on this approach, a framework named Cloudroid Swarm was developed to transparently migrate multirobot applications to cloud servers. A framework, CoRoL, for on-floor COBOT resource sharing supports complex ML task execution resource efficiency and manages the potential presence of malicious entities in heterogeneous manufacturing setup [17].

Mazdin *et al.* (2019) presented a method with swarm based COBOT for simultaneous coverage of surfaces [18]. The robots are assigned to different viewpoints in order to allow effective 3D reconstruction of object. Many leading organizations are using the power of human-machine collaboration for the transformation of their business and improve the bottom line.

Swarm robotics uses the concepts of collective intelligence to make it possible for cooperative robots to function as a unit. This field of study focuses on creating algorithms that make coordination and cooperation easier. Notable algorithms that can improve the performance of robot swarms are particle swarm optimization and ant colony optimization [24].

The main way to get into the swarm robotics research field for a new edge is through the work of Dorigo *et al.* [25]. The work explains every aspect of swarm robotics, starting with its foundations, innate qualities, and intriguing ownerships.

Garattoni *et al.* [26] addressed the description of an engineering process for the creation, analysis, and preservation of robotic swarms. The writers then present swarm robotics from an engineering perspective and discuss related works that support the advancement of swarm robotics as a field of engineering. One more stubborn defiance in swarm robotics relates to formal modelling of systems that can simulate robotic swarms in an efficient manner.

It traverses the potential applications of swarm robotics and scientific involvements, linking some of the guiding principles of ongoing research projects and relating on numerous aspects of design, analysis, and emerging behaviors, before posing some unanswered questions about swarm robotics in practical applications.

The concept of decentralized control is fundamental to swarm intelligence. Individual robots interact with peers to accomplish group objectives and make decisions based on available local information. Microservices can help with the robust communication mechanisms that are required for this [27, 28]. In a decentralized swarm, autonomous agents act on local information rather than directives from an imposing dominion that work together to generate complex behaviour [29, 30].

16.3 Objective and Focus of the Work

The objective of the proposed work is to evaluate the performance of the MS that contains respective intelligence for collaborative knowledge pool of COBOT while deploying. The collaborative execution of multiple MS and their performance impacts on the server resources. It can highlight important key aspects for different services of COBOT. As such, the objective of the proposed work is to evaluate the overall service performance of MS as well as SI pool and COBOT communication.

The focus of the work highlights the programming logic of COBOT. The programming logic for the COBOT is to deliver the ability to evaluate the state of the problem and carry the action that helps in collaborative execution. Traditionally, for a robotic program, a human being is involved in offline mode only. The programs that are involved in this execution are not flexible enough or human aware that cannot be changed during program runtime.

The functional method of a robot works in deterministic environment. In human computer interaction (HCI) method, the operator can add additional functionality to the environment. In the program of COBOT, the involvement of human goes beyond the traditional role programmers. The explicit and implicit involvement of an operator can be experienced for affecting the execution of the COBOT program. The explicit involvement of operator reflects the direct communication where the operator can instruct the COBOT for functional execution. The implicit involvement of operator reflects the indirect communication where the COBOT realizes the state of the human being and executes the functional policies accordingly.

The functional policy can be generated based on the pre conditioned data set of the problem or it can be set by the programmer as per probability of the problem execution. Based on the involvement of operational mode, the proposed work identifies three vital areas where the programming logic can be developed and deployed for flexible execution of the collaborative work in supportive as well as in sequential execution of HCI methods. They are (1) learning (2) optimization, and (3) communication.

In learning stage, the COBOTS learn from the operator the activities as well as the skills that are primarily required for problem solution. The activities may include trial and error, observation, and evaluation of answers through the questionnaires. In this stage, the programmer develops the learning model from which the COBOTS learn the basic instructions. It could be in pattern of iterations with trial-and-error policy, testing and training data, demonstration of work, etc. During runtime, the operator could explicitly affect the working policy of COBOT by providing explicit parameter such as environment parameter, feedback and question answers, and personalized working process. The operator can also help in prior while developing the probabilistic knowledge of the learning algorithm based on the practical environment.

In the communication stage, the operator communicates the COBOT through different verbal and non-verbal instructions. The communication through non-verbal instruction includes user interface, gestures, haptics, etc. The programmer plans all programming logic and actions based on the possible functionality for which the COBOT may perform against the verbal and non-verbal instructions. In this stage, the operator may trigger the COBOT's functionality that is predefined in the system.

In the optimization stage, the algorithms that are developed are optimized to minimize the functional execution of the COBOT. The mathematical models are developed to reduce the cost of workload, consumption of energy and time. The optimization also maximizes the quality of the product as well as the physical comfort to achieve the desirable performance. During runtime of COBOT's program, the human being can impact the performance of COBOT, as it implicitly affects the functional cost. The optimization can provided better output than a human being in the production.

16.4 Experimental Arrangement

The experimental arrangement for the proposed work is discussed in this section. Figure 16.1 shows the block diagram of the experimental arrangement. Here, the user interface (UI) of COBOT usually contains the interactive presentation of the system. It captures the command from the end user or the operator of COBOT. The UI program communicates the parent MS which contains the intelligence. The intelligence of parent MS is deployed through the features of AIML.

The AIML file contains all relevant information for respective operations related to communication, optimization, and learning service. Based on functional requirement, the parent MS passes the instruction to the respective sub-MS. The sub-MS also contains the relevant intelligence for the respective core logical function. If required, the sub-MS can interact with other MSs for collaborative execution of intelligence. As such, the service of SI can be generated through collaborative MSs. All sub-MS can deliver the functional output to the parent MS for service execution. The quality aspects of MS execution are evaluated through performance analysis. The flowchart of assessment framework for evaluating the performance of MS is shown in Figure 16.2.

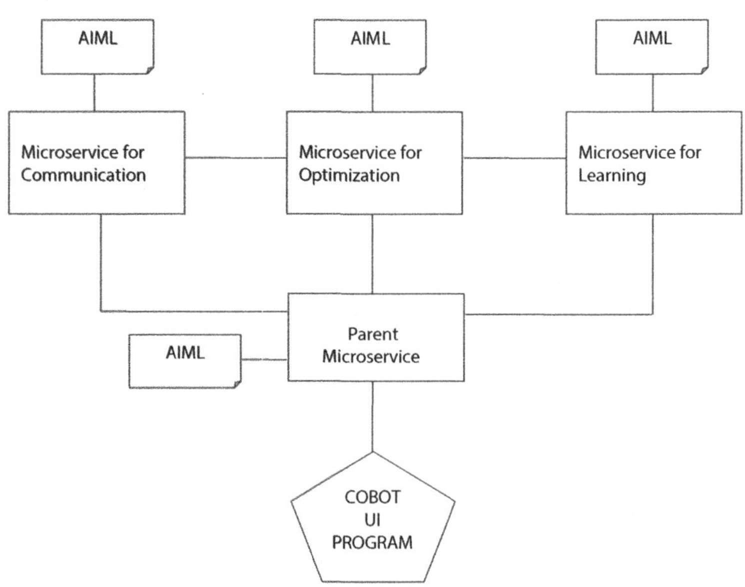

Figure 16.1 Block diagram of experimental arrangement.

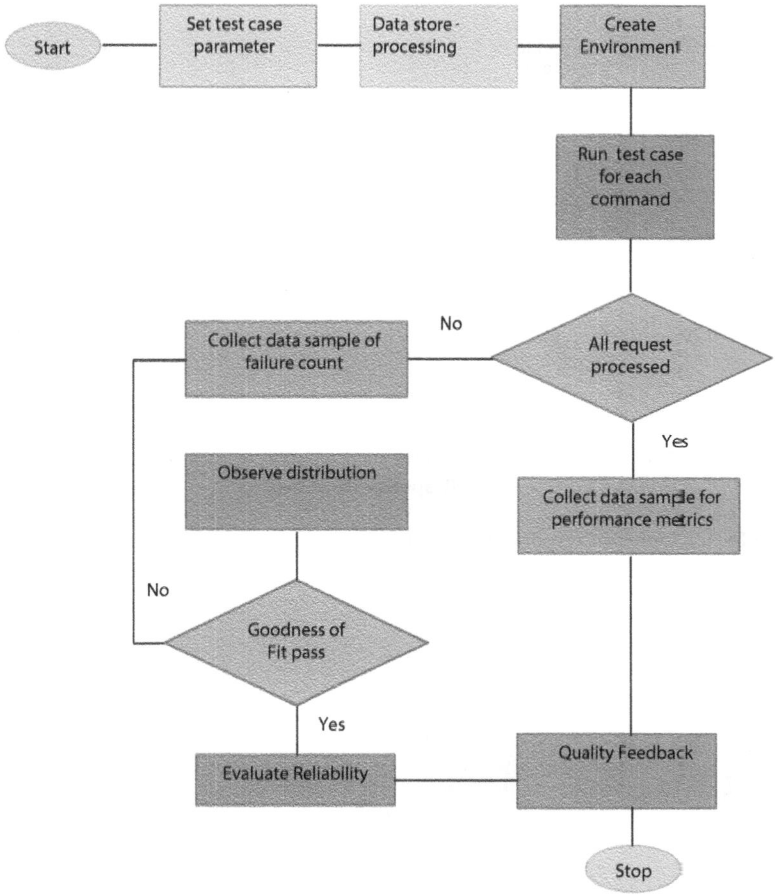

Figure 16.2 Flowchart of assessment framework for evaluating the performance of microservice.

The consecution execution for 100 accesses of COBOT UI is evaluated, and the performance metric response time is observed. The response time further evaluated with sample analysis. The distribution of response time is evaluated through data visualization technique. Figures 16.3 and 16.4 show the probability plot of normality and histogram with single peak value. As such, it can be concluded that the data sample is normal in distribution.

However, the consecution execution up to 1,800 accesses of COBOT UI is seen without failure in command execution. However, beyond that, the MS execution fails in service. More fault count but with highest stability of above deployment methodology is observed beyond 1,800 accesses. As such, a

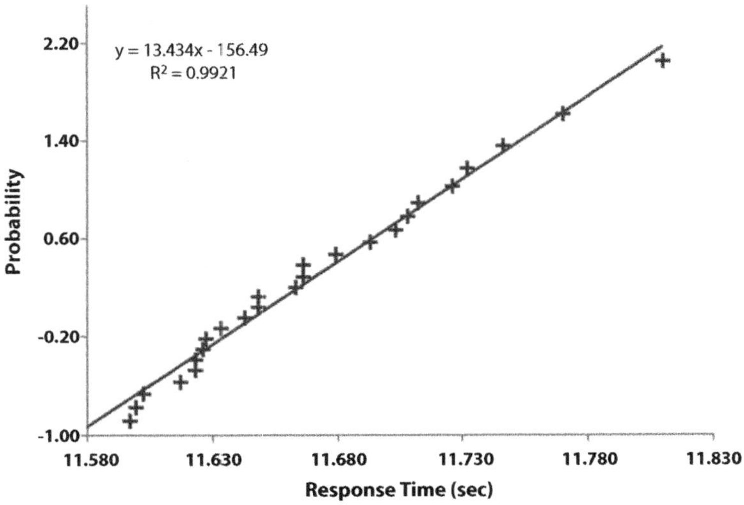

Figure 16.3 NP plot for response time of the service.

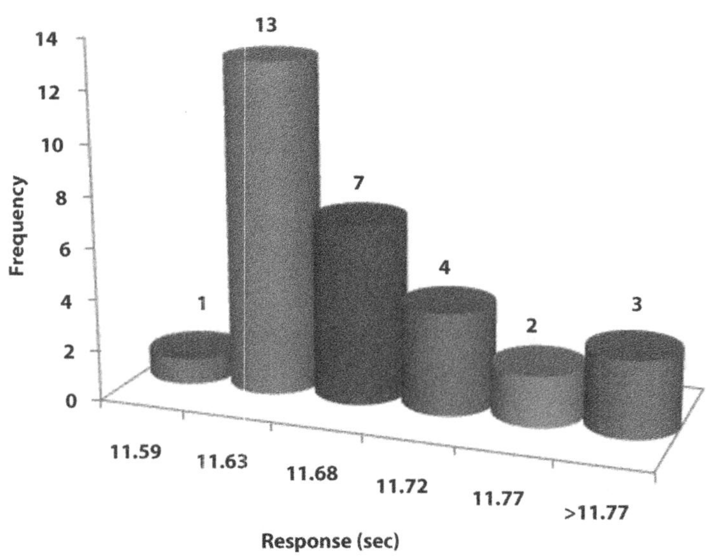

Figure 16.4 HG plot for response time of the service.

further study is carried out to observe the reliability of the proposed deployment beyond its stability stage. For stress level of 1,800 accesses, the failure records are observed. The data plot of failure count for the service is shown in Figure 16.5. Table 16.1 shows the failure count ranges.

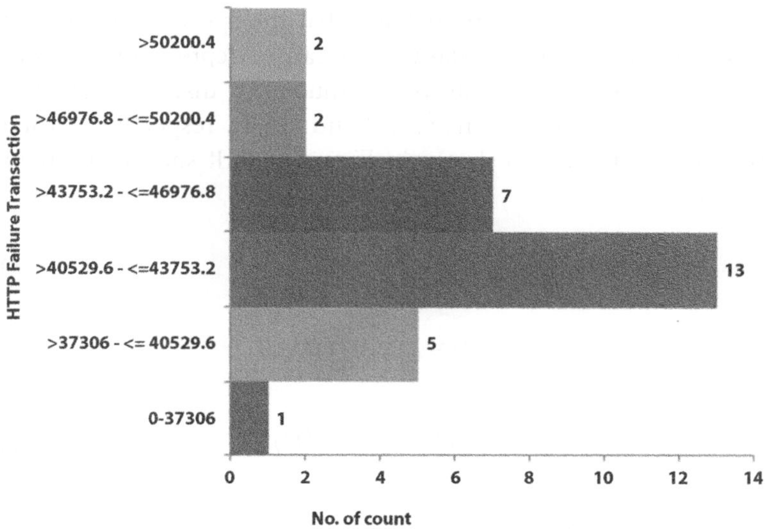

Figure 16.5 Data plot of failure count for the service.

Table 16.1 Failure count range and frequency.

Range of failure count	Frequency
>0–≤37306	1
>37,306–≤40529.6	5
>40,529.6–≤43753.2	13
>43,753.2–≤46976.8	7
>46,976.8–≤50200.4	2
>50,200.4	2

The Weibull Cumulative Distribution Function (CDF) for the recorded fault count of proposed deployment is evaluated using Equation (16.1) as discussed elsewhere at Chandra *et al.* and Coles [19–21].

$$\text{CDF} = 1 - \exp\{(\text{-Fc/SC})^{\text{SH}}\} \tag{16.1}$$

The CDF represents the probability that the specific quality attribute holds a value less than or equal to x. It can be represented as Equation (16.2). If the distribution follows continuous or discrete, then the $f(x)$ can be expressed as Equations (16.3) and (16.4), respectively. The CDF of observed fault count is shown in Figure 16.6. It shows continuous in distribution.

$$f(x)=P[X\leq x]=\alpha \tag{16.2}$$

$$f(x)=\int_{-\infty}^{x} f(\mu)d\mu \tag{16.3}$$

$$f(x)=\sum_{i=0}^{x} f(i) \tag{16.4}$$

The observation of the test is further evaluated using Anderson-Darling test (ADT). The expression for ADT test is defined as A^2, and Equations (16.5) and (16.6) are discussed elsewhere at [22, 23]. Whereas, for small data sample, the modified ADT can be expressed as using Equation (16.7). The intermediate assessment of ADT for normality is shown in Table 16.2.

$$A^2 = -N - S \tag{16.5}$$

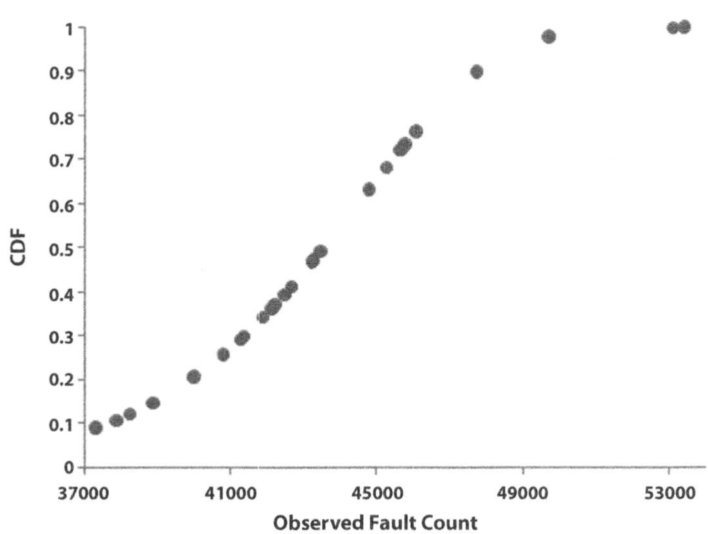

Figure 16.6 Normal CDF plot of observed fault count.

Table 16.2 Assessment of case III with ADT for test of normality.

Row count	**F(y)**	**1 – F(y)**	**1 – $F(y_{n+1-i})$**	**ln F(y)**	**ln(1 – $F(y_{n+1-i})$)**
1	0.07	0.92	0.03	−2.61	−3.41
2	0.08	0.91	0.03	−2.49	−3.40
3	0.09	0.91	0.03	−2.40	−3.24
4	0.10	0.89	0.07	−2.24	−2.65
5	0.14	0.85	0.16	−1.90	−1.82
6	0.15	0.84	0.19	−1.89	−1.65
7	0.19	0.80	0.19	−1.61	−1.62
8	0.23	0.76	0.20	−1.44	−1.60
9	0.24	0.75	0.20	−1.41	−1.59
10	0.29	0.70	0.24	−1.22	−1.40
11	0.32	0.67	0.30	−1.13	−1.13
12	0.32	0.67	0.49	−1.11	−0.69
13	0.33	0.66	0.52	−1.10	−0.64
14	0.36	0.63	0.53	−1.01	−0.63
15	0.38	0.61	0.53	−0.94	−0.63
16	0.46	0.53	0.61	−0.76	−0.49
17	0.46	0.53	0.63	−0.75	−0.45
18	0.47	0.52	0.66	−0.74	−0.41
19	0.50	0.49	0.67	−0.68	−0.39
20	0.69	0.30	0.67	−0.36	−0.38
21	0.75	0.24	0.70	−0.28	−0.34
22	0.79	0.20	0.75	−0.22	−0.27
23	0.79	0.20	0.76	−0.22	−0.26
24	0.80	0.19	0.80	−0.21	−0.22

(Continued)

Table 16.2 Assessment of case III with ADT for test of normality. (*Continued*)

Row count	F(y)	1 – F(y)	1 – F(y_{n+1-i})	ln F(y)	ln(1 – F(y_{n+1-i}))
25	0.81	0.19	0.84	−0.21	−0.16
26	0.83	0.16	0.85	−0.17	−0.16
27	0.93	0.07	0.89	−0.07	−0.11
28	0.96	0.03	0.91	−0.04	−0.09
29	0.96	0.03	0.91	−0.0	−0.08
30	0.96	0.03	0.92	−0.03	−0.07

$$S = \sum_{i=1}^{n} \frac{(2i-1)}{N}[\ln F(Y_i) + \ln(1 - F(Y_{N+1-i}))] \quad (16.6)$$

$$A^{2^*} = (1 + 0.2/\sqrt{n})A^2 \quad (16.7)$$

The AD test expresses the null hypothesis as the collected data sample of SoC for measurement of PV module is from the specified distribution. The alternate hypothesis assumes that the data sample of Soc is not from the specified distribution. The A^{2*} and p-values are evaluated to be 0.582 and 0.129, respectively. As the evaluated A^{2*} and p-value is greater than 0.05, the null hypothesis can be accepted. As such, the evidence is achieved that there is no difference exists in between the normal data and the collected data sample of MS execution for COBOT UI program.

16.5 Reliability Assessment

The Weibull distribution reliability metric (R) is evaluated for COBOT UI by using Equation (16.8) [18]. The population mean of CDF is evaluated using Equation (16.9).

$$R = (1\text{-CDF}) \quad (16.8)$$

$$M_P = \text{Mean} \pm T.S_{Dev} / \sqrt{N} \quad (16.9)$$

In this case, Equation (16.8) is evaluated as 1 – (CDF Mean). The CDF mean of Case III is observed to be 0.49. As such the R is evaluated to be 0.51. It means that, even during observation of fault count, the reliable service delivery is 51%. At 95% confidence level, the $T_{(0.05, 29)}$ and S_{Dev} are evaluated to be 2.045 and 0.277, respectively. The M_p lies in between 0.59 to 0.39. As such using Equation (16.9), the population reliability lies in between 0.41 and 0.61 that is the during high stress level the reliability of the service monitoring lies between 41% and 61%.

16.6 Overall Discussion

The overall observation is shown in Table 16.3. The stability of the MS for SI execution for COBOT is observed stable for access against 1,800. However, beyond that, the deployment of COBOT UI throws errors. The distribution of performance metrics is evaluated through statistical analysis and observed to normal in its distribution. The COBOT UI experiences a failure response while executing the SI for a consecutive access beyond 1,800. The reliability is evaluated for accesses of beyond 1,800, and it is in the range of 41% to 61%. That is, the COBOT UI is stable beyond its stress limit with a probability of failure, and beyond that, the reliability of COBOT UI can degrade gradually. The MS can work collaboratively to

Table 16.3 Overall observation.

Observed parameter	Results
Stability of COBOT, MS, and AI	Observed up to 1,800 consecutive access
Stability beyond expectation level COBOT, MS, and AI	Throws error
Distribution of data sample	Normal
Reliability	Strong upto 1,800 consecutive access
Reliability beyond 1,800 consecutive access	Lies in between 41% and 61%.
GoF (goodness of fit) test	Observed
CDF of fault count	Continuous distribution

access the intelligence in swarm pattern. It can deliver the response at 95% confidence level.

16.7 Conclusion

The MS as well as the SI can be deployed for service delivery through COBOT UI. The collaborative MS can access the intelligence for functional execution of the respective operation of COBOT. The COBOT UI is stable upto a respective consecutive access point. Beyond that, the COBOT UI may experience failure in service as well as degradation in reliability may beexperienced. Beyond the stress limit of 1,800 consecutive commands, the reliability lies in the range of 41% to 61%. The stability of the COBOT UI is observed at 95% confidence level. As part of future work, we propose to observe the COBOT UI with load balancing web server and find the correlation of metrics for evaluating the efficiency of SI of MS in COBOT UI program.

References

1. Zaki, J., Islam, S.M.R., Alghamdi, N.S., Abdullah-Al-Wadud, M., Kwak, K.-S., Introducing Cloud-Assisted Micro-Service-Based Software Development Framework for Healthcare Systems. *IEEE Access*, 10, 33332–33348, 2022, doi: 10.1109/ACCESS.2022.3161455.
2. Blinowski, G., Ojdowska, A., Przybyłek, A., Monolithic vs. Microservice Architecture: A Performance and Scalability Evaluation. *IEEE Access*, 10, 20357–20374, 2022, doi: 10.1109/ACCESS.2022.3152803.
3. Lyu, Z., Wei, H., Bai, X., Lian, C., Microservice-Based Architecture for an Energy Management System. *IEEE Syst. J.*, 14, 4, 5061–5072, Dec. 2020, doi: 10.1109/JSYST.2020.2981095.
4. Raj, V. and Ravichandra, S., Microservices: A perfect SOA based solution for Enterprise Applications compared to Web Services. *2018 3rd IEEE International Conference on Recent Trends in Electronics, Information & Communication Technology (RTEICT)*, Bangalore, India, pp. 1531–1536, 2018, doi: 10.1109/RTEICT42901.2018.9012140.
5. Ma, S.-P., Lu, T.-W., Li, C.-C., Migrating Monoliths to Microservices based on the Analysis of Database Access Requests. *2022 IEEE International Conference on Service-Oriented System Engineering (SOSE)*, Newark, CA, USA, pp. 11–18, 2022, doi: 10.1109/SOSE55356.2022.00008.
6. Jiang, Y., Zhang, N., Ren, Z., Research on Intelligent Monitoring Scheme for Microservice Application Systems. *2020 International Conference on*

Intelligent Transportation, Big Data & Smart City (ICITBS). pp. 791–794, 2020, doi: 10.1109/ICITBS49701.2020.00173.

7. Wan, F., Wu, X., Zhang, Q., Chain-Oriented Load Balancing in Microservice System. *2020 World Conference on Computing and Communication Technologies (WCCCT)*, pp. 10–14, 2020, doi: 10.1109/WCCCT49810.2020.9169996.
8. Aktaş, K., Hakan Kilinç, H., Arica, N., Microservice Interaction Prediction in Communication Platform as a Service. *2022 30th Signal Processing and Communications Applications Conference (SIU)*, pp. 1–4, 2022, doi: 10.1109/SIU55565.2022.9864867.
9. Bicchi, A., Peshkin, M.A., Colgate, J.E., Safety for Physical Human–Robot Interaction, in: *Springer Handbook of Robotics*, Siciliano, B., Khatib, O. (Eds.), Springer, Berlin, Heidelberg, 2008, https://doi.org/10.1007/978-3-540-30301-5_58.
10. Universal Robotics, Our history, 2018, [Online], Available: https: //www.universalrobots.com/about-universal-robots/our-history/.
11. Robotiq, Cobots EBook, 2018, [Online], Available: https: //blog.robotiq.com/collaborative-robotebook.
12. Robots and robotic devices - Collaborative robots, ISO Standard ISO/TS 15066:2016, 2016.
13. Muller, R., Vette, M., Mailahn, O., Process-oriented task assignment for assembly processeswith human-robot interaction. *Procedia CIRP*, 44, 210–215, 2016.
14. Peternel, L., Kim, W., Babic, J., Ajoudani, A., Towards ergonomic control of human-robot comanipulation and handover, in: *2017 IEEE-RAS 17th International Conference on Humanoid Robotics (Humanoids)*, IEEE, 2017.
15. Yin, Z., Liu, J., Chen, B., Chen, C., A Delivery Robot Cloud Platform Based on Microservice. *J. Rob.*, 2021, 10, 2021, Article ID 6656912, https://doi.org/10.1155/2021/6656912.
16. Zhai, Y., Ding, B., Zhang, P., Luo, J., Cloudroid Swarm: A QoS-Aware Framework for Multirobot Cooperation Offloading. *Wireless Commun. Mobile Comput.*, 2021, 18, 2021, Article ID 6631111, https://doi.org/10.1155/2021/6631111.
17. Bharti, S. and McGibney, A., CoRoL: A Reliable Framework for Computation Offloading in Collaborative Robots. *IEEE Internet Things J.*, 9, 19, 18195–18207, 2022.
18. Mazdin, P. and Rinner, B., Coordination of Mobile Agents for Simultaneous Coverage, in: *Volume 11873 of Lecture Notes in Computer Science*, pp. 170–185, Springer International Publishing, Cham, 2019, doi: 10.1007/978-3-030-33792-6_11.
19. Chandra, M., Singpurwalla, N.D., Stephen, M.A., Kolmogorov statistics for tests of fit for the extreme value and Weibull distribution. *J. Am. Stat. Assoc.*, 76, 375, 729–731, 1981, Taylor & Francis.

20. Coles, S.G., On goodness-of-fit tests for the two-parameter Weibull distribution derived from the stabilized probability plot. *Biometrika*, 76, 3, 593–598, 1989.
21. Medhi, S., Bora, A., Bezboruah, T., Security Impact on e-ATM Windows Communication Foundation Services using Certificate based Authentication and Protection: An implementation of Message Level Security based on .NET Technique. *Int. J. Inf. Retr. Res.*, 6, 3, 37–51, 2016, DOI :10.4018/IJIRR.2016070103.
22. Military handbook - MIL-HDBK-17-1F. Polymer matrix composites guidelines for characterization of structural materials, in: *Composite Materials Handbook*, vol. 1, Department of Defense, U.S., 2002.
23. Medhi, S., Bora, A., Bezboruah, T., Investigations On Some Aspects of Reliability of Content Based Routing SOAP based Windows Communication Foundation Services. *Int. J. Inf. Retr. Res.*, 7, 1, 17–31, 2017, doi: DOI: 10.4018/IJIRR.2017010102.
24. Bonani, M. *et al.*, Swarm Robotics: A Review from the Swarm Engineering Perspective. *Swarm Intell.*, 4, 1, 1–41, 2010.
25. Dorigo, M. *et al.*, Swarm Intelligence in Robotics. *Proc. IEEE*, 91, 9, 1464–1478, 2004.
26. Garattoni, L. and Birattari, M., Swarm robotics, in: *Wiley Encyclopedia of Electrical and Electronics Engineering*, Wiley-Interscience, Hoboken, NJ, USA, 2016.
27. Gómez, J.M. *et al.*, Towards Microservices in Robotics. *IEEE Rob. Autom. Lett.*, 4, 2, 2280–2287, 2019.
28. Brambilla, M. *et al.*, Swarm Intelligence for Self-Organizing and Self-Assembling Robots. *IEEE Trans. Rob.*, 29, 2, 224–238, 2013.
29. Correll, N. *et al.*, Analysis and Observations from the First Amazon Picking Challenge. *IEEE Trans. Autom. Sci. Eng.*, 14, 1, 172–188, 2017.
30. Janson, S., Merkle, D., Middendorf, M., A decentralization approach for swarm intelligence algorithms in networks applied tomulti swarm PSO. *Int. J. Intell. Comput. Cybern.*, 1, 25–45, 2008.

17

Security Issues and Trends of Industrial Robots and Cobots

Samikshya Sarangi[1], Aswani Kumar Cherukuri[1*], S. Chandramouliswaran[1], Annapurna Jonnalgadda[2] and Firuz Kamalov[3]

[1]School of Information Technology and Engineering, Vellore Institute of Technology, Vellore, India
[2]School of Computer Science and Engineering, Vellore Institute of Technology, Vellore, India
[3]School of Engineering, Canadian University Dubai, Dubai, United Arab Emirates

Abstract
The market for cyber-physical systems, which includes collaborative robots (also known as cobots), is continuing to expand at a rapid rate. This is mostly caused by megatrends, such as smart factories, which involve the connection of various types of machinery as well as interaction between humans and robots. On the one hand, the implementation of these new technologies results in a greater degree of automation and adaptability in manufacturing. On the other hand, however, it raises demand for the safety and security of the systems, as the potential for such collaborative situations to cause harm to human beings is enhanced. Hence, when dealing with a cobot, it is recommended to have a heightened knowledge of the many areas of cyber security and how they relate to the safety of the system. This chapter is about the security of industrial robots and cobots. The chapter starts with an introduction to the fourth industrial revolution and the various aspects of it. It discusses the various vulnerabilities and attacks that are faced by industrial robots and cobots.

***Keywords*:** Attack surface, attack tree, cobots, Industry 4.0, industrial robots, security, vulnerability

**Corresponding author*: cherukuri@acm.org

V. Ramasamy, S. Balamurugan and Sheng-Lung Peng (eds.) Intelligent Robots and Cobots: Industry 5.0 Applications, (355–376) © 2025 Scrivener Publishing LLC

17.1 Introduction

Both the "Industry 4.0" and the underlying digital transformation are making rapid strides forward. As a consequence of the digital revolution, people's lives and places of employment are undergoing significant transformations; despite this, the general public is optimistic about the possibilities that Industry 4.0 may bring [1]. Robotics, particularly industrial robots, collaborative robots, and other types of robots are becoming increasingly important in the context of Industry 4.0. Workers will need to actively engage with robots in an environment that is interconnected as a result of the new industrial paradigm known as Industry 4.0. This will present them with a new set of obstacles [2, 3]. On the one hand, while they have been incredibly helpful to humans in terms of reducing burdens and increasing productivity, there is still an issue that needs to be addressed on the other. The problem at hand is making certain that industrial robots and cobots are safe to use. To program them, a large amount of data is fed as the training data, testing data, and eventually as input. These data may contain sensitive information that, if it were to become public knowledge, may put the operation of a company or an industrial process at risk. The process of moving from automation to true collaboration with robots in manufacturing will create a new area of research that will require the use of a practice lens [4]. In this work, an attempt is made to comprehend the numerous attack surfaces and vulnerabilities that are present in robots and cobots, as well as a look at the preventative measures that can be taken. In the latter portion of this piece, we will also investigate industrial controls, security mechanisms, as well as standards and policies connected to them.

Industrial robots have been utilized for several years now for work that is repetitive and requires a high level of precision [16]. They are never allowed to interact with people because the risks are too great. However, new collaborative robots (cobots), which do not require protective walls and so permit direct interaction with humans, have become a trend [17]. Cobots are robots that work together with humans. From a total of 355,000 robots last year, 18,000 (4.8%) cobots have already been sold, with growth rates of as much as 30% annually [18]. Additionally, an increase in overall market share of up to 30% is predicted by 2027 [17]. As a direct consequence of this, a market that is expanding and in which universal robots have been the forerunner is becoming an increasingly appealing target for new competitors [17]. The main contributions of this work are summarized as follows:

- A summary of industrial bots and cobots.
- A detailed analysis of security issues, vulnerabilities, attack tree and attack surface of industrial bots, and cobots.
- Further, we also provide mitigation strategies, security mechanisms, and policies.

17.2 Cobots and Industrial Bots

17.2.1 Industrial Robots

Industrial robots are mechanical devices made for carrying out activities in hazardous or difficult-to-reach places, such as the metalworking or automotive industries. They first appeared in the post–World War II era, when there was a severe manpower shortage in industries and firms had to heavily invest in their production facilities. The first industrial robot in the country was created by inventor George Devol in 1961 and installed in the plants of General Motors. This vacuum tube-powered robot, known as Unimate, was designed to transport and weld hazardous die castings. Later, rivals from Europe and Japan copied this concept for their markets. For instance, the first Japanese computer-controlled robot, the Kawasaki-Unimate 2000, saw use for the first time in 1968.

A cobot collaborates with human workers to complete tasks as opposed to a robot, which works alone. Because there are no fences around the cobot, operators can operate securely close to it unlike with industrial robots. It then becomes possible to control it manually or collaborate with another person.

17.2.2 Industry 4.0 and Beyond

The Fourth Industrial Revolution, or Industry 4.0, is the incorporation of cutting-edge technologies and data analytics into the production process. In order to improve manufacturing procedures and boost efficiency, this involves the use of cyber-physical systems, the Internet of Things, cloud computing, artificial intelligence, and big data analytics. People's lives and places of employment are changing significantly as a result of the digital revolution, yet, despite this, the general public is hopeful about the opportunities that Industry 4.0 may bring for the advancement of sustainability [1].

Beyond Industry 4.0, there are emerging technologies such as quantum computing, 5G, and blockchain that are expected to revolutionize manufacturing even further. These technologies have the potential to enable even greater connectivity, speed, and accuracy in data collection and analysis, which can drive the creation of new business models and disrupt traditional supply chains.

As Industry 4.0 and beyond continue to evolve, manufacturers will need to adapt to new technologies and embrace the concept of "smart factories" that leverage data-driven insights to optimize operations and improve product quality. This will require a focus on upskilling and reskilling the workforce, as well as collaboration across industries to ensure that emerging technologies are developed responsibly and sustainably.

The German initiative known as "Industry 4.0" has been well-known globally for the past 10 years. Other countries have created similar strategic initiatives, and much research has gone into developing and deploying various Industry 4.0 technology. The European Commission unveiled Industry 5.0 10 years after the launch of Industry 4.0. Although Industry 5.0 is thought to be value-driven, Industry 4.0 is thought to be technology-driven. The coexistence of two Industrial Revolutions raises issues, which call for debates and explanations [4]. This work will explore a specific area of Industry 4.0, that is the domain of robots and cobots.

17.2.3 Cobots

Cobots stands for "collaborative robots." Robots had increased productivity in industries after decades of automation, but they were still constrained by their work environment and closed constitution. Robots could not handle more flexible and developing procedures, and workers could not quickly access them to help and reprogram them [6]. Early in the 1990s, a group of engineers, mathematicians, and computer scientists began to consider the possibility of creating robots that are more sensitive, intelligent, and safe. They aimed to create a robotic hand that could manipulate objects with ease, just like a human hand. But they faced two significant issues. First, how can a robot hand be made to move like a human one? Second, how can a robot be made to feel human-like?

Cobots are low-cost industrial robots that can operate close to humans without putting them in risk. Cobots can be utilized for a variety of tasks, including machine maintenance, welding, packaging, and palletizing because they are more versatile than normal or standard industrial robots. The majority of cobot implementations take place in manufacturing and industrial settings, although cobots can be employed in a wide range of industries, from pharma to medical to agriculture.

The goal of Industry 4.0 is to enable the factory of the future, which will have more flexible collaborative robots known as "cobots," new types of intelligent information systems, and automation. According to experts, complete automation—or removing the human from production, is not regarded as viable. Instead, a cooperative effort between humans and various forms of intelligent machines is the desired outcome. However, integrating true human–robot collaboration into assembly lines is trickier than simply automating specific procedures. Human–computer interaction (HCI) and human–robot interaction (HRI) research have so far concentrated on how users engage with Industry 4.0, automation, and cobots [4]. Kopp *et al.* [7] have developed an empirically driven framework to identify the success factors for introducing industrial HRI in practice. Through case studies and interviews with 29 companies in various industries, they identified eight key success factors, including clear HRI goals, employee involvement, and appropriate HRI technology selection. The authors found that successful HRI implementation requires a combination of technical and non-technical factors and emphasized the importance of addressing the concerns and needs of employees who will be working alongside robots. Overall, the framework can guide companies looking to introduce HRI in their operations [21].

Industrial robots have mechanical arms with many directions of motion that may be programmed to do a variety of tasks in varied environments. These business robots can do the following:

- In various settings, including vacuum and high-pressure chambers.
- In both spotless workspaces and abjectly filthy places.
- In hazardous locations where there is a risk of explosions, diseases, radiation, or other extremely harmful hazards for people.

Robotic arms may include grippers designed specifically for handling delicate and breakable things, whereas other robotic arms may have grippers capable of lifting and grabbing weights weighing several tonnes.

17.2.4 Other Bots

Articulated robots are versatile machines used in manufacturing that closely resemble the human arm in both appearance and function.

These robots are typically mounted on rotating bases and can have between two and 10 axes of motion, with the most common type having four to six axes. Some articulated robots with six axes incorporate an additional axis, often referred to as a seventh axis or robot transfer unit, which allows the robot to move along a predetermined path and perform more complex motions. This axis is extremely useful for tasks like assembly, arc welding, material handling, machine tending, and packaging because it may move the robot to various terminals along a line or across adjacent lines.
Cartesian robots, also known as gantry robots or linear robots, are a type of industrial robot that operates on a three-axis Cartesian coordinate system (X, Y, and Z). These robots use linear motion to move along a fixed path, rather than the rotational motion used by articulated robots. In tasks requiring extreme precision, like pick-and-place operations, assembly, and material handling, cartesian robots are often used. They are frequently used for wafer handling and inspection in the semiconductor sector. Cartesian robots differ from other industrial robot types by their simplicity and accuracy. They are a common choice for many manufacturing applications because they are frequently simpler to maintain and program. However, their linear motion system means that they are not as versatile as articulated robots and may not be suitable for tasks that require a greater range of motion.
SCARA robots, which stand for "selected compliance assembly robot arms," function using a three-axis system that is analogous to that of cartesian robots. Because they rotate, they can first expand to fit into more confined areas and then contract again as required. This may be useful when loading or unloading workstations, as well as moving components between work cells. SCARA robots are widely used for a variety of tasks, including assembly, painting, and palletizing, in addition to their applications in the biomedical manufacturing industry.

Detla robots, have three arms that are positioned in a triangle pattern on a type of industrial robot. It is often referred to as a parallel robot or a spider robot. The motors that move the arms in unison are coupled to a base and in control of the arms. Due of their ability to move with such great speed and accuracy, robots are perfect for tasks like pick-and-place operations, packaging, and assembly. Delta robots are renowned for their blazing speed, pinpoint accuracy, and careful handling of fragile materials. They are often used in industries such as food and beverage, pharmaceuticals, and electronics manufacturing. One of the key advantages of Delta robots is their ability to maintain high levels of accuracy even when operating at high speeds. They are also able to work in confined spaces, making them a popular choice for applications that require a small footprint. However, due to their specialized design, Delta robots may not be suitable for all manufacturing applications and may require more specialized training and maintenance.

17.2.5 Potential Use Cases

Cobots and industrial bots have the potential to revolutionize the manufacturing and industrial sectors in numerous ways. Cobots, or collaborative robots, are designed to work safely alongside human workers, whereas industrial robots are typically larger, more powerful robots that can perform a variety of tasks in manufacturing, assembly, and packaging.

Cobots may be used to assist with jobs that are too risky or challenging for people to complete. Cobots, for instance, can be employed in locations with dangerous chemicals or high temperatures, lowering the chance of worker injury. To free up human workers to concentrate on more difficult duties, they can also be utilized to handle large objects or carry out monotonous activities.

Industrial robots, on the other hand, can be used for a wide range of tasks, from welding and painting to assembly and packaging. They can work quickly and efficiently, reducing production times and increasing output. They can also be programmed to perform tasks with a high level of precision, reducing errors and waste.

The potential use cases for cobots and industrial bots are vast and varied, and as the technology continues to improve, we can expect to see them being used in more and more industries. From automotive and aerospace to food and pharmaceuticals, the benefits of cobots and industrial bots are clear, and they have the potential to transform the way we work and live. The most common use cases have been mentioned below:

Industrial robots are versatile machines that can be used in a wide range of applications across various industries. Some of the most common use cases of industrial robots include the following:

Assembly: Industrial robots can be programmed to perform repetitive assembly tasks with high precision and speed. They can assemble complex parts and components with ease, making them an ideal choice for manufacturing industries.

Welding: Industrial robots can be used for spot welding, arc welding, and other types of welding processes. They can work in hazardous environments and produce high-quality welds consistently.

Painting: Industrial robots can be used for painting large surfaces or intricate details with high precision. They can also work with different types of paints and coatings.
Packaging: Industrial robots can be used for packaging products in different shapes and sizes. They can also handle delicate and fragile items with care, reducing the risk of damage during the packaging process.
Material handling: Industrial robots can be used for material handling tasks such as loading and unloading parts, transporting materials, and palletizing products. They can work in harsh environments and lift heavy loads safely and efficiently.
Inspection and quality control: Industrial robots can be used for inspecting products and components for defects, ensuring consistent quality control throughout the production process.
Cutting and machining: Industrial robots can be used for cutting and machining different materials such as metal, wood, and plastic. They can perform these tasks with high precision and speed, reducing the time and cost of manufacturing.
Cleaning and maintenance: Industrial robots can be used for cleaning and maintaining industrial equipment and machinery. They can reach tight spaces and perform tasks that are difficult or dangerous for humans.

Collaborative robots, also known as cobots, are designed to work alongside humans and can be programmed to perform a wide range of tasks. Recent research also tells about enhancing the quality of support by leveraging the strengths of different bots, such as their knowledge bases and language models, to provide a more comprehensive and personalized response to user queries. This outperforms the traditional single-bot systems [20]. Some of the common use cases of collaborative robots include the following:

Assembly: Collaborative robots can work alongside humans to perform repetitive and intricate assembly tasks. They can also handle small and delicate components, reducing the risk of damage and improving the overall efficiency of the assembly process.
Pick and place: Collaborative robots can be used for pick and place tasks such as packing and palletizing. They can work alongside humans to pick up items from a conveyor belt and place them in a designated location, reducing the strain on human workers and improving productivity.
Quality control: Collaborative robots can be used for quality control tasks such as inspecting products and components for defects. They can work

alongside humans to identify defects and ensure consistent quality control throughout the production process.

Material handling: Collaborative robots can be used for material handling tasks such as transporting materials and parts. They can work alongside humans to move heavy loads, reducing the risk of injury and improving efficiency.

Welding and cutting: Collaborative robots can be used for welding and cutting tasks, such as spot welding and laser cutting. They can work alongside humans to perform these tasks, improving the overall speed and quality of the process.

Packaging: Collaborative robots can be used for packaging tasks such as filling and sealing containers. They can work alongside humans to package products, reducing the risk of injury and improving productivity.

Inspection: Collaborative robots can be used for inspection tasks such as checking for defects and inconsistencies. They can work alongside humans to inspect products and ensure consistent quality control.

Healthcare: Collaborative robots can be used in healthcare settings to assist with patient care and medication delivery. They can work alongside healthcare professionals to improve efficiency and reduce the risk of injury.

Overall, collaborative robots can be used in a wide range of industries and applications. They are designed to work alongside humans and can help to improve efficiency, productivity, and safety in many different settings.

17.3 Security of Cobots and Other Bots

The unanticipated alteration of cobot parameters by a cyber assault is a common impact that might occur as a result of the effect that safety has on security [8]. If this occurs during a human–machine interaction, then the cobot (or its workpiece/payload) may clamp or impact a section of the worker's body, which may result in severe damage [19]. Having said that, this form of reciprocal engagement also operates in the opposite direction. External safety devices helps to reduce the likelihood of collisions occurring. These sensors, on the other hand, call for increased network traffic and, as a result, increase the system's vulnerability to attack. These examples demonstrate the connection between safety and security, as well as the necessity of approaching both issues simultaneously to achieve optimal results.

As with any technology, there are security concerns associated with the use of cobots and other robots [10–12]. Some of the key security concerns include the following:

- **Physical safety:** Cobots work alongside humans, and there is a risk of physical harm if they malfunction or behave unexpectedly. It is important to ensure that cobots are designed with safety features such as collision detection and emergency stop buttons.
- **Cybersecurity:** Cobots are connected to networks and other devices, which can create vulnerabilities that hackers can exploit. It is important to ensure that cobots have strong passwords, are regularly updated with security patches, and are protected by firewalls and other security measures.
- **Data privacy:** Cobots collect and transmit data, which can include sensitive information such as trade secrets and personal information. It is essential that the data is encrypted, and also access is restricted to authorized personnel.
- **Malware:** Cobots can be infected with malware, which can cause them to malfunction or behave unexpectedly. It is important to ensure that cobots are regularly updated with security patches and protected by antivirus software.
- **Integration with legacy systems:** Cobots may need to be integrated with legacy systems, which can create security vulnerabilities if those systems are not properly secured. It is important to ensure that legacy systems are updated with security patches and protected by firewalls and other security measures.

It is important to ensure that cobots and other robots are designed, installed, and maintained with security in mind. This includes implementing physical safety measures, cybersecurity protocols, and data privacy protections.

17.3.1 Threat Landscape

Hollerer *et al.* [5] presented a security assessment of a collaborative robot (cobot) to identify potential vulnerabilities and threats. The assessment was carried out on a specific cobot model, and various attack scenarios were simulated, including network attacks and physical attacks. The results showed that the cobot was susceptible to several types of attacks, including those that could compromise safety and privacy. The authors emphasize the importance of considering security in the design and operation of cobots and suggest the development of standardized security guidelines and testing methodologies for cobots. There is a need for retraining and upskilling of workers to effectively collaborate with robots, the potential displacement of workers due to automation, the high costs of implementing collaborative robots, and the need for improved safety standards and regulations for collaborative robots [9]. The threat landscape for cobot security is constantly evolving as new technologies are developed and attackers find new ways to exploit vulnerabilities. Some of the key threats to cobot security include the following:

Malware: Malware can infect cobots and other robots, causing them to behave unexpectedly or maliciously. Attackers can use malware to steal data, disrupt operations or even cause physical harm to humans.
Unauthorized access: Attackers may be able to gain unauthorized access to cobots and other robots, either by exploiting vulnerabilities in the devices themselves or by stealing login credentials. Once they have access, attackers can steal data, disrupt operations, or cause physical harm to humans.
Phishing: To deceive users into disclosing login credentials or other sensitive information, cybercriminals may use phishing emails or other social engineering techniques. Attackers may be able to use this to access other systems, networks, cobots, and other robots.
Denial-of-service attacks: Attackers may use denial-of-service attacks to flood cobots and other robots with traffic, causing them to slow down or stop working altogether. This can disrupt operations and cause physical harm to humans.
Supply chain attacks: Attackers may target the supply chain for cobots and other robots, either by introducing malware or other malicious code into components or by compromising the devices during the manufacturing process.
Insider threats: Insiders, including employees and contractors, may intentionally or unintentionally introduce vulnerabilities into cobots and other robots. This can include using weak passwords, failing to update software, or ignoring security protocols. Figure 17.1 shows the categorization of various attacks.

The threat landscape for cobot security is complex and constantly evolving. Organizations need to implement robust security measures to protect against these threats, including regular security audits, vulnerability scans, and employee training on security best practices.

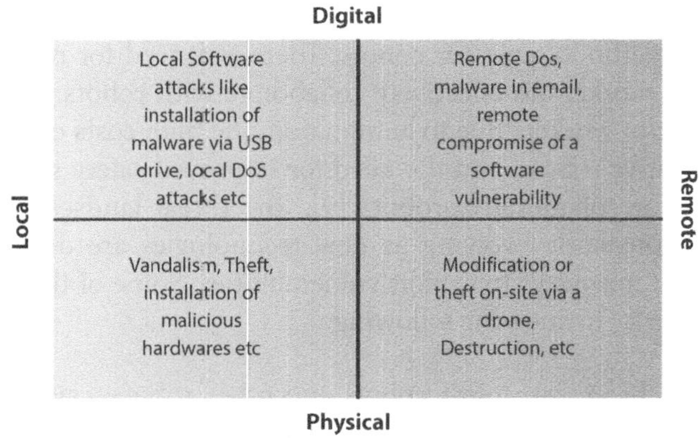

Figure 17.1 IT security attack categorization.

17.3.2 Attack Surface

Like any technology, cobots also have their attack surface and potential vulnerabilities that need to be addressed. The attack surface of cobots includes the hardware and software components of the robot, as well as the communication channels and interfaces between the robot and other devices in the network. Potential vulnerabilities of cobots include insecure communication protocols, weak authentication and authorization mechanisms, and software vulnerabilities that can be exploited by attackers to gain unauthorized access to the robot or the network. In addition, cobots can also be physically manipulated or tampered with by attackers, which can pose a significant risk to human safety [5]. Implementing appropriate security measures, including as encryption, powerful authentication and authorization systems, and regular updates and patches, is crucial to reducing these threats. In addition, it is essential to teach human employees how to spot and report any suspicious behaviors or other anomalies relating to the behavior of the cobot. The attack surface in cobots refers to the various points of entry that attackers can use to gain unauthorized access, compromise, or exploit the system. These points of entry can include the following:

Network connectivity: Cobots are often connected to networks, which can be a potential entry point for attackers. This includes both wired and wireless connections, as well as any other networked devices that the cobot interacts with.

Control software: Cobots are typically controlled by software that can be vulnerable to attack. Attackers can exploit vulnerabilities in the software to gain control of the cobot, modify its behavior, or steal data.
Peripheral devices: Cobots may interact with auxiliary equipment like sensors or cameras, which can serve as an attack vector. If these gadgets are not properly secured, then they may be open to attack.
Operating system: The operating system that the cobot runs on can be vulnerable to attack if it is not properly secured. Attackers can exploit vulnerabilities in the operating system to gain control of the cobot or steal data.
Physical access: Physical access to the cobot can also be a potential entry point for attackers. This includes accessing the control panel, USB ports, or other physical components.
Supply chain: The supply chain for cobots can also be vulnerable to attack, as attackers may be able to introduce malware or other malicious code during the manufacturing or distribution process.
Human–machine interface: The interface through which humans interact with the cobot can be targeted by attackers to perform social engineering attacks or steal sensitive information.

The attack surface for cobots can be extensive, and it is important to implement robust security measures to protect against potential attacks. This includes implementing network security protocols, regularly updating software and operating systems, securing peripheral devices, limiting physical access to the cobot, and working with trusted vendors and suppliers to secure the entire supply chain.

17.3.3 Attack Tree for Cobots

An attack tree is a visual representation of the various steps that an attacker might take to compromise a system. The attack tree for cobots can be broken down into several different levels, each representing a different stage in the attack process. Figure 17.2 shows the attack tree of cobots. Here is an example of a possible attack tree for cobots.

Level 1: Gain access to the network

- Exploit a vulnerability in a connected device (such as a sensor or camera)
- Exploit a vulnerability in the network infrastructure (such as a router or firewall)
- Phishing or social engineering to gain access to login credentials

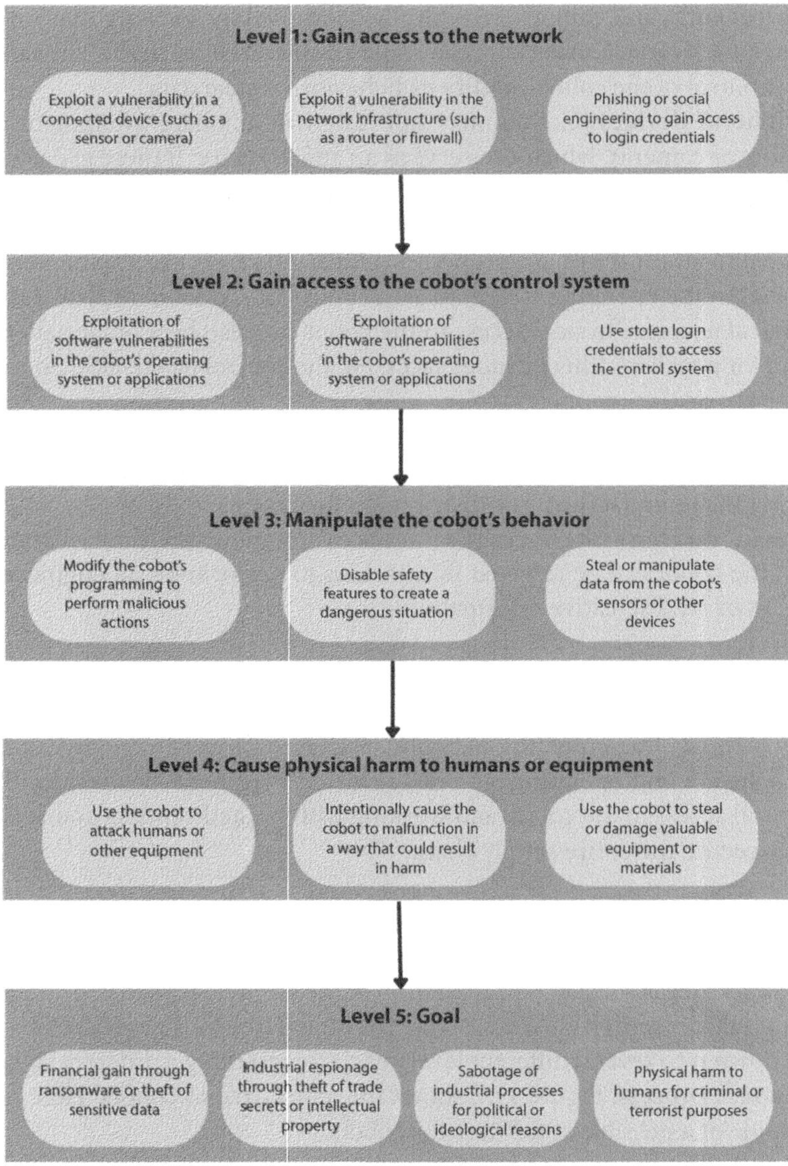

Figure 17.2 Illustration of an attack tree for cobot attack.

Level 2: Gain access to the cobot's control system

- The exploitation of software vulnerabilities in the cobot's operating system or applications;
- The exploitation of network vulnerabilities in the communication protocols used by the cobot;
- Use stolen login credentials to access the control system.

Level 3: Manipulate the cobot's behavior

- Modify the cobot's programming to perform malicious actions;
- Disable safety features to create a dangerous situation;
- Steal or manipulate data from the cobot's sensors or other devices.

Level 4: Cause physical harm to humans or equipment

- Use the cobot to attack humans or other equipment;
- Intentionally cause the cobot to malfunction in a way that could result in harm;
- Use the cobot to steal or damage valuable equipment or materials.

Level 5: Goal

- Financial gain through ransomware or theft of sensitive data;
- Industrial espionage through theft of trade secrets or intellectual property;
- Sabotage of industrial processes for political or ideological reasons;
- Physical harm to humans for criminal or terrorist purposes.

The attack tree for cobots can be complex and multifaceted, with attackers potentially using a combination of techniques to compromise the system. It is important to implement multiple layers of security to protect against these attacks, including network security protocols, software updates, physical access controls, and employee training on security best practices [13–15].

17.4 Mitigation Strategies

When it comes to industrial bots and cobots, there are various controls and regulations in place to ensure their safe operation and compliance with industry standards. These controls can be categorized into several areas, including safety standards, risk assessments, programming and configuration controls, human–machine interface controls, maintenance and inspection controls, and environmental controls [24].

Firstly, safety standards play a critical role in ensuring the safe design and use of industrial bots and cobots in industrial environments. These standards include guidelines such as ISO 10218-2 and ISO/TS 15066, which provide detailed guidance on safe robot and collaborative robot operations. Compliance with these safety standards is essential to minimize the risk of accidents and ensure the safety of human workers who interact with these machines.

Secondly, risk assessments are another crucial aspect of the industrial bot and cobot controls. Before deploying an industrial robot or cobot, a thorough risk assessment must be conducted to identify potential hazards and establish appropriate safety measures. This ensures that any potential risks are identified and mitigated before the machine is put into operation.

Programming and configuration controls are also essential when it comes to industrial bot and cobot safety. These controls ensure that the robot or cobot is programmed and configured in a way that promotes safe operation. This includes setting up proper emergency stops, limiting the speed and force of robot movements, and defining safe zones.

Human–machine interface controls are also critical to ensure the safe operation of industrial bots and cobots. These controls focus on designing intuitive and user-friendly interfaces for operators to control the machines. This includes providing clear visual feedback and easy-to-use programming interfaces to minimize the risk of human error.

Maintenance and inspection controls are also necessary to ensure the continued safe operation of industrial bots and cobots. Regular maintenance schedules must be established, and regular inspections must be conducted to identify any potential safety hazards.

Lastly, environmental controls are essential to minimize the impact of industrial bots and cobots on the environment. This includes proper disposal of waste materials, minimizing energy consumption, and reducing emissions.

Industrial robots and cobots are subject to a range of controls and regulations to ensure their safe operation and compliance with industry standards.

These controls are crucial in minimizing the risk of accidents and ensuring the safety of human workers who interact with these machines [25].

17.4.1 Security Mechanisms

Collaborative robots, or cobots, are increasingly being deployed in industrial settings, and, as with any connected device, they are vulnerable to cyber-attacks. To protect against such attacks, several security mechanisms are deployed for cobot protection [24, 25].

Firstly, access control mechanisms limit access to the robot's control system to authorized personnel only. This can involve the use of passwords or biometric authentication such as fingerprint or facial recognition. This helps prevent unauthorized users from accessing or manipulating the robot's controls.

Secondly, encryption is used to protect data transmitted between the cobot and its control system, making it difficult for hackers to intercept and manipulate robot commands.

Thirdly, firewalls are used to block unauthorized access to the robot's control system from the internet, preventing external attacks.

Fourthly, intrusion detection systems are used to detect and respond to unauthorized access attempts, whereas anti-malware software is deployed to protect the cobot from malware infections.

Finally, physical security measures such as tamper-proofing and anti-tampering mechanisms are deployed to prevent unauthorized access to the robot's internal components and systems.

By deploying a combination of these security mechanisms, cobots can be protected from cyber-attacks, ensuring their safe and secure operation in industrial settings.

17.4.2 Security Standards

Industrial bots and cobots have become an integral part of manufacturing and other industrial processes, performing various tasks ranging from assembly, packaging, and material handling, among others. As these machines are connected to the internet and communicate with other systems, they are prone to cybersecurity threats that can potentially harm workers, damage machines, and lead to substantial financial losses.

To address these risks, various security standards have been developed to ensure the safe and secure operation of industrial bots and cobots.

These standards are comprehensive guidelines that cover a range of areas, including design, installation, operation, and maintenance. They provide manufacturers, operators, and other stakeholders with a framework for creating and maintaining a secure operating environment.

The ISO 13482 standard is one of the most crucial security standards for industrial bots and cobots. This standard provides guidelines for the safety of personal care robots, ensuring that these devices are safe for use in healthcare settings. The standard covers aspects such as risk assessment, safety requirements, and user instructions, and is designed to ensure that the devices are safe for users and meet industry standards.

Another essential security standard for industrial robots is the ISO 10218 standard. The standard covers safety requirements for industrial robots, including robot design, installation, and operation. The standard aims to ensure that industrial robots operate safely in their intended environments and can provide workers with a safe and secure working environment.

In addition to these safety standards, several cybersecurity standards are relevant to industrial bots and cobots. For instance, the International Electrotechnical Commission (IEC) 62443 standard provides guidelines for industrial cybersecurity, whereas the National Institute of Standards and Technology (NIST) Cybersecurity Framework outlines best practices for managing and reducing cybersecurity risks.

Finally, there are industry-specific standards that are relevant to certain types of industrial bots and cobots. For example, the ANSI/RIA R15.06-2012 standard covers safety requirements for industrial robots used in manufacturing, while the American Society for Testing and Materials F3196 standard provides guidelines for the safety of collaborative robots.

In conclusion, compliance with these security standards is critical for ensuring the safe and secure operation of industrial bots and cobots. These standards provide manufacturers, operators, and other stakeholders with a framework for creating and maintaining a secure operating environment, protecting both the devices themselves and the people who work with them.

17.4.3 Security Policies

Industrial robots and collaborative robots (cobots) are increasingly being used in a variety of industries, from manufacturing to healthcare. As these robots become more prevalent in the workplace, it is important to ensure they are secure and not vulnerable to attacks that could compromise their functionality or cause harm to workers. Here are some security policies that can help ensure the security of industrial bots and cobots:

- Implement access controls: Access to industrial bots and cobots should be limited to authorized personnel only. This can be achieved through physical controls, such as locked doors or keycard access, as well as software controls, such as password protection.
- Conduct regular vulnerability assessments: Regular assessments of the bots and their associated systems should be performed to identify any vulnerabilities. These assessments should include both physical and software components of the bots, as well as any network connections or interfaces [22].
- Implement software and firmware updates: Regular software and firmware updates should be applied to the bots to ensure that they are running the latest, most secure versions. Updates should be tested thoroughly before deployment to ensure they do not introduce new vulnerabilities.
- Implement network security measures: Any network connections or interfaces used by the bots should be secured with firewalls and other network security measures. Network traffic should be monitored for signs of suspicious activity.
- Implement physical security measures: Physical security measures, such as cameras and alarms, should be implemented to protect the bots from theft or damage. Bots should also be physically secured when not in use.
- Train employees on security best practices: All employees who interact with industrial bots and cobots should be trained on security best practices, such as how to identify and report suspicious activity or potential security breaches.
- Have an incident response plan in place: In the event of a security breach or other incident, there should be a plan in place for responding to the incident. This plan should include procedures for isolating the affected bots, assessing the extent of the damage, and restoring functionality.

By implementing these security policies, organizations can help ensure the security of their industrial bots and cobots, protecting both the robots themselves and the workers who interact with them [22–25].

17.5 Conclusions

In conclusion, Industry 4.0 is rapidly transforming the manufacturing landscape with the introduction of advanced technologies such as robots and cobots. These technologies are poised to revolutionize the manufacturing sector, improving productivity and efficiency while reducing labor costs. However, with the benefits of these technologies come security concerns that cannot be ignored.

The potential use cases of cobots and robots are extensive and range from simple pick-and-place tasks to more complex and intelligent manufacturing processes. As these technologies become more sophisticated, it is critical to address the associated security concerns. The threat landscape for industrial bots and cobots is complex, with various types of attacks possible, including network-based attacks, physical attacks, and social engineering attacks.

To mitigate these security concerns, industrial controls such as access controls and vulnerability assessments can be implemented. Additionally, security mechanisms such as encryption and intrusion detection systems can be utilized to protect against attacks. Compliance with standards and policies, such as the IEC 62443 and the NIST Cybersecurity Framework, can provide a solid foundation for security best practices.

The potential benefits of robots and cobots in Industry 4.0 are significant, but they must be accompanied by robust security measures to prevent attacks and protect workers. As these technologies continue to evolve and become more widespread, organizations must prioritize security and implement best practices to ensure the safe and effective use of industrial bots and cobots.

References

1. Ghobakhloo, M., Industry 4.0, digitization, and opportunities for sustainability. *J. Cleaner Prod.*, 252, 119869, 2020.
2. Bragança, S., Costa, E., Castellucci, I., Arezes, P.M., A brief overview of the use of collaborative robots in industry 4.0: human role and safety, in: *Occupational and environmental safety and health*, pp. 641–650, 2019.
3. Xu, X., Lu, Y., Vogel-Heuser, B., Wang, L., Industry 4.0 and Industry 5.0—Inception, conception and perception. *J. Manuf. Syst.*, 61, 530–535, 2021.
4. Weiss, A., Wortmeier, A.-K., Kubicek, B., Cobots in industry 4.0: A roadmap for future practice studies on human–robot collaboration. *IEEE Trans. Hum.-Mach. Syst.*, 51, 4, 335–345, 2021.

5. Hollerer, S., Fischer, C., Brenner, B., Papa, M., Schlund, S., Kastner, W., Fabini, J., Zseby, T., Cobot attack: a security assessment exemplified by a specific collaborative robot. *Procedia Manuf.*, 54, 191–196, 2021.
6. El Zaatari, S., Marei, M., Li, W., Usman, Z., Cobot programming for collaborative industrial tasks: An overview. *Rob. Auton. Syst., 116*, 162–180, 2019.
7. Kopp, T., Baumgartner, M., Kinkel, S., Success factors for introducing industrial human-robot interaction in practice: an empirically driven framework. *Int. J. Adv. Manuf. Technol.*, 112, 685–704, 2021.
8. Kirschgens, L.A., Ugarte, I.Z., Uriarte, E.G., Rosas, A.M., Vilches, V.M., Robot hazards: from safety to security. arXiv preprint arXiv:1806.06681, 2018.
9. Galin, R. and Meshcheryakov, R., Automation and robotics in the context of Industry 4.0: the shift to collaborative robots. *IOP Conf. Ser.: Mater. Sci. Eng.*, 537, 3, 032073, 2019, IOP Publishing.
10. Yaacoub, J.-P.A., Noura, H.N., Salman, O., Chehab, A., Robotics cyber security: Vulnerabilities, attacks, countermeasures, and recommendations. *Int. J. Inf. Secur.*, 21, 1, 115–158, 2022.
11. Bi, Z.M., Chen, B., Xu, L., Wu, C., Malott, C., Chamberlin, M., Enterline, T., Security and safety assurance of collaborative manufacturing in industry 4.0. *Enterp. Inf. Syst.*, 16, 12, 2008512, 2022.
12. Castillo, J.F., Ortiz, J.H., Velásquez, M.F.D., Saavedra, D.F., COBOTS in industry 4.0: Safe and efficient interaction, in: *Collaborative and humanoid robots*, p. 3, 2021.
13. Dutta, V. and Zielińska, T., Cybersecurity of robotic systems: Leading Challenges and robotic system design methodology. *Electronics* 10, 22, 2850, 2021.
14. Three pillars of successful cobot-deployment: safety, quality, and flexibility from a single source Published on 2019-11-04 14:52:30 UTC in Flexible Manufacturing, blog link: https://industrial.omron.eu/en/solutions/blog/three-pillars-cobots.
15. Clim, A., Cyber security beyond the Industry 4.0 era. A short review on a few technological promises. *Inf. Econ.*, 23, 2, 34–44, 2019.
16. Vogel-Heuser, B., Bauernhansl, T., ten Hompel, M., *Handbook Industry 4.0 [Handbuch Industrie 4.0]*, Springer Nature, Springer Vieweg Berlin, Heidelberg, 2017.
17. Xiao, M., The collaborative robot market–2019, Interact Analysis, 2020, available at https://www. interactanalysis. com/the-collaborative-robot-market-2019/(accessed 7th December, 2021).
18. International Federation of Robotics, World Robotics Report 2020, 2020, URL: https://ifr.org/ifr-press-releases/news/record2.7-million-robots-work-in-factories-around-the-globe.
19. Tang, G. and Webb, P., Human–robot shared workspace in aerospace factories, in: *Human–Robot Interaction*, pp. 72–79, Imprint Chapman and Hall/CRC, New York, 2019.

20. Subramaniam, S., Aggarwal, P., Dasgupta, G.B., Paradkar, A., Cobots-a cognitive multi-bot conversational framework for technical support, in: *Proceedings of the 17th International Conference on Autonomous Agents and MultiAgent Systems*, pp. 597–604, 2018.
21. Khalid, A., Kirisci, P., Ghrairi, Z., Thoben, K.D., Pannek, J., Towards implementing safety and security concepts for human-robot collaboration in the context of Industry 4.0, in: *39th International MATADOR Conference on Advanced Manufacturing*, vol. 2, pp. 55–63, 2017.
22. Hollerer, S., Chabrová, M., Sauter, T., Kastner, W., Combined Modeling Techniques for Safety and Security in Industrial Automation: A Case Study, in: *2022 15th International Conference on Security of Information and Networks (SIN)*, pp. 1–4, IEEE, 2022.
23. Lichte, D. and Wolf, K.-D., Use case-based consideration of safety and security in cyber physical production systems applied to a collaborative robot system, in: *Safety and Reliability–Safe Societies in a Changing World*, pp. 1395–1401, CRC Press, London, 2018.
24. Peshkin, M., & Colgate, J.E., Cobots. *Ind. Robot: Int. J.*, 26, 5, 335–341, 1999.
25. Gleirscher, M., Johnson, N., Karachristou, P., Calinescu, R., Law, J., Clark, J., Challenges in the safety-security co-assurance of collaborative industrial robots, in: *The 21st Century Industrial Robot: When Tools Become Collaborators*, pp. 191–214, 2022.

18

Blockchain Technology for the Cobot's Cybersecurity Issues

B. Manikandan[1], G. Ganesh Kumar[1*], S. Kanakaprabha[2], R. Vijaya Kumar Reddy[3] and R. Janani[4]

[1]*Department of Information Technology, Hindusthan Institute of Technology, Coimbatore, Tamil Nadu, India*
[2]*Department of Computer Science and Engineering, Rathinam Technical Campus, Anna University, Coimbatore, India*
[3]*Department of Computer Science and Engineering, K L Education Foundation, Vaddeswaram, India*
[4]*Department of Information Technology, Sri Krishna College of Engineering and Technology, Coimbatore, Tamil Nadu, India*

Abstract
Cobots, or collaborative robots, have emerged as a promising technology that combines human capabilities with robotic systems to enhance productivity and efficiency in various industries. However, the integration of cobots into interconnected networks and their reliance on digital communication channels raise significant cybersecurity concerns. Traditional security measures such as firewalls and encryption mechanisms may not be sufficient to safeguard cobots against evolving cyber threats. This paper explores the potential contribution of blockchain technology in addressing cybersecurity issues associated with cobots. Blockchain, as a decentralized and immutable ledger, offers several inherent security features that can benefit cobots. Firstly, blockchain's distributed nature eliminates the need for a central authority, reducing the risk of single points of failure and unauthorized access. This decentralization enhances the resilience of cobots' communication networks, preventing potential cyber-attacks from compromising the entire system.

Secondly, the transparency and immutability of blockchain records provide an effective mechanism for verifying and validating cobots' actions and transactions. By recording every interaction and transaction on the blockchain, it becomes

**Corresponding author*: gkumarcse@gmail.com

V. Ramasamy, S. Balamurugan and Sheng-Lung Peng (eds.) Intelligent Robots and Cobots: Industry 5.0 Applications, (377–400) © 2025 Scrivener Publishing LLC

difficult for malicious actors to manipulate or tamper with data. This feature ensures the integrity of critical cobot operations and supports auditability, which is crucial for compliance and forensic investigations.

Furthermore, blockchain-based smart contracts can play a vital role in securing interactions between cobots and their human operators or other entities. Smart contracts enable the establishment of predefined rules and protocols, ensuring that cobots execute only authorized actions. The decentralized execution of smart contracts eliminates the need for intermediaries, reducing the risk of unauthorized modifications or tampering.

Additionally, blockchain technology can facilitate secure identity management for cobots. Through the use of cryptographic keys and digital signatures, cobots can establish their unique identities on the blockchain, ensuring authentication and authorization of their actions within the network. This prevents unauthorized cobots from gaining access to sensitive resources or compromising the integrity of the system.

However, challenges exist in implementing blockchain for cobots' cybersecurity. These challenges include scalability issues, energy consumption, and the integration of legacy systems with blockchain technology. Overcoming these challenges requires further research and development to optimize blockchain protocols and explore interoperability solutions.

***Keywords*:** Cobots, cobot's cybersecurity, blockchain

18.1 Introduction

The rapid advancement of robotics and automation has led to the development of collaborative robots, commonly known as cobots. These intelligent machines work alongside humans, enhancing productivity and efficiency in various industries. However, as cobots become integrated into interconnected networks and rely on digital communication channels, they face significant cybersecurity challenges. The potential vulnerabilities in cobot systems pose risks such as unauthorized access, data tampering, and malicious attacks that can disrupt operations and compromise safety.

Traditional cybersecurity measures, such as firewalls and encryption, have been effective in protecting traditional computer systems. However, the unique characteristics of cobots, including their real-time collaboration with humans and their ability to interact with multiple devices, require innovative approaches to ensure their security. Blockchain technology has emerged as a promising solution that can address these cybersecurity issues and contribute to the safe and reliable operation of cobots.

Blockchain, often linked with cryptocurrencies like Bitcoin, functions as a decentralized and distributed ledger, documenting transactions across numerous computers or nodes. It boasts essential traits such as decentralization, transparency, immutability, and smart contracts, which can notably fortify the cybersecurity stance of cobots. Utilizing these attributes, blockchain offers a resilient and secure structure to oversee and safeguard digital interactions and data within cobots.

This paper aims to explore the possible contribution of blockchain technology in addressing the cybersecurity challenges faced by cobots. It will delve into the specific security features offered by blockchain and discuss how they can be leveraged to ensure the integrity, confidentiality, and obtainability of cobot systems. Additionally, the paper will highlight the benefits of blockchain in areas such as secure communication, data integrity, identity management, and smart contract execution for cobots [6].

Furthermore, the paper will also discuss the challenges and limitations associated with implementing blockchain technology in the context of cobots. These challenges may include scalability issues, energy consumption concerns, and the integration of existing systems with blockchain protocols. Understanding these obstacles is crucial for devising practical and effective solutions to harness the full potential of blockchain in securing cobots.

By examining the contribution of blockchain technology to cobots' cybersecurity issues, this paper aims to provide insights and recommendations that can guide researchers, industry professionals, and policymakers in ornamental the safety and trustworthiness of cobot systems. Ultimately, the adoption of blockchain technology can pave the way for widespread deployment of cobots across industries, driving innovation, efficiency, and productivity while mitigating potential cybersecurity risks [9].

The topics included in the contribution of blockchain technology for cobots' cybersecurity issues can be as follows:

Cybersecurity Challenges in Cobots: An overview of the unique cybersecurity challenges that cobots face, such as unauthorized access, data tampering, interception of communications, and potential disruption of operations.

Introduction to Blockchain Technology: A brief introduction to blockchain technology, explaining its decentralized and distributed nature, cryptographic principles, and the concept of a decentralized ledger.

Security Features of Blockchain: Exploring the inherent security features of blockchain, including decentralization, transparency, immutability, consensus mechanisms, cryptographic hashing, and cryptographic signatures.
Data Integrity and Auditability: Examining how blockchain can ensure the integrity of cobots' data by recording transactions and interactions on an immutable ledger, enabling auditability, traceability, and forensic investigations.
Secure Identity Management: Exploring how blockchain-based identity management can provide secure authentication and authorization mechanisms for cobots, ensuring that only authorized entities can access and interact with the system.
Smart Contracts for Secure Interactions: Discussing the role of blockchain-based smart contracts in defining and enforcing predefined rules and protocols for cobots' interactions, enabling secure and automated execution of authorized actions.
Challenges and Limitations: Addressing the challenges and limitations associated with implementing blockchain in the context of cobots, including scalability issues, energy consumption concerns, and the integration with existing systems.
Case Studies and Examples: Providing real-world examples and case studies where blockchain technology has been successfully applied to enhance cobots' cybersecurity, highlighting the benefits and outcomes achieved.
Future Directions and Recommendations: Offering insights into future research directions, potential improvements in blockchain protocols for cobots, and recommendations for stakeholders, researchers, and policymakers to maximize the contribution of blockchain technology to cobots' cybersecurity.

18.2 Literature Review

Soori, M., Dastres, R., Arezoo, B., & Jough, F. K. G. (2024) [1]. Intelligent robotic systems in Industry 4.0. Journal of Advanced Manufacturing Science and Technology. Editorial Committee of JAMST DOI: 10.51393/j.jamst.2024007 2709-2135©2024 JAMS). In this paper, the security aspects of Industry 4.0 focus on the specific challenges and considerations related to collaborative robots. It serves as a resource for researchers, professionals, and policymakers interested in understanding and implementing robust security measures for cobots within the Industry 4.0 paradigm. By highlighting best practices and emerging technologies, the chapter contributes to the advancement of secure collaboration and the protection of cyber-physical systems in the context of Industry 4.0.

Li *et al.* (2020) [2]. A Survey on the Security of Robotics and Automation Systems: Threats, Solutions, and Challenges. IEEE Access, 8, 90169–90187: In this paper, a comprehensive overview of the security considerations in robotics and automation systems. It serves as a valuable resource for researchers, practitioners, and policymakers seeking to understand the threats, solutions, and challenges associated with securing these systems. By synthesizing existing knowledge and highlighting emerging technologies, the survey aims to facilitate further research and development efforts in enhancing the security of robotics and automation systems

Jerbi and Besnard (2019) [3]. Cybersecurity for Industrial Robots: Challenges and Solutions. IEEE Robotics & Automation Magazine, 26(4), 66–76: This paper provides valuable insights into the cybersecurity considerations for industrial robots. It serves as a reference for researchers, industrial practitioners, and policymakers seeking to safeguard industrial automation systems against cyber threats. The proposed solutions and best practices contribute to the development of a resilient cybersecurity framework for industrial robots, ensuring their safe and secure operation within modern manufacturing environments.

Strobel *et al.* (2020) [11]: Blockchain technology ensures the security of robot swarms: An examination of consensus protocols and their resilience to byzantine robots proposes a system where robots within a swarm must demonstrate their integrity to their peers by exchanging cryptographic proofs to collaborate on mission completion. This approach's implications are showcased in two distinct swarm robotics missions: foraging and maze formation. In both tasks, swarm robots effectively cooperated and executed sequential actions without explicit knowledge of the missions' overarching objectives. The findings presented in this study illustrate the practicality of utilizing Merkle trees as a cooperation mechanism in swarm robotics systems, evident in both simulated environments and real-robot experiments. This bears significance for forthcoming decentralized robotics applications where security holds paramount importance.

Jia *et al.* (2020) [4]: This work's primary contribution lies in formally describing a networked robotic control system through the agent (module) concept. This description proves beneficial in two key aspects: (a) aiding in specifying the secure linking of control and interface components, and (b) facilitating high-level control development through formal approaches like finite state machines. This approach serves to prevent logical errors, thereby bolstering the system's security.

18.3 Introduction to Cybersecurity

Safeguarding computer systems, networks, software, and data against unauthorized access, attacks, and theft is the essence of cybersecurity. In today's highly digital and interconnected landscape, its significance cannot be emphasized enough. Cybersecurity involves a broad spectrum of tactics and plans dedicated to shielding information technology infrastructure and resources from a diverse array of threats, spanning hackers, cyber-criminals, viruses, malware, and other malicious entities.

Why is Cybersecurity Important?

The rapid expansion of the internet and the widespread adoption of digital technologies have opened up new avenues for communication, commerce, and innovation. However, these advancements have also exposed vulnerabilities that can be exploited by cyber threats. Figure 18.1 shows the advantages of cybersecurity. Some reasons why cybersecurity is crucial include the following:

Preserving Trust: Businesses and individuals rely on digital services for everyday activities. Maintaining strong cybersecurity measures helps build trust between users and service providers.
Preventing Financial Loss: Cyber-attacks can lead to fiscal losses for dealings due to theft of assets, loss of revenue, and costly downtime during recovery.
Safeguarding Critical Infrastructure: Cybersecurity is essential for safeguarding dangerous infrastructure, such as control grids, transportation

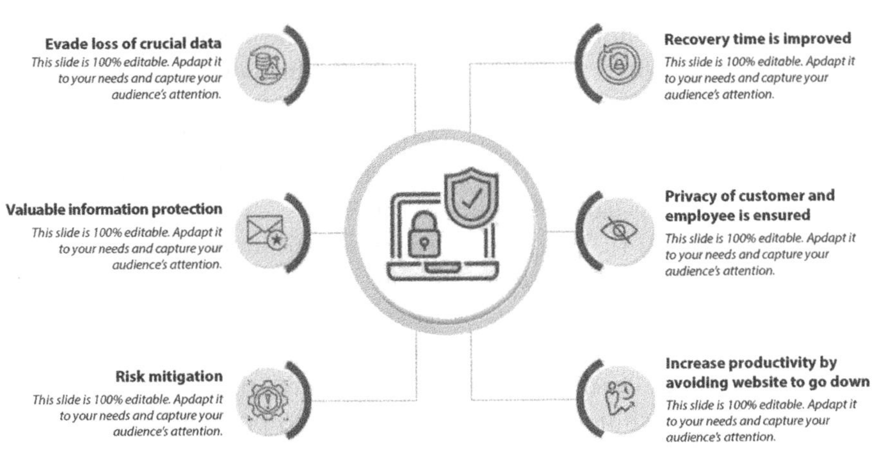

Figure 18.1 Advantages of cybersecurity.

systems, and healthcare facilities, as their disruption could have devastating consequences.
National Security: Nations rely heavily on digital infrastructure, making them vulnerable to cyber-attacks. Strong cybersecurity measures are essential for safeguarding a country's security interests.

Core Principles of Cybersecurity:
Integrity: Guaranteeing the accuracy and reliability of data and systems, preventing unauthorized modifications.
Availability: Ensuring that information and systems are available and operational when needed, and protecting them from disruptions.
Authentication: Verifying the identity of users and devices to prevent unauthorized access.
Authorization: Defining and enforcing access privileges to limit user access to appropriate resources.
Non-Repudiation: Ensuring that the origin and receipt of data or communications cannot be denied by the sender or recipient.

18.4 Cybersecurity Challenges

Cybersecurity faces numerous challenges in today's fast-paced and interconnected digital world. Some of the most significant challenges include the following:

Sophisticated Cyber-Attacks: Cybercriminals and hackers are continuously developing advanced attack techniques, making it challenging for cybersecurity defenses to keep up.
Insider Threats: Malicious or negligent actions by employees or individuals with access to sensitive information pose a significant risk to organizations.
Rapidly Evolving Threat Landscape: The emergence of new threats, such as zero-day vulnerabilities and polymorphic malware, requires constant adaptation and innovation in cybersecurity strategies.
Supply Chain Risks: Cyber-attackers often target third-party vendors and suppliers to gain access to larger organizations, exploiting weaknesses in the supply chain [14].
Shortage of Skilled Professionals: There is a global shortage of cybersecurity professionals, making it difficult for organizations to find and retain skilled experts.

Data Breaches and Privacy Concerns: High-profile data breaches erode public trust, and stringent data privacy regulations require organizations to handle personal information with care.
Legacy Systems and Outdated Software: Older systems and software that are no longer supported or updated can be vulnerable to known exploits.
Cloud Security: Migrating to the cloud introduces new security challenges, and misconfigurations or inadequate access controls can lead to data breaches [5].
Balancing Security and Usability: Implementing robust security measures can sometimes lead to inconvenience or decreased user experience, creating a challenge in finding the right balance.
Attribution and Cybercrime Jurisdiction: Tracing the origin of cyberattacks and holding attackers accountable across international boundaries can be complex and difficult.
Security Compliance and Regulations: Organizations must comply with an ever-increasing number of cybersecurity regulations, which can be complex and costly to implement.
Emerging Technologies: The integration of emerging technologies like artificial intelligence, quantum computing, and 5G introduces both new possibilities and potential security vulnerabilities.

Addressing these challenges requires a holistic and proactive approach to cybersecurity, involving a combination of technology, processes, policies, and a well-trained workforce. It also requires cooperation between

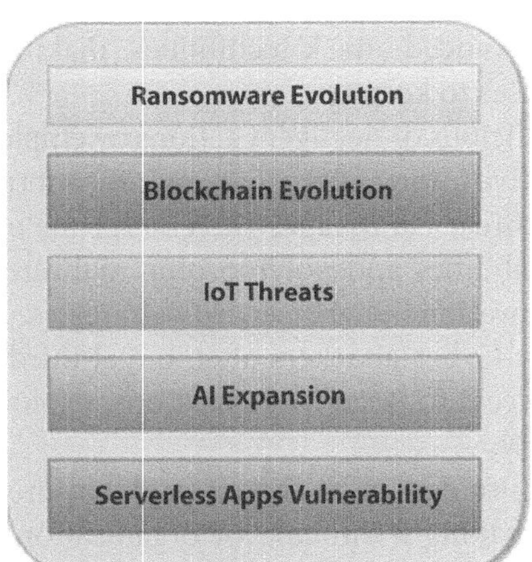

Figure 18.2 Cybersecurity challenges in cobots.

public and private sectors to share threat intelligence and collaborate on finding effective solutions. As the digital landscape evolves, the cybersecurity industry must continue to adapt to ensure a safer and more secure cyberspace. Figure 18.2 shows the challenges of cybersecurity.

18.4.1 Cybersecurity Challenges in Cobots

Cobots, or collaborative robots, introduce a unique set of cybersecurity challenges due to their integration into interconnected networks and their collaboration with humans. The following are some key cybersecurity challenges specific to cobots:

Unauthorized Access: Cobots connected to a network are vulnerable to unauthorized access by malicious actors. Unauthorized access can lead to unauthorized control, manipulation, or sabotage of cobots, posing risks to the safety of humans working alongside them and potentially causing damage to equipment or processes. It looks like you are mentioning "Unauthorized Access Cobots Connected." If you are referring to a situation involving unauthorized access to collaborative robots (cobots) that are connected to a network or system, here are some steps you could consider taking:

Isolate the Cobots: If you suspect unauthorized access, then disconnect the cobots from the network or isolate them from the rest of your systems to prevent further potential compromise.
Assess the Situation: Determine the extent of the unauthorized access. What actions were taken or could potentially be taken by the unauthorized party?
Incident Response Plan: Follow your organization's incident response plan, if one is in place. If not, consider creating one to outline steps for addressing cybersecurity incidents.
Forensics and Investigation: Conduct a thorough investigation to understand how the unauthorized access occurred, what systems were affected, and what data might have been compromised.
Patch and Update: Ensure that the cobots' software and firmware are up to date. Vulnerabilities in software could be exploited, so keeping everything patched is crucial.
Change Credentials: Change passwords, access keys, and other credentials used to access and control the cobots.

Implement Access Controls: Strengthen access controls by restricting access to authorized personnel only. Implement strong authentication methods.
Network Security: Review and enhance network security measures. Firewalls, intrusion detection systems, and other security technologies can help detect and prevent unauthorized access.
Communication: Inform relevant stakeholders about the incident, including IT personnel, security teams, and management.
Preventive Measures: Once the incident is resolved, focus on preventing similar incidents in the future. Continuously monitor and assess your cybersecurity posture.

Data Tampering: Cobots generate and rely on vast amounts of data, including sensor readings, operational parameters, and human interactions. Malicious actors may attempt to tamper with this data, leading to incorrect decisions or actions by cobots. For example, altering sensor readings could cause cobot to make incorrect judgments or perform dangerous actions.

Data tampering involving cobots can be a serious security concern. Data tampering refers to unauthorized changes or alterations made to data, which can lead to inaccurate information, faulty decisions, and potential safety risks when it comes to cobots.

Communication Interception: Cobots often communicate and exchange data with other devices, such as human operators, other robots, or central control systems. Interception of these communications can expose sensitive information or allow attackers to eavesdrop on commands, potentially leading to unauthorized control or theft of valuable intellectual property.

Insider Threats: The collaboration between cobots and human operators introduces the risk of insider threats. A malicious or disgruntled employee with access to cobots may intentionally manipulate their behavior or compromise their security, leading to unauthorized actions or compromising sensitive information.

Vulnerabilities in Software and Firmware: Cobots rely on software and firmware for their operations, which may contain vulnerabilities that can be exploited by attackers. These vulnerabilities can be in the cobot's operating system, control software, or third-party applications, providing opportunities for unauthorized access or control.

Vulnerabilities in the software and firmware of cobots can pose significant risks to their operation, data integrity, and even safety. It is crucial to address these vulnerabilities to ensure the security and proper functioning of your cobots. Here is how you can handle vulnerabilities in cobot software and firmware:

Regular Updates and Patching: Keep the cobot software and firmware up to date by regularly applying patches and updates provided by the manufacturer. This helps address known vulnerabilities and security issues.

Vulnerability Assessment: Perform regular vulnerability assessments and penetration testing on your cobot systems to identify potential weaknesses. This proactive approach can help you discover vulnerabilities before they are exploited.

Vendor Security Information: Stay informed about security advisories and alerts issued by the cobot manufacturer. Subscribe to their security notification services to receive timely information about vulnerabilities and patches.

Secure Development Lifecycle: Encourage the cobot manufacturer to follow a secure development lifecycle when creating software and firmware. This includes security assessments during the design, coding, and testing phases.

Third-Party Components: Cobots often use third-party components (libraries, modules, etc.). Ensure that these components are also regularly updated and free from vulnerabilities.

User Training: Train personnel who interact with the cobots on security best practices. Educate them about the risks of downloading unauthorized software, clicking on suspicious links, or using weak passwords.

Secure Configuration: Ensure that cobot software and firmware are configured securely. Disable unnecessary services, change default passwords, and implement security settings recommended by the manufacturer.

Secure Supply Chain: Verify the security practices of your cobot suppliers and manufacturers. Ensure that they follow secure coding practices and supply chain integrity to prevent tampering.

Collaboration With Experts: If you lack the expertise, then consider involving cybersecurity professionals who specialize in industrial control systems and robotics. They can provide insights and guidance on securing cobot systems.

Regulatory Compliance: Ensure that your cobot systems comply with relevant industry regulations and standards for cybersecurity and safety.

Lack of Secure Authentication: Cobots need to authenticate and verify the identity of human operators or other entities they interact with. Weak or inadequate authentication mechanisms can result in unauthorized access, allowing attackers to compromise the cobot's functionality or gain control over it.

Physical Security Risks: Physical access to cobots can also pose cybersecurity risks. Unauthorized physical access to cobot may allow an attacker to manipulate or tamper with its hardware components, install malicious devices, or extract sensitive data stored within the system.

Addressing these cybersecurity challenges requires a comprehensive approach that combines secure network architecture, robust access controls, encryption mechanisms, intrusion detection systems, regular software/firmware updates, security awareness training, and continuous monitoring and response capabilities. Additionally, the adoption of advanced technologies like blockchain can contribute to enhancing the security and resilience of cobots' operations.

Figure 18.3 shows the cybersecurity challenges in cobots.

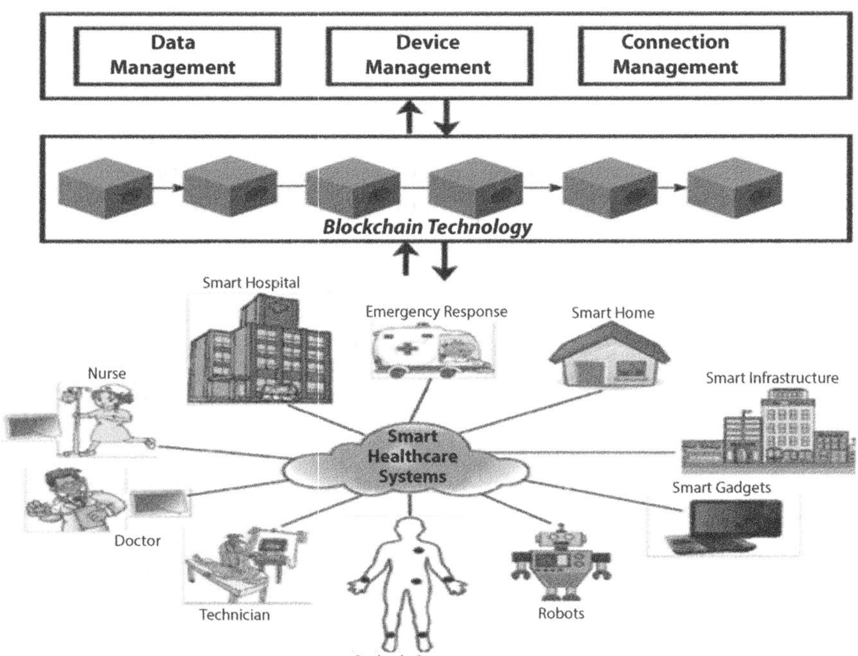

Figure 18.3 Cybersecurity challenges in cobots.

18.5 Introduction to Blockchain Technology

Blockchain technology, initially developed as the foundation for cryptocurrencies like Bitcoin, has emerged as a transformative innovation capable of revolutionizing numerous industries and sectors. Evolving beyond its origins, blockchain has become a versatile tool with applications extending beyond digital currencies. Its decentralized and transparent characteristics, combined with robust cryptographic principles, open new avenues for secure and reliable digital transactions and interactions.

At its core, a blockchain is a distributed and decentralized ledger that records transactions across multiple computers or nodes. Instead of relying on a central authority or intermediary, blockchain enables peer-to-peer transactions and consensus mechanisms that validate and secure the data stored within the system. This decentralized structure provides several key advantages, including transparency, immutability, and enhanced security.

Transactions within a blockchain are organized into blocks, interconnected through cryptographic links, forming a chain. Each block comprises a timestamp, a distinct identifier (hash), and a pointer to the preceding block, establishing an uninterrupted sequence of past transactions. Upon addition to the blockchain, altering or meddling with the data within a block becomes exceedingly difficult due to the implementation of cryptographic hashing algorithms and the decentralized structure of the network [12]. Figure 18.4 illustrates the cybersecurity challenges in cobots.

One of the notable features of blockchain technology is transparency. The entire transaction history, from the genesis block to the latest block, is visible and accessible to all participants in the network. This transparency ensures accountability and helps prevent fraudulent activities by providing a comprehensive and auditable record of transactions.

Immutability is another crucial characteristic of blockchain. Once a transaction is recorded and confirmed by the network consensus, it cannot be modified or deleted. This immutability ensures the integrity of the data stored on the blockchain, making it highly resistant to tampering or unauthorized changes [13].

The security of blockchain is maintained through cryptographic mechanisms. Each transaction is secured using cryptographic algorithms, ensuring the confidentiality and integrity of the data exchanged between participants. Digital signatures and cryptographic keys enable secure authentication and authorization, preventing unauthorized access or tampering with the data.

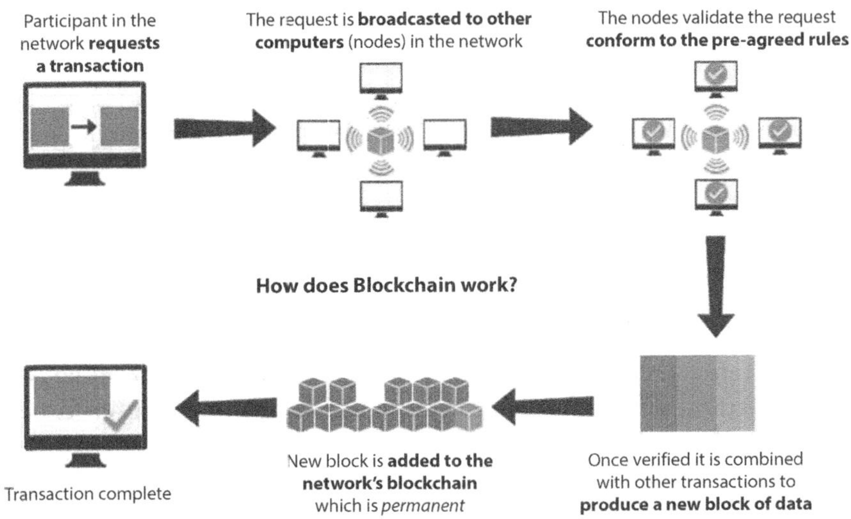

Figure 18.4 Cybersecurity challenges in cobots.

18.6 Security Features of Blockchain in Cobot's Cybersecurity

Blockchain technology offers several inherent security features that can contribute to cobots' cybersecurity. Figure 18.5 shows the improvising Cybersecurity with blockchain. These features include the following:

Decentralization: Blockchain operates on a decentralized network of nodes, eliminating the need for a central authority. This decentralized structure reduces the risk of single points of failure and unauthorized control. In the context of cobots, decentralization enhances the resilience of communication networks, preventing potential cyber-attacks from compromising the entire system.

Transparency: Blockchain provides transparency by maintaining a complete and immutable record of all transactions and interactions. This transparency enables stakeholders to verify and validate the integrity of cobots' actions and data. Any changes or tampering attempts are immediately noticeable, promoting accountability and discouraging malicious activities.

Consensus Mechanisms: Blockchain relies on consensus mechanisms to validate and confirm transactions. These mechanisms ensure that all

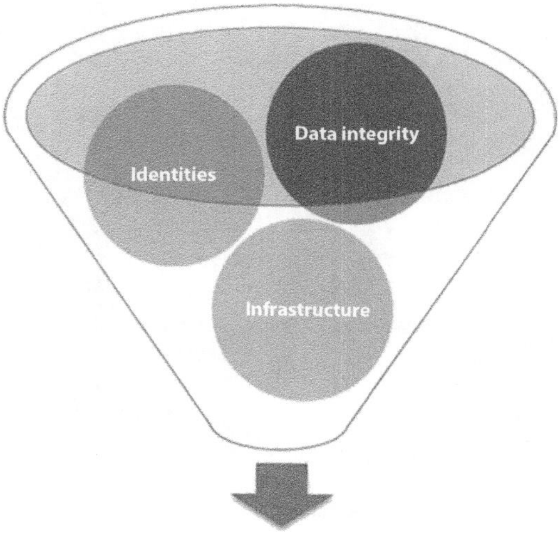

Figure 18.5 Cybersecurity with blockchain.

participants in the network agree on the state of the blockchain. By achieving consensus, blockchain mitigates the risk of fraudulent or malicious transactions and provides a secure foundation for cobots' interactions and data exchange.

Cryptographic Security: Blockchain utilizes cryptographic algorithms to secure transactions and protect data. Cryptographic techniques such as hashing and digital signatures ensure the confidentiality, integrity, and authenticity of cobots' communications and transactions. These mechanisms provide robust protection against unauthorized access, tampering, and data interception.

Secure Identity Management: Blockchain technology enables secure identity management for cobots. Through the use of cryptographic keys and digital signatures, cobots can establish their unique identities on the blockchain. This ensures authentication and authorization of their actions within the network, preventing unauthorized cobots from gaining access to sensitive resources or compromising the integrity of the system. Figure 18.6 shows the Cybersecurity with blockchain.

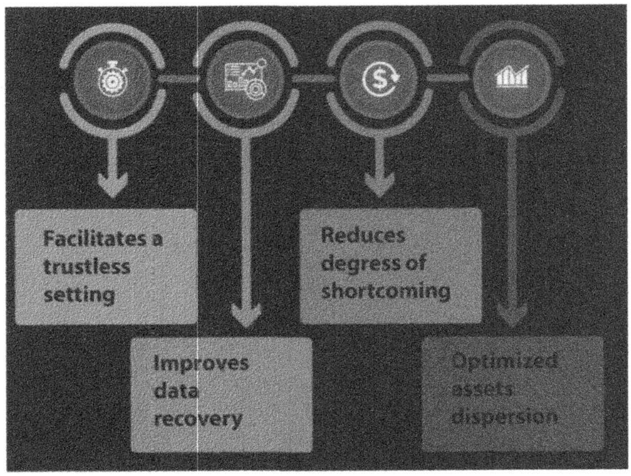

Figure 18.6 Cybersecurity with blockchain.

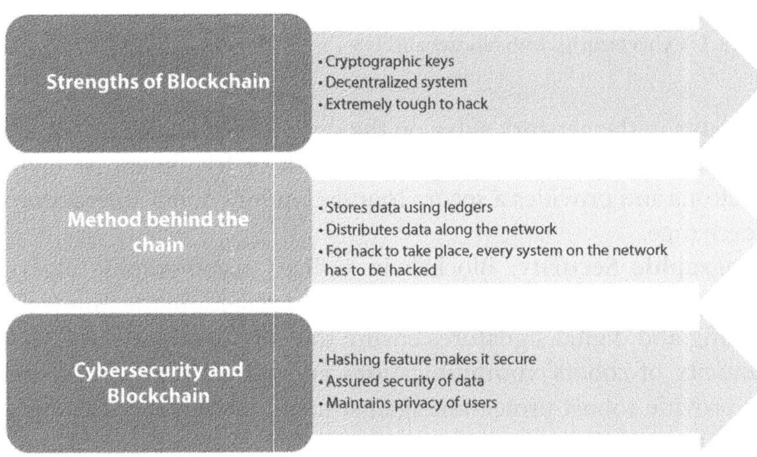

Figure 18.7 Blockchain technology is going to improve the way cybersecurity.

Blockchain technology has the potential to significantly improve the way cybersecurity is implemented and managed. Figure 18.7 shows the blockchain technology is going to improve the way cybersecurity.

By leveraging these security features, blockchain technology can enhance the integrity, confidentiality, and availability of cobot systems. These features provide a robust foundation for securing communication channels, ensuring data integrity, facilitating secure identity management, and enforcing authorized actions through smart contracts. However, it is important to consider the specific implementation and integration

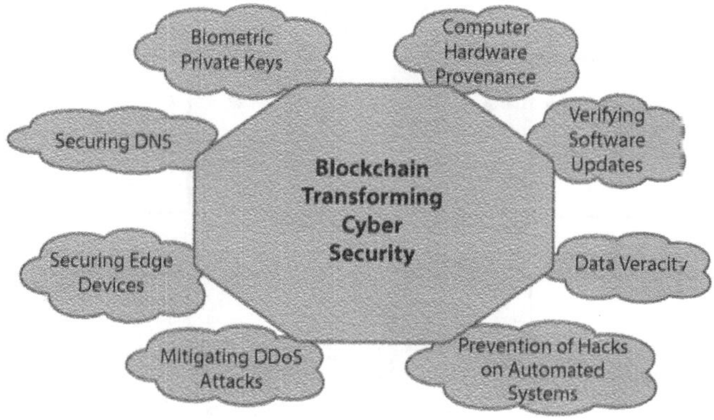

Figure 18.8 Blockchain technology transforming cybersecurity.

challenges that may arise when applying blockchain technology to cobots' cybersecurity [8].

Blockchain technology has the potential to transform cybersecurity in several ways. It introduces new paradigms that can enhance security and address various challenges in the digital world. Here are some key aspects of how blockchain is impacting cybersecurity. However, it is essential to recognize that blockchain technology is not a silver bullet for all cybersecurity challenges. While it offers significant advantages, it also has its limitations, such as scalability and potential vulnerabilities in specific implementations. Therefore, the successful integration of blockchain technology into cybersecurity requires careful consideration of the specific use cases, network design, and a comprehensive approach to overall security measures. Figure 18.8 shows the blockchain technology transforming cybersecurity.

18.7 Application Areas

There are several important applications of cybersecurity for cobots, aiming to protect them from potential threats and ensure the safety and integrity of their operations. Figure 18.9 shows the Cobots' multipurpose uses. Some key applications of cobot's cybersecurity include the following [15, 16]:

Access Control: This involves authentication and authorization protocols to verify the identity and permissions of users accessing the cobot system. Robust access control prevents unauthorized individuals from manipulating or compromising the cobots' functionality or data.

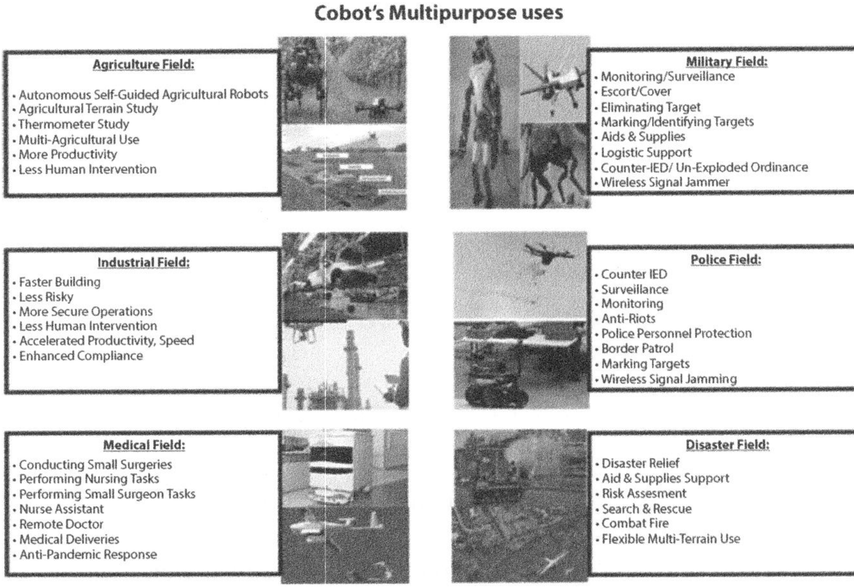

Figure 18.9 Cobots' multipurpose uses.

Secure Communication: Cobots often communicate with humans, other robots, or central control systems. Implementing secure communication protocols, such as encryption and secure channels, protects the confidentiality and integrity of data exchanged between cobots and other entities. Secure communication prevents eavesdropping, data tampering, or interception of sensitive information.

Securing the communication of cobots is crucial to prevent unauthorized access, data interception, and tampering. Here are steps you can take to ensure secure communication for your cobots:

Virtual Private Networks (VPNs): Use VPNs to establish encrypted tunnels for remote access to cobots. This helps protect data transmitted over public networks and ensures secure communication between remote operators and cobots.

Secure Protocols: Use secure communication protocols such as HTTPS (SSL/TLS) for web interfaces and SSH for remote access. These protocols encrypt data to prevent eavesdropping.

Data Integrity and Auditability: Ensuring the integrity of data generated and processed by cobots is crucial. Implementing mechanisms to detect and prevent data tampering, such as cryptographic hashes or digital signatures, helps maintain the integrity of cobots' operational data. Additionally, maintaining an audit trail of cobots' actions and interactions enables forensic analysis, compliance monitoring, and accountability [10].

18.8 Challenges and Considerations Cobot's Cybersecurity

Securing the cybersecurity of cobots involves several challenges and considerations that need to be addressed. Some of the key challenges and considerations for cobot's cybersecurity are as follows:

Interconnected Networks and Systems: Cobots often operate within interconnected networks, communicating with various devices and systems. The complexity of these networks increases the attack surface and introduces vulnerabilities that can be exploited. Securing the entire network infrastructure and ensuring the integrity of all interconnected components is a challenge.

Lack of Security by Design: Cobots are often designed with a primary focus on functionality and productivity, with security considerations as an afterthought. Retrofitting security measures onto existing cobots or integrating security into the design of new cobots poses challenges, as it requires a holistic and systematic approach to ensure effective cybersecurity.

Human Factors and User Awareness: Humans play a crucial role in cobot operations, from programming and maintenance to interacting with cobots directly. Human error, lack of cybersecurity awareness, and unintentional actions can introduce vulnerabilities or bypass security controls. Educating and raising awareness among cobot operators and users about cybersecurity best practices is essential.

Legacy Systems and Compatibility: Integrating cobots into existing legacy systems or environments may pose compatibility challenges. Legacy systems may have outdated or incompatible security measures, increasing the risk of vulnerabilities. Ensuring compatibility between cobots and existing systems while maintaining adequate security is a consideration in cobot cybersecurity [17].

Scalability and Interoperability: As the number of deployed cobots increases, managing and scaling cybersecurity measures across a large fleet of cobots becomes challenging. Ensuring interoperability and consistency in security measures across different cobot models and manufacturers is crucial for maintaining a secure and cohesive ecosystem.

Threat modeling is a critical process in cybersecurity that helps identify potential security threats and vulnerabilities in a system or application. When dealing with collaborative robots, also known as cobots, threat modeling becomes essential to ensure their safe and secure operation. Here are some steps to perform cobot threat modeling:

Identify Assets and Data: Begin by identifying the assets and data that the cobot interacts with or has access to. This could include physical assets, such as machinery and sensors, as well as digital assets, like control software, configuration data, and sensitive information [18].
Define the System Scope: Clearly define the boundaries and components of the cobot system. This includes the cobot itself, any peripheral devices, communication channels, and interfaces with other systems.
Create an Attack Surface: Identify all the entry points through which an attacker might try to compromise the cobot or gain unauthorized access to its functions, data, or control.
Identify Threats: Consider potential threats that could exploit vulnerabilities in the cobot system. This might include physical attacks, cyber-attacks, malware, unauthorized access, and insider threats.
Analyze Vulnerabilities: Assess the weaknesses and vulnerabilities in the cobot system that could be exploited by the identified threats. These vulnerabilities could be related to software, hardware, communication protocols, or the physical environment.
Determine Impact: Understand the potential impact of successful attacks on the cobot system. Consider safety risks to human operators, damage to equipment, data breaches, financial losses, or disruption of critical processes [19].
Apply Mitigations: Develop and implement appropriate mitigations for the identified risks. This might include software patches, security updates, access controls, encryption, network segmentation, physical barriers, and training for human operators.
Testing and Validation: Test the cobot system against the identified threats and evaluate the effectiveness of the implemented mitigations. Penetration testing, vulnerability assessments, and red team exercises can help validate the security posture of the cobot system.

It is essential to involve stakeholders from different domains, such as robotics engineers, software developers, security experts, and end users, in the threat modeling process [20]. Collaboration ensures a comprehensive understanding of the cobot system and improves the quality of threat identification and mitigation strategies. Figure 18.10 shows the cobot threat modeling life cycle.

Figure 18.10 Cobot threat modeling life cycle.

18.9 Conclusion

In conclusion, the cybersecurity of cobots is of paramount importance to ensure their safe and reliable operation within interconnected networks. As cobots become increasingly integrated into industries [7], addressing cyber threats and protecting against potential vulnerabilities is crucial. Cobot cybersecurity presents unique challenges due to their collaboration with humans, reliance on interconnected systems, and the need for real-time decision-making. However, leveraging appropriate cybersecurity measures can mitigate these challenges and establish a secure environment for cobots.

The transparency and immutability of blockchain technology hold promise for enhancing cobot cybersecurity. Blockchain's decentralized nature and smart contracts can provide secure communication channels, data integrity, secure identity management, and enforce authorized actions. These features contribute to establishing trust, accountability, and resilience in cobot systems.

Overcoming challenges such as evolving threats, legacy system integration, limited computing resources, and human factors requires a holistic and collaborative approach. Continuous monitoring, vulnerability management, and user training plans are essential mechanisms of an effective cobot cybersecurity strategy.

It is crucial for stakeholders, including cobot manufacturers, system integrators, cybersecurity experts, and policymakers, to work together to develop industry-wide standards, best practices, and regulations that address the specific cybersecurity needs of cobots.

References

1. Soori, M., Dastres, R., Arezoo, B., Jough, F.K.G., Intelligent robotic systems in Industry 4.0: A review. *J. Adv. Manuf. Sci. Technol.*, 2709–2135, January 2024, Editorial Committee of JAMST, DOI: 10.51393/j.jamst.2024007.
2. Li, M., Wang, X., Xu, J., Zhou, Z., A Survey on the Security of Robotics and Automation Systems: Threats, Solutions, and Challenges. *IEEE Access*, 8, 90169–90187, 2020.
3. Jerbi, M. and Besnard, F., Cyber security for Industrial Robots: Challenges and Solutions. *IEEE Rob. Autom. Mag.*, 26, 4, 66–76, 2019.
4. Jia, C., Li, X., Chen, C., Xie, X., Xu, L., Cyber security Challenges in Smart Manufacturing: From Industrial Control Systems to Collaborative Robots. *IEEE Trans. Ind. Inf.*, 16, 8, 5410–5419, 2020.
5. Roman, R., Zhou, J., Lopez, J., On the Features and Challenges of Security and Privacy in Distributed Internet of Things. *Comput. Netw.*, 57, 10, 2266–2279, 2013.
6. DeMarinis, N., Tellex, S., Kemerlis, V.P., Konidaris, G., Fonseca R., Scanning the internet for ros: A view of security in robotics research, in: *2019 international conference on robotics and automation (ICRA)*, pp. 8514–8521, 2019.
7. Ferrer, E.C., Hardjono, T., Pentland, A., Dorigo, M., Secure and secret cooperation in robot swarms. *Sci. Rob.*, 6, 56, eabf1538, 2021, doi: 10.1126/scirobotics.abf1538.
8. Ferrer, E.C., Jiménez, E., Luis Lopez-Presa, J., Martín-Rueda, J., Following leaders in byzantine multi robot systems by using blockchain technology. *IEEE Trans. Rob.*, 38, 1101–1117, 2022.
9. Mayoral-Vilches, V., Carbajo, U.A., Gil-Uriarte, E., Industrial robot ransomware: Akerbeltz, in: *2020 fourth IEEE international conference on robotic computing (IRC)*, pp. 432–435, 2020.
10. Queralta, J.P., Li, Q., Ferrer, E.C., Westerlund, T., Secure encoded instruction graphs for end-to-end data validation in autonomous robots. *IEEE Internet Things J.*, 9, 18, 18028–18040, 2022, doi: 10.1109/jiot.2022.3164545.
11. Strobel, V., Ferrer, E.C., Dorigo, M., Blockchain technology secures robot swarms: A comparison of consensus protocols and their resilience to byzantine robots. *Front. Rob. AI*, 7, 54, 2020, doi: 10.3389/frobt.2020.00054.
12. Dorigo, M., Trianni, V., Şahin, E., Groß, R., Labella, T.H., Baldassarre, G. *et al.*, Evolving self-organizing behaviors for a swarm-bot. *Auton. Rob.*, 17, 223–245, 2004, 10.1023/B:AURO.0000033973.24945.f3.

13. Nishida, Y., Kaneko, K., Sharma, S., Sakurai, K., Suppressing chain size of blockchain-based information sharing for swarm robotic systems, in: *Proceedings of the Sixth International Symposium on Computing and Networking Workshops (CANDARW 2018)*, IEEE Press, Los Alamitos, CA, pp. 524–528, 2018.
14. Rubenstein, M., Ahler, C., Hoff, N., Cabrera, A., Nagpal, R., Kilobot: a low cost robot with scalable operations designed for collective behaviors. *Robot. Auton. Syst.*, 62, 966–975, 2014, 10.1016/j.robot.2013.08.006.
15. Sargeant, I. and Tomlinson, A., Maliciously manipulating a robotic swarm, in: *Proceedings of ESCS'16 - The 14th International Conference on Embedded Systems, Cyber-Physical Systems, & Applications*, CSREA Press, Bogart, GA, pp. 122–128, 2016.
16. Khalid, A., Kirisci, P., Khan, Z.H., Ghrairi, Z., Thoben, K.D., Pannek, J., Security framework for industrial collaborative robotic cyber-physical systems. *Comput. Ind.*, 97, 132–145, 2018. [Google Scholar] [CrossRef] [Green Version].
17. Sunder, S., Foundations for innovation in cyber-physical systems, in: *Proceedings of the NIST CPS Workshop*, Chicago, IL, USA, 13–14 March 2012, vol. 13, pp. 1–12. [Google Scholar].
18. Dutta, V. and Zielinska, T., Networking technologies for robotic applications, arXiv 2015, arXiv:1505.07593. [Google Scholar].
19. Dreyer, I. and Stang, G., Foresight in governments-practices and trends around the world, in: *Yearbook of European Security*, vol. 1368, pp. 1–26, EU Institute for Security Studies, Paris, France, 2013. [Google Scholar].
20. Hon, W.K., Millard, C., Singh, J., Twenty legal considerations for clouds of things, in: *Queen Mary School of Law Legal Studies Research Paper*, vol. 216, pp. 1–47, SSRN, Rochester, NY, USA, 2016. [Google Scholar].

Index

Also of Interest

Check out these published and forthcoming titles in the "Industry 5.0 Transformation Applications" series from Scrivener Publishing

Generative Artificial Intelligence
Concepts and Applications
Edited by R. Nidhya, D. Pavithra, Manish Kumar, A. Dinesh Kumar and S. Balamurugan
Forthcoming 2025. ISBN 978-1-394-20922-4

Optimized Computational Intelligence Driven Decision Making
Theory, Application and Challenges
Edited by Hrudaya Kumar Tripathy, Sushruta Mishra, Minakhi Rout, S. Balamurugan and Samaresh Mishra
Forthcoming 2025. ISBN 978-1-394-24253-5

Cyber-Physical Systems for Innovating and Transforming Society 5.0
Edited by Tanupriya Choudhury, Abhijit Kumar, Ravi Tomar, S. Balamurugan and Ankit Vishnoi
Forthcoming 2025. ISBN 978-1-394-19771-2

Intelligent Robots and Cobots
Concepts and Applications for Industry 5.0 Transformation
Edited by Ramasamy V., S. Balamurugan and Sheng-Lung Peng
Forthcoming 2025. ISBN 978-1-394-19817-7

Welding Practices for Industry 5.0
Edited by Syed Quadir Moinuddin, Shaik Himam Saheb, Ashok Kumar Dewangan, Muralimohan Cheepu and S. Balamurugan
Published 2024. ISBN 978-1-394-17241-2

Metaverse and Immersive Technologies
An Introduction to Industrial, Business and Social Applications
Edited by Chandrashekhar A., Shaik Himam Saheb, Sandeep Kumar Panda, S. Balamurugan and Sheng-Lung Peng
Published 2023. ISBN 978-1-394-17454-6

www.scrivenerpublishing.com

Printed and bound by CPI Group (UK) Ltd, Croydon, CR0 4YY
17/09/2026
14964280-0005